Mathematical Understanding for Secondary Teaching: A Framework and Classroom-Based Situations

Mathematical Understanding for Secondary Teaching: A Framework and Classroom-Based Situations

Edited by

M. Kathleen Heid and Patricia S. Wilson, with Glendon W. Blume

Copublished by:
NCTM (#15128)
&

INFORMATION AGE PUBLISHING, INC.
Charlotte, NC • www.infoagepub.com

Library of Congress Cataloging-in-Publication Data

Mathematical understanding for secondary teaching : a framework and classroom-based situations / edited by M. Kathleen Heid and Patricia S. Wilson, with Glendon W. Blume.
    pages cm
  Includes bibliographical references.
  ISBN 978-1-68123-113-6 (pbk.) -- ISBN 978-1-68123-114-3 (hardcover) -- ISBN 978-1-68123-115-0 (ebook) 1. Mathematics--Study and teaching (Secondary) 2. Mathematics teachers--Training of. 3. High school teachers--Training of. I. Heid, Mary Kathleen, editor. II. Wilson, Patricia S., editor. III. Blume, Glendon W., editor.
  QA11.2.M27745 2015
  510.71'2--dc23

                              2015019679

NCTM stock #15128.

CONTENTS

PREFACE

Knowing mathematics and being able to use it is critical to success in a large range of professions, especially for those who teach mathematics to students in PreK–12 settings. Increasingly, the attention of mathematics education researchers and scholars has turned toward characterizing the nature of mathematical knowledge for teaching (MKT). Although much of the initial work in this area (as well as the MKT acronym) was based on investigation of mathematical knowledge for teaching elementary mathematics, the work reported in this volume is focused on mathematical understanding for secondary teaching. This mathematical understanding is characterized in the Mathematical Understanding for Secondary Teaching Framework (MUST).

Descriptions of mathematics for secondary teaching have typically been described through lists of required mathematics courses or majors. However, characterizing mathematics for secondary teaching cannot be fully accomplished through lists of courses or majors, or even through lists of mathematical concepts, procedures, or schemas, because such lists suggest a static nature to mathematics. By focusing on the more dynamic phrase, *mathematical understanding*, rather than on *mathematical knowledge*, we were more able to reflect the nature of mathematical understanding as growing and evolving.

Although one might construct a description of the mathematical understanding needed for teaching by conducting a careful thought experiment, such an ex-

Mathematical Understanding for Secondary Teaching: A Framework and Classroom-Based Situations, pages xi–xiv.
Copyright © 2015 by Information Age Publishing

periment might not accurately reflect the realities of the classroom. To keep our work firmly grounded in practice, we committed to developing our framework based on mathematical opportunities that we have seen arise in the practice of teaching mathematics. From those incidents, we developed descriptions of the mathematics that a teacher in that setting could productively use, and from those descriptions we developed our framework for Mathematical Understanding for Secondary Teaching.

The first six chapters of the volume describe the MUST Framework as well as its development and uses. Chapters 1 and 2 describe the Framework, the end product of our work. The basis for the Framework was the set of Situations, and chapter 3 explains the structure of a Situation. Chapter 4 explains how we arrived at the Framework from the Situations, and chapter 5 discusses contexts in which mathematics educators might engage in creating novel Situations. Chapter 6 describes the myriad ways (proposed or actualized) in which the Situations and the Framework might be used in the context of teacher preparation or professional development.

Chapters 7 through 49 are the Situations. Each Situation (a) describes a practice-based incident we witnessed that presented an opportunity for a teacher to draw on his or her mathematical understanding, and (b) elaborates on several areas of mathematics that the teacher in this setting could productively call upon. The Situations are intended to describe mathematical understandings that a teacher could productively use in the designated settings and not to prescribe the pedagogy that one might use in these settings. The Situations are drawn from topics in secondary school mathematics that merit some mathematical attention. The content of the 43 Situations in this volume span, but do not cover, the secondary school mathematics curriculum. Very loosely categorized, chapters 7 through 15 address areas of number and operations, including division involving 0, products of two negative numbers, absolute value, and selected properties of complex numbers. Chapters 16 through 24 fit broadly into the category of algebraic properties and equivalent forms, and include discussion of inverses, exponents, and inequalities. Chapters 25 through 29 continue the algebra discussion with Situations related to graphs and quadratic functions, and chapters 30 and 31 elaborate on algebraic transformations. Chapters 32 through 34 are situated at the intersection of algebra and geometry, dealing with such topics as parametric drawings and loci of moving points. Chapters 35 to 40 focus more specifically on topics usually assigned to geometry, such as similarity of figures, circumscription of polygons, and areas of plane figures. The Situations are rounded out in chapters 41 through 49 with discussion of trigonometric relationships, statistical concepts, and mathematical induction.

We envision the MUST Framework providing an opportunity for people to cross borders and form working groups that focus on improving the mathematical understanding of those who teach mathematics at the secondary level. We would encourage individuals to use the MUST Framework, but we see great potential in groups containing mathematicians, teachers, curriculum designers, and mathematics educators working together to promote mathematical understanding.

ACKNOWLEDGEMENTS

Throughout the volume, we acknowledge the many contributions from school mathematics teachers, mathematics education leaders, mathematics teacher educators, mathematics education researchers, and mathematicians to the creation and development of the Situations and the Framework. We thank them for how generously they shared their insights about mathematical understanding for secondary teaching. A list of those who participated in our three national conferences appears in the Appendix.

Additional individuals made specific, important contributions to bringing this volume to fruition. First, we thank the authors. Seldom do so many mathematics education faculty members and graduate students work together across different universities, and this volume is testimony to our success in this long-term collaboration. Among the authors were the many graduate students who not only generated Prompts for the Situations and developed Mathematical Foci but also contributed to the generation of the Framework. The Situations underwent a plethora of successive revisions, and authors of a Situation include (to the best of our knowledge and record keeping), in the order of participation, all the individuals who participated in writing the Situation.

Our thanks also go out to Tom Banchoff, who lent his mathematical eye to our initial set of 10 Situations, and to John Lediaev and Walter Seaman, two mathematicians who poured over the entire set of Situations and offered specific advice on the mathematics we were describing. Their suggestions were insightful and highly useful, and we know that the volume is all the better for their work. We also thank Susan Peters and Dick Askey for responding to specific content-related questions. Finally, we thank Younhee Lee and Fernanda Bonafini, two Penn State mathematics education graduate students who, although they entered the program too late to serve as authors, took on the task of formatting the edited manuscripts. Their careful look at the mathematical content of those chapters also improved the content of the chapters.

Finally, we would like to thank Spud Bradley, Janice Earl, and their colleagues at the National Science Foundation for their tenacious support for the Centers for Learning and Teaching in Mathematics and Science as well as for collaboration across Centers. The work reported in this book was supported in part by National Science Foundation Grant No. ESI-0227586 for the Center for Proficiency in Teaching Mathematics (CPTM) at the University of Georgia and from National Science Foundation Grant No. ESI-0426253 for the Mid-Atlantic Center for Mathematics Teaching and Learning (MACMTL) at the Pennsylvania State University. Any opinions, findings, or conclusions and recommendations expressed herein are those of the authors and do not necessarily reflect the views of the National Science Foundation.

It is our hope that mathematics professional developers, mathematics teacher educators, school mathematics leaders and teachers, mathematics education researchers, and mathematicians find the Framework and Situations valuable as they work on understanding and enhancing mathematical understanding for secondary teaching.

CHAPTER 1

BACKGROUND FOR THE MATHEMATICAL UNDERSTANDING FRAMEWORK

Jeremy Kilpatrick

Secondary school mathematics comprises far more than facts, routines, and strategies. It includes a vast array of interrelated mathematical concepts, ways to represent and communicate those concepts, and tools for solving all kinds of mathematical problems. It requires reasoning within the confines of a mathematical system and creativity, providing learners with mathematical competence while also laying a foundation for further studies in mathematics and other disciplines.

To facilitate the learning of secondary school mathematics, teachers need a particular kind of understanding. Mathematical understanding for teaching at the secondary level is the mathematical expertise and skill a teacher has and uses for the purpose of promoting students' understanding of, expertise with, and appreciation for mathematics. It requires that teachers not only know more mathematics than they teach but also know it more deeply and in a more connected and abstract manner.

As educators of prospective mathematics teachers of secondary school mathematics, we realized several years ago that our efforts to develop the understanding they need would be enhanced if we had some means of characterizing that under-

Mathematical Understanding for Secondary Teaching: A Framework and Classroom-Based Situations, pages 1–7.

standing. We decided to construct a framework—Mathematical Understanding for Secondary Teaching—that would enable teachers and teacher educators to analyze secondary mathematics from the perspective of the understanding that any teacher of that mathematics ought to have.

We began by identifying mathematical events that had occurred in the context of teaching secondary mathematics, asking what understanding a teacher might productively use in order to take advantage of the mathematics entailed in the event. Our interest in characterizing mathematical understanding was influenced by the work of Deborah Ball and her colleagues at the University of Michigan (e.g., Ball, 2003; Ball & Bass, 2003; Ball & Sleep, 2007a, 2007b; Ball, Thames, & Phelps, 2008). In particular, Ball et al. partitioned mathematical knowledge for teaching (MKT) into components that distinguish between subject matter knowledge and pedagogical content knowledge (Shulman, 1986). Our focus was on mathematical understanding. As we worked on developing our own framework, we considered other attempts to characterize secondary teachers' mathematical understandings (e.g., Adler & Davis, 2006; Cuoco, 2001; Cuoco, Goldenberg, & Mark, 1996; Even, 1990; Ferrini-Mundy, Floden, McCrory, Burrill, & Sandow, 2005; McEwen & Bull, 1991; Peressini, Borko, Romagnano, Knuth, & Willis-Yorker, 2004; Tatto et al., 2008). Our intention has been to add to the work in this area, which continues to expand. We believe that our approach brings something new to the conversation about teachers' mathematical knowledge.

A NEW FRAMEWORK: MATHEMATICAL UNDERSTANDING FOR SECONDARY TEACHING

Mathematical Understanding for Secondary Teaching (MUST) is related to but not identical to mathematical knowledge for teaching (MKT). In examining the work that others have done in developing frameworks for MKT, we became increasingly convinced that whatever framework we developed should reflect a more dynamic view than that implied by *mathematical knowledge. Knowledge* suggests a static condition, whereas *understanding* connotes a process with a constantly evolving state. In our approach, understanding is revealed in the observable and evolving application of a teacher's knowledge. Furthermore, MUST is related to, but different from, pedagogical content knowledge (PCK) (Shulman, 1986). Like PCK, MUST focuses on mathematics, but unlike PCK, MUST does not attempt to describe pedagogical knowledge or pedagogical proficiency. We wanted to keep our focus on the mathematical understanding that teachers of secondary mathematics need quite apart from what they do in their teaching to make use of that understanding. That focus allows our framework to be used to analyze the mathematics in a classroom situation without getting into the specifics of what a teacher and his or her students might do with that mathematics.

Our framework has been developed out of classroom practice focused on mathematics at the secondary level. A distinguishing characteristic of our framework is the variety of classroom contexts from which we have drawn examples.

We have observed the work of practicing teachers, preservice teachers, and mathematics educators and have used episodes from teaching settings to examine and characterize MUST, as described in the following discussion of our development of Situations.

SITUATIONS

Starting our pursuit of a framework from the bottom up, we developed a collection of sample *Situations*[1] as a way of capturing classroom practice. Each Situation portrays an incident that occurred in the context of teaching secondary mathematics in which some mathematical point is at issue. (For details of our approach, see chapter 3.) Looking across Situations, we attempted to characterize the knowledge and understanding of mathematics that secondary school mathematics teachers could productively use but that other users of mathematics may not necessarily need or use frequently.

Each Situation begins with a *Prompt*—an episode that has occurred in the context of teaching mathematics (usually in a mathematics classroom) and raises issues that illuminate the mathematics understanding that would be beneficial for secondary teachers. The Prompt may be a question raised by a student, an interesting response by a student to a teacher's question, a student error, or some other stimulating event. We then outline, in descriptions called *Mathematical Foci*, mathematics that is relevant to the Prompt. The set of Foci for a particular Situation is not meant to be an exhaustive accounting of the mathematics a teacher might draw upon, but we believe the Foci include key points to be considered. For each Situation, each Focus associated with it describes a different mathematical idea, and the set of Foci constitute the bulk of the Situation. There is no offer of pedagogical advice or comment about what mathematics the teacher should actually discuss in a class in which such an episode may occur. Rather, we describe the mathematics itself and leave it to the teacher or mathematics educator to decide what to use and how to do so. Along with the Foci, each Situation includes an opening paragraph, called a *Commentary*, to set the stage for the Mathematical Foci. The Commentary gives an overview of what is contained in the Foci and serves as an advance organizer for the reader. Some Situations also include a *Postcommentary* to include extensions of the mathematics addressed in the Situation.

Throughout the process of writing and revising the Situations, we identified and used aspects of what we would come to include in our MUST Framework. For example, various representations helped us to think about the mathematics in the Prompt. Perhaps there was a geometric model that was helpful or a graph or numerical representation to provide insight or clarification. At times a particular analogy was pertinent to the Prompt. We were not interested in making every Situation follow a particular format in which the same representations (such as analytical, graphical, verbal) were used again and again. We wanted to emphasize representations that we perceived as particularly helpful or relevant in relation to the Prompt.

Another example of our use of aspects of mathematical understanding in writing and revising Situations was the use of connections to other mathematical ideas, or extensions to concepts beyond those currently at hand. For example, if a Prompt addressed sums of integers, we described (though not in great detail) sums of squares of integers. Topics like this that extended ideas in the Situation would be discussed in a Postcommentary at the end of the Situation. Another way to extend a mathematical Focus is to adjust the assumptions. For example, in a Situation involving fractional exponents greater than 0, one could consider the implications of working with exponents that are negative or that are not real numbers.

Our use of these aspects of what would eventually constitute the MUST Framework drew our attention to what we believed were pertinent elements of mathematical understanding for teaching. This process helped us construct and clarify the framework and grounded our development of the elements of the framework.

Evolution of the Framework

By examining Mathematical Foci for more than 40 Situations, we developed a framework characterizing MUST for secondary school mathematics. In the Situations, we could see the need, for example, for a teacher to be skilled in such tasks as using multiple representations of a mathematical concept, making connections between concepts, proving mathematical conjectures, determining the mathematics in a student comment or error, understanding the mathematics that comes before and after the task at hand, or discerning when students' questions raise mathematical issues that should be explored given the time available. In seeking to develop and improve the framework, we have responded to comments and suggestions given by experts (mathematicians, mathematics educators, teacher educators, and teacher leaders [e.g., department heads, mathematics supervisors]) in the field of mathematics education. We gathered this input at two Situations Development Conferences at The Pennsylvania State University.[2]

The purpose of the first conference was to engage a group of mathematicians and mathematics educators, each of whom had done work on the mathematics of secondary teachers, in providing feedback on our conceptualization of secondary teachers' mathematical understanding as embodied in 10 of the Situations that we had developed. At that point we had not developed a framework; rather, we were at the stage of writing and revising Situations with the goal of being able to characterize mathematical *knowledge* (then *proficiency* and finally *understanding*) for teaching at the secondary level and constructing a framework for doing so. We received input from the experts about the Situations themselves, and that input challenged us to continue to refine our work and to consider some additional mathematical ideas that we had not included in the Foci of the 10 Situations we shared. We also sought advice from the participants about what they considered to be key aspects of mathematical knowledge for teaching at the secondary level. A few of the ideas arising from that discussion were analysis of student thinking and student work, mathematical reasoning, mathematical connections, and mathematical habits of mind. We went back to work on the framework, trying to incorporate advice we had received at the conference, in order to continue the process of characterizing MKT (later, MUST). We began to build lists of items (content and processes) to be included in the framework (e.g., entities such as mathematical connections and representations, and actions such as choosing appropriate mathematical examples).

During the second conference, we presented a version of our framework to a group of mathematicians, mathematics educators, and teacher leaders for the purpose of seeking feedback and advice, as well as to discuss ideas about how the Framework and Situations could be used and disseminated. We received positive responses from participants regarding how they envisioned using the Situations in their work with prospective or practicing teachers. The feedback we received on the framework document included comments about both the content and the format of the document. In small-group and large-group sessions, we discussed ideas for improving the framework—what to change or clarify, what to leave out, and what to add. Following the conference, we carefully considered this input as we worked to improve our framework document.

At different times over the course of our work, we have focused on the Situations, the Framework, or both. Working on these two parts of our project in parallel has been helpful in keeping them both in view, particularly as our development of the Situations has informed our construction of the Framework. We believe that the Framework now can be used to better interpret the Situations, to write new Situations, and to further our understanding of mathematical understanding for teaching.

CONCLUSION

The work of describing MUST continues, but we believe this framework is already an important contribution to the mathematics education community. The core mathematical understanding needed by teachers is different from that needed in other professions. A teacher's work requires general mathematical knowledge as well as expertise in the kinds of tasks described in this framework: accessing the mathematical thinking of learners, developing multiple representations of a mathematical concept, knowing how to use the curriculum in a way that will help further the mathematical understanding of students, and so on.

A distinctive feature of our work is that we have chosen to develop a framework for mathematical understanding for secondary teaching (MUST) that, rather than seeking primarily to identify the knowledge and specific understandings of mathematics useful in secondary teaching, would highlight the dynamic nature of secondary teachers' mathematical understandings. Mathematical understanding for teaching should grow and deepen over the course of a teacher's career, and the lenses that comprise our framework characterize the nature of the mathematical proficiencies, actions, and contexts that set those understandings apart. Just as the mathematical understanding of teachers evolves over time, so will the field's conceptualization of MUST, and the project in which we have engaged is and will continue to be a work in progress.

A second distinctive feature of the framework is that we focus solely on mathematical understanding at the secondary school level. We believe that this focus is essential to the profession's conversation about teacher knowledge. Secondary mathematics differs from elementary school mathematics in its breadth, rigor, abstraction, and explicitness of mathematical structure required. Therefore, teaching at the secondary level requires mathematical understanding that advances the nature and level of reasoning in which students engage.

Finally, we believe we bring a distinctive perspective in that our framework has arisen from the practice of secondary mathematics teaching in a wide variety of settings including courses for prospective teachers, high school classes taught by practicing teachers, and classes taught by student teachers.

Just as we have sought the input of many mathematicians, mathematics teachers, and teacher educators during construction of this framework, we welcome comments on our final product from those in the field. Furthermore, we would like to gain further insight from others into MUST, perhaps by building on the ideas presented here.

NOTES

1. We reserve the term *Situations* to refer to the specific products of our work.
2. A third conference was conducted at University of Georgia to generate ideas about uses of the Situations and Framework. Chapter 6 describes ideas generated at that conference.

REFERENCES

Adler, J., & Davis, Z. (2006). Opening another black box: Researching mathematics for teaching in mathematics teacher education. *Journal for Research in Mathematics Education, 36,* 270–296.

Ball, D. L. (2003, February). *What mathematical knowledge is needed for teaching mathematics?* Paper presented at the Secretary's Summit on Mathematics, U.S. Department of Education, Washington, DC.

Ball, D. L., & Bass, H. (2003). Toward a practice-based theory of mathematical knowledge for teaching. In B. Davis & E. Simmt (Eds.), *Proceedings of the 2002 Annual Meeting of the Canadian Mathematics Education Study Group* (pp. 3–14). Edmonton, AB: CMESG/GCEDM.

Ball, D. L., & Sleep, L. (2007a, January). *How does mathematical language figure in the work of teaching? How does this shape Mathematical Knowledge for Teaching (MKT) and teacher education?* Paper presented at the Association of Mathematics Teacher Educators annual meeting, Irvine, CA.

Ball, D. L., & Sleep, L. (2007b, January). *What is mathematical knowledge for teaching, and what are features of tasks that can be used to develop MKT?* Presentation at the Center for Proficiency in Teaching Mathematics presession at the meeting of the Association of Mathematics Teacher Educators, Irvine, CA.

Ball, D. L., Thames, M. H., & Phelps, G. (2008). Content knowledge for teaching: What makes it special? *Journal of Teacher Education, 59,* 389–407. doi:10.1177/0022487108324554

Cuoco, A. (2001). Mathematics for teaching. *Notices of the AMS, 48,* 168–174.

Cuoco, A., Goldenberg, E. P., & Mark, J. (1996). Habits of mind: An organizing principle for mathematics curricula. *Journal of Mathematical Behavior, 15,* 375–402.

Even, R. (1990). Subject matter knowledge for teaching and the case of functions. *Educational Studies in Mathematics, 21,* 521–544. doi:10.1007/BF00315943

Ferrini-Mundy, J., Floden, R., McCrory, R., Burrill, G., & Sandow, D. (2005). *A conceptual framework for knowledge for teaching school algebra.* Unpublished manuscript, Michigan State University. Retrieved from http://www.educ.msu.edu/kat/papers.htm

McEwen, H., & Bull, B. (1991). The pedagogic nature of subject matter knowledge. *American Educational Research Journal, 28,* 316–344. doi:10.3102/00028312028002316

Peressini, D., Borko, H., Romagnano, L., Knuth, E., & Willis-Yorker, C. (2004). A conceptual framework for learning to teach secondary mathematics: A situative perspective. *Educational Studies in Mathematics, 56,* 67–96. doi:10.1023/B:EDUC.0000028398.80108.87

Shulman, L. (1986). Those who understand: Knowledge growth in teaching. *Educational Researcher, 15*(2), 4–14. doi:10.3102/0013189X015002004

Tatto, M. T., Schwille, J., Senk, S., Ingvarson, L., Peck, R., & Rowley, G. (2008). *Teacher Education and Development Study in Mathematics (TEDS-M): Conceptual framework.* East Lansing: Teacher Education and Development International Study Center, College of Education, Michigan State University.

CHAPTER 2

MATHEMATICAL UNDERSTANDING FOR SECONDARY TEACHING: A FRAMEWORK[1]

Jeremy Kilpatrick, Glendon Blume, M. Kathleen Heid,
James Wilson, Patricia Wilson, and Rose Mary Zbiek

Mathematical understanding for secondary teaching (MUST) must be tailored to the work of teaching. The core mathematical understanding needed for teaching is different from the mathematical understanding needed for engineering, accounting, or the medical professions. It is even different from the mathematical understanding a mathematician needs. For example, a mathematician may prove a theorem, and an architect may perform geometric calculations. For these users of mathematics, it is sufficient that they have the skills and abilities for the task at hand. But a teacher's work includes these tasks as well as interpreting students' mathematics, knowing where students are on the path of mathematical understanding, developing multiple representations of mathematical concepts tailored to students' understandings, using their advanced mathematical understanding to craft tasks and examples with a specific set of characteristics, and so on.

Mathematical Understanding for Secondary Teaching: A Framework and Classroom-Based Situations, pages 9–30.

A teacher's mathematical understanding for secondary teaching is continuously varying. We make a distinction between knowledge and understanding. Knowledge may be seen as static and something that cannot be directly observed, whereas understanding can be viewed as the use of the knowledge one has. Understanding can be observed in a teacher's actions and the decisions he or she makes. Also, because of its nature, a teacher's understanding grows and deepens in the course of his or her career.

The focus of our framework is on mathematics for *secondary school* teachers. That is, we seek to characterize the mathematical understanding that is useful to secondary teachers as distinct from the understanding needed by elementary school mathematics teachers. We believe that MUST is different from mathematical understanding for elementary school teaching because of the nature of the mathematical content involved and the level of cognitive development of the students: (a) There is a wider range of mathematics content (i.e., more topics are studied); (b) there is a greater emphasis on formality, axiomatic systems, and rigor in regard to mathematical proof; (c) there is more explicit attention to mathematical structure and abstraction (e.g., identities, inverses, domain, and undefined elements); and (d) the cognitive development of secondary students is such that they can reason differently from elementary school children about concepts and relationships in mathematical systems as well as about conceptually challenging mathematical ideas such as proportionality, probability, and mathematical induction.

Our framework has been developed out of *classroom practice,* and we have drawn examples from a wide variety of contexts related to the teaching and learning of secondary mathematics. We have examined episodes occurring in the work of prospective and practicing secondary mathematics teachers and mathematics educators at the college level. From this collection, we have determined elements of mathematical understanding that would be beneficial to secondary teachers. We describe a wide sample, as opposed to a comprehensive catalog, of mathematical understanding for teaching that comes from our analyses of these practice-based episodes.

Mathematical understanding for teaching is not the same as understanding in pedagogy, nor is it equivalent to pedagogical content knowledge (Shulman, 1986). Being equipped with the understanding described in our MUST Framework is not simply a matter of "knowing the mathematics" adjoined to "knowing how to teach." The task of teaching mathematics cannot be partitioned into such simple categories.

A FRAMEWORK FOR MUST

Mathematical understanding for secondary teaching (MUST) can be viewed as having three different perspectives on the same entity: Mathematical Proficiency, Mathematical Activity, and Mathematical Context of Teaching (Figure 2.1). Just as differently positioned spotlights cast shadows that emphasize different aspects

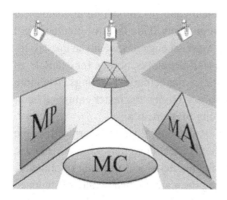

MP   Mathematical Proficiency
MA   Mathematical Activity
MC   Mathematical Context

FIGURE 2.1. Mathematical Understanding for Secondary Teaching Viewed From Three Perspectives.

of the phenomenon being viewed, so the different perspectives of MUST bring to light different aspects of mathematical understanding for secondary teaching. Each perspective emphasizes a different view of MUST. Moreover, MUST, the phenomenon that is being investigated, is not a static entity or endpoint, but rather a developing quality.

Characterizing the mathematical understanding needed by secondary teachers requires understanding the overall mathematical capacities (Mathematical Proficiency) needed by teachers, the specific mathematical actions that teachers need to carry out (Mathematical Activity), and the settings in which teachers need to bring these mathematical capacities and actions into play (Mathematical Context).

Mathematical Proficiency includes aspects of mathematical knowledge and ability, such as conceptual understanding and procedural fluency, that teachers need themselves and that they seek to foster in their students. The mathematical proficiency that teachers need goes well beyond what one might find in secondary students or even in the average educated adult. Students' development of mathematical proficiency is arguably heavily influenced by their teachers' Mathematical Proficiency. Secondary teachers of mathematics need proficiency not only with the mathematics that they are teaching their students but also with the mathematics their students should have learned in elementary school, and they need proficiency with the mathematics their students may encounter when taking mathematics and related subjects in college.

Engaging in *Mathematical Activity* can be thought of as "doing mathematics." Whereas the Mathematical Proficiency perspective describes the general types of mathematical understandings that a teacher should have and use, the Mathematical Activity perspective describes specific mathematical actions in which the teacher should be able to engage. Teachers need a conscious, elaborated command of their nature and particulars. Examples include representing mathematical objects and operations, connecting mathematical concepts, modeling mathematical phenomena, and justifying mathematical arguments. This facet of mathematical understanding for teaching is on display as teachers engage students in the day-to-day study of mathematics. Teachers need a well-developed ability, for example, to notice mathematical structure and to generalize mathematical findings. The more adept a teacher is in meaningful Mathematical Activity, the better equipped he or she will be to facilitate the learning and doing of mathematics.

Engaging in the *Mathematical Context of Teaching* diverges sharply from the mathematical understanding needed in other professions requiring mathematics. One of its aspects is an understanding of the mathematical thinking of students, which may include, for example, recognizing the mathematical nature of students' errors and misconceptions. Teachers need to be able to decide whether a proof might be circular or incomplete, how well a proposed solution satisfies the conditions of a problem, and whether an alternative definition is equivalent to one already proposed. Another aspect of the mathematical work of teaching is knowledge of and expertise in the mathematics that comes before and after what is being studied currently. A teacher benefits from knowing what students have learned in previous years so that he or she can help them build upon that prior knowledge. The teacher also needs to provide a foundation for the mathematics they will be learning later, which requires knowing and understanding the mathematics in the rest of the curriculum.

The three perspectives of MUST—Mathematical Proficiency, Mathematical Activity, and Mathematical Context of Teaching—together form a robust picture of the mathematics required of a teacher of secondary mathematics. It is not enough for a teacher to know the mathematics that students are learning. Teachers must also possess a depth and extent of mathematical understanding and the ability to enact that understanding that will equip them to foster their students' mathematical proficiency. The three perspectives of MUST are interactive. Mathematical Activity and the Mathematical Context of Teaching emerge from, and depend upon, the teacher's Mathematical Proficiency. Mathematical Proficiency is enacted through Mathematical Activity, which is tailored to the Mathematical Context of Teaching.

AN EXAMPLE OF MUST USE

In responding to the following situation, no matter how it is handled pedagogically, the teacher needs to make use of all facets of his or her MUST:

> In an Algebra II class, students had just finished reviewing the rules for exponents. The teacher wrote $x^m \cdot x^n = x^5$ on the board and asked the students to make a list of values for m and n that made the statement true. After a few minutes, one student asked, "Can we write them all down? I keep thinking of more."

To decide whether the student's question is worth pursuing, to frame additional questions appropriately, and to know how to proceed from there, the teacher needs conceptual understanding and productive disposition (two strands of Mathematical Proficiency). The concept of an exponent is more complicated than might initially be apparent. Does the rule $x^m \cdot x^n = x^{m+n}$ always apply? Must the domain of x be restricted? Must the domain of m and n be restricted? These are questions that the teacher needs sufficient mathematical proficiency to address. With respect to mathematical activity, the teacher's skill in representing exponents, knowing constraints that may be helpful in dealing with them, and making connections between exponents and other mathematical phenomena are all crucial to teaching the concept successfully. What are the advantages of a graphical representation of an exponential function as opposed to a symbolic representation? How is the operation of exponentiation connected to the operation of multiplication? Does an exponent always indicate repeated multiplication? With respect to the mathematical work of teaching, it is critical that the teacher knows and understands the mathematics that typically comes before and after the point in the curriculum at which a problem like the one involving the rule for $x^m \cdot x^n$ is addressed. For example, if this problem is being discussed in a beginning algebra course, it is important to realize that students have probably had limited exposure to exponents and may think about them only in terms of the repeated multiplication of natural numbers. And to lay a good foundation for later studies of exponential functions, the teacher needs to know that there may be discontinuity in the graph of x^n depending on the domain of both the base and the exponent.

ELABORATION OF THE MUST PERSPECTIVES

The philosopher Gilbert Ryle (1949) claimed that there are two types of knowledge: The first is expressed as "knowing that," sometimes called *propositional* or *factual* knowledge, and the second as "knowing how," sometimes called *practical* knowledge. Because we wanted to capture this distinction and at the same time to enlarge the construct of *mathematical knowledge for teaching* to include such mathematical aspects as reasoning, problem solving, and disposition, we have adopted the term *understanding* throughout this document instead of using the term *knowledge*. An outline of our framework for the three perspectives of MUST is shown in Figure 2.2. In the sections that follow, we amplify each perspective in turn.

MATHEMATICAL PROFICIENCY

A principal goal of secondary school mathematics is to develop all facets of the learners' mathematical proficiency, and the teacher of secondary mathematics needs to be able to help students with that development. Such expertise on the teacher's part requires that the teacher not only understand the substance of secondary school mathematics deeply and thoroughly but also know how to guide

1. **Mathematical Proficiency**
   Conceptual Understanding
   Procedural Fluency
   Strategic Competence
   Adaptive Reasoning
   Productive Disposition
   Historical and Cultural Knowledge

2. **Mathematical Activity**
   Mathematical noticing
      Structure of mathematical systems
      Symbolic form
      Form of an argument
      Connect within and outside mathematics
   Mathematical reasoning
      Justifying/proving
      Reasoning when conjecturing and generalizing
      Constraining and extending
   Mathematical creating
      Representing
      Defining
      Modifying/transforming/manipulating

3. **Mathematical Context of Teaching**
   Probe mathematical ideas
   Access and understand the mathematical thinking of
   learners
   Know and use the curriculum
   Assess the mathematical knowledge of learners
   Reflect on the mathematics of practice

FIGURE 2.2. Framework for Mathematical Understanding for Secondary Teaching (MUST).

students toward greater proficiency in mathematics. We recognized that the multifaceted nature of teachers' Mathematical Proficiency could be described in much the same way as it is used in *Adding It Up* (Kilpatrick, Swafford, & Findell, 2001) to describe the Mathematical Proficiency of students (see Figure 2.2). We found that the five strands of Mathematical Proficiency described in *Adding It Up* worked well to describe the Mathematical Proficiency needed by teachers when supplemented with a strand that included historical and cultural knowledge of mathematics.

There is a range of proficiency in each strand, and a teacher may become increasingly proficient in the course of his or her career. At the same time, certain categories may involve greater depth of mathematical knowledge than others. For example, *conceptual understanding* involves a different kind of knowledge than *procedural fluency,* though both are important. For the most part, only rote knowledge is required in order to demonstrate procedural fluency in mathematics. Conceptual understanding, however, involves (among other things) knowing *why* the procedures work.

Conceptual understanding

Conceptual understanding is sometimes described as the "knowing *why*" of mathematical proficiency. A person may demonstrate conceptual understanding by such actions as deriving needed formulas without simply retrieving them from memory, evaluating an answer for reasonableness and correctness, understanding connections in mathematics, or formulating a proof. Conceptual understanding for teachers is not only knowing concepts but also being able to use that knowledge to promote student understanding.

Examples of conceptual understanding include the following:

1. Knowing and understanding where the quadratic formula comes from (including being able to derive it).
2. Seeing the connections between right triangle trigonometry and the graphs of trigonometric functions.
3. Understanding how the introduction of an outlying data point can affect mean and median differently.

Procedural fluency

A person with procedural fluency knows some conditions for when and how a procedure may be applied and can apply it competently. Procedural fluency alone, however, would not allow one to independently derive new uses for a previously learned procedure, such as completing the square to solve $ax^6 + bx^3 = c$. Procedural fluency can be thought of as part of the "knowing *how*" of mathematical proficiency. Such fluency is useful because the ability to quickly recall and accurately execute procedures significantly aids in the solution of mathematical problems. Procedural fluency is not only being able to apply mathematical procedures but also knowing when to apply it.

The following are examples of calling on procedural fluency:

1. Recalling and using the algorithm for long division of polynomials.
2. Sketching the graph of a linear function.
3. Finding the area of a polygon using a formula.

Strategic competence

Strategic competence requires procedural fluency as well as a certain level of conceptual understanding. Demonstrating strategic competence requires the ability to generate, evaluate, and implement problem-solving strategies. That is, a person must first be able to generate possible problem-solving strategies (such as utilizing a known formula, deriving a new formula, solving a simpler problem, trying extreme cases, or graphing), and then must evaluate the relative effectiveness of those strategies. The person must then accurately implement the chosen strategy. Strategic competence is different from procedural fluency in that it requires creativity and flexibility because problem-solving strategies cannot be reduced to mere procedures.

Examples of strategic competence include the following:

1. Recognizing problems in which the quadratic formula is useful (which goes beyond simply recognizing a quadratic equation or function).
2. Figuring out how to partition a variety of polygons into "helpful" pieces so as to find their areas.
3. Investigating a special case as a way to approach a problem whose solution for the general case is not immediately apparent.

Adaptive reasoning

A teacher or student with adaptive reasoning is able to recognize current assumptions and adjust to changes in assumptions and conventions. Adjusting to these changes involves comparing assumptions and working in a variety of mathematical systems. For example, because they are based on different assumptions, Euclidean and spherical geometries are structurally different. A person with adaptive reasoning, when introduced to spherical geometry, would consider the possibility that the interior angles of a triangle do not sum to 180°. Furthermore, he or she would be able to construct an example of a triangle, within the assumptions of spherical geometry, whose interior angles sum to more than 180°.

Examples of adaptive reasoning include the following:

1. Recognizing that division by an unknown is problematic.
2. Working with both common definitions for a trapezoid.
3. Operating in more than one coordinate system.

Productive disposition

Those with a productive disposition believe that they can benefit from engaging in mathematical activity and are confident that they can succeed in mathematical endeavors. They are curious and enthusiastic about mathematics and are therefore motivated to see a problem through to its conclusion, even if that involves thinking about the problem for an extended period of time so as to make progress. People with a productive disposition are able to notice mathematics in the world around them and apply mathematical principles to situations outside the mathematics classroom.

Examples of productive disposition include the following:

1. Noticing symmetry in the natural world.
2. Persevering through multiple attempts to solve a problem.
3. Taking time to write and solve a system of equations for a real-world application such as comparing phone-service plans.

Historical and cultural knowledge

Having knowledge about the history of mathematics is beneficial for many reasons. One prominent benefit is that a person with such knowledge will likely have a deeper understanding of the origin and significance of various mathematical conventions, which in turn may increase his or her conceptual understanding of mathematical ideas. For example, knowing that the integral symbol ∫ is an elongated *s*, from the Latin *summa* (meaning *sum* or *total*) may remind a person of what the integral function is. Some other benefits of historical knowledge include an awareness of which mathematical ideas have proven to be the most useful in the past, an increased ability to predict which mathematical ideas will likely be of use to students in the future, and an appreciation for current developments in mathematics.

Cross-cultural knowledge (i.e., awareness of how people in various cultures or even in various disciplines conceptualize and express mathematical ideas) may have a direct impact on a person's mathematical understanding. For example, a teacher or student may be accustomed to defining a rectangle in terms of its sides and angles, but people in some non-Western cultures define a rectangle in terms of its diagonals. Being able to conceptualize both definitions can strengthen one's mathematical proficiency.

The following are additional examples of historical and cultural knowledge:

1. Being familiar with the historical progression from Euclidean geometry to multiple geometric systems.
2. Being able to compare mathematicians' convention of measuring angles counterclockwise from horizontal with the convention (used by pilots, ship captains, etc.) of indicating directions in terms of degrees clockwise from North.
3. Understanding similarities and differences in algorithms typically taught in North America and those taught elsewhere.
4. Knowing that long-standing "open problems" in mathematics continue to be solved and new problems posed.

MATHEMATICAL ACTIVITY

Through a Mathematical Activity perspective, we focus on the ever-evolving mathematical actions within mathematical understanding. The perspective of Mathematical Activity organizes the verbs of doing mathematics—the actions one takes with mathematical objects. The three strands—mathematical noticing, mathematical reasoning, and mathematical creating—intertwine in Mathematical Activity.

Mathematical noticing

The first strand of Mathematical Activity, mathematical noticing,[2] involves recognizing similarities and differences in structure, form, and argumentation both in mathematical settings and in real-world settings. Mathematical noticing requires identifying mathematical characteristics that are particularly salient for the purpose at hand and focusing on those characteristics in the presence of other available candidates for foci.

Structure of mathematical systems. Noticing structure is foundational to making mathematical conclusions. Examples of the structure on which one focuses can be the definitions and axioms that govern a mathematical system. Noticing and using the structure of mathematical systems underpins other mathematical activities such as deriving properties of a system. Whereas many users of mathematics rely on these system, form, and argumentation structures, teachers need to notice similarities and differences among the structures in varied mathematical settings.

As students proceed through secondary school mathematics, the rate of introduction of new mathematical systems increases. Although the new systems use similar operations on similar objects, teachers need to be constantly vigilant regarding the constraints under which each system operates. Teachers need to notice invariant as well as changing aspects of mathematical structure as the curriculum moves from the study of rational numbers to the study of real and complex numbers, variables, polynomials, matrices, and functions.

Examples of noticing mathematical structure include the following:

1. Noticing the effects on a geometry when the parallel postulate is not assumed.
2. Being aware that familiar operations do not have the same meaning when applied to different mathematical objects and structures, and hence knowing not to generalize properties of multiplication over the set of real numbers to multiplication over the set of matrices.
3. Recognizing the entities of inverses and compositions across a broad range of mathematical settings.

4.   Noticing connections between (and features of) different methods for solving problems (e.g., noticing the structural similarities between the Euclidean algorithm and the long division algorithm).

5.   Noticing differences between the same objects in different systems (e.g., noticing the difference in solutions when solving an equation in the real number system and in the complex number system).

6.   Noticing differences in algebraic structure (e.g., noticing properties of a system such as the field properties, properties of equivalence relations, and properties of equality) and applying the knowledge of this structure to algebraic transformations.[3]

Symbolic form. Recognizing algebraic symbolic forms allows teachers to identify and use potential symbolic rules with those forms.

Examples of noticing symbolic forms include the following:

1.   Being aware that that the truth of $f(a) + f(b) = f(a + b)$ depends on the nature of the function f, and that students tend to apply this "student's distributive property"[4] indiscriminately.

2.   Noticing differences and similarities in notation and distinguishing among the meanings of notations that are similar in appearance (e.g., noticing differences in the uses of familiar notation such as the superscript -1—as in x^{-1} and f^{-1}—depending on context, and being able to identify and explain the conditions under which specific meanings for the notation are appropriate).

Form of an argument. Secondary teachers have a particular need to notice the specific form of mathematical arguments, whether advanced in a textbook or by a student. Noticing the form of a mathematical argument allows teachers to identify missing elements or redundant portions of the argument. Two examples of argument forms that secondary teachers find to be difficult for students are arguments by contradiction and arguments by mathematical induction.

Connect within and outside mathematics. Connecting within mathematics requires teachers to extract the characteristics and structure of the mathematics they are teaching and notice those characteristics and structure in other areas of mathematics. Teachers who notice connections between mathematical representations of the same entity and between mathematical entities and their properties can provide rich and challenging environments for their students. Such teachers are able to move smoothly from question to question, both fielding student questions and posing challenges that require students to connect mathematical ideas.

Examples of noticing connections within mathematics include the following:

1.   Noticing different manifestations and representations of the same mathematical system (e.g., noticing that paper-folding, symmetries of a triangle, paper-and-pencil games, and Escher-type drawings are all venues

for studying transformations such as reflections, rotations, translations, and glide reflections, and that transformations can be represented and manipulated through matrix operations as well as through mappings on a plane).

2. Recognizing relationships between alternative algorithms, student-generated algorithms, and standard algorithms (e.g., noticing that Peasant multiplication and standard multiplication algorithms can be derived using the field properties).

3. Noticing the affordances of the different representations and that different representations highlight different strengths and weaknesses for conveying information about what is being represented.

Connecting to areas outside of mathematics requires teachers to have a disposition to notice mathematics outside of their classroom and to seek mathematical explanations for real-world quantitative relationships. The point is not that there is some ordained list of applications that a teacher needs to know, but rather that there are intriguing topics that teachers can explore with their students by applying secondary school mathematics and that teachers should be willing and able to seek out the resources to investigate these topics. Connecting within and outside mathematics means looking for and noticing applications of mathematics as well as circumstances from which to extract mathematics, while at the same time recognizing the constraints that the context places on a mathematical result. Not every teacher may need to know something about a particular connection, but all teachers need to know the properties of the mathematical entities about which they are teaching well enough to recognize an application when they see it. This recognition involves seeing the properties of the mathematical entities well enough to match them to the situation (Zbiek & Conner, 2006).

Examples of noticing connections to the world outside of mathematics include the following:

1. Noticing the slope of lines comparing growth in linear relationships described in economics.

2. Noticing the impact of outliers in describing climate change.

3. Noticing the mathematics of decision-making that underpins video games.

4. Noticing the mathematics in models used in the trucking industry to distribute food or packages throughout the country.

5. Noticing the ways in which mathematics underpins today's electronic technology (e.g., noticing that video games employ matrix operations to animate images on the screen through geometric transformations).

6. Noticing ways in which mathematics underpins different industries (e.g., noticing that designers of automobiles use Bezier curves to render pictures of new designs for cars).

Mathematical reasoning

The second strand of Mathematical Activity is mathematical reasoning. Mathematical reasoning includes activities such as justifying and proving as well as reasoning in the context of conjecturing and generalizing. Mathematical reasoning results in the production of a mathematical argument or a mathematical rationale that supports the plausibility of a conjecture or generalization.

Justifying/proving. Teaching mathematics well requires justifying mathematical claims through logically deduced connections among mathematical ideas. Formal justification, or proof, requires basing arguments on a logical sequence of statements supported by definitions, axioms, and known theorems, whereas informal arguments involve reasoning from empirically derived—but often limited—data, reasoning by analogy, establishing plausibility based on similar instances, and the like. When creating formal or informal arguments, teachers need to be on the alert for special cases that they need in order to recognize or generate an exhaustive list of cases, and they need to recognize the limitations of reasoning from diagrams.

Teachers of secondary mathematics need a different type of justification ability from that of other users of mathematics because they are required to formulate and structure arguments across a range of appropriate levels. Teachers need to be comfortable with a range of strategies for mathematical justification, including both formal justification and informal arguments. Secondary school mathematics teachers need to be able to understand and formulate different levels and types of mathematically and pedagogically viable justifications and proofs (e.g., proof by contradiction and proof by induction). They also need to recognize the need to specify assumptions in an argument, and they must be able to state assumptions on which a valid mathematical argument depends. Teachers' arguments often need not be as elegant as those for which mathematicians typically strive, and teachers need to be able to create proofs that explain as well as proofs that convince (Hanna, 1989; Hersh, 1993).

Examples of justifying/proving include the following:

1. Constructing an array of justifications for why the sum of the first n natural numbers is $\frac{n(n+1)}{2}$, including appealing to cases, making strategic choices for pairwise grouping of numbers, and appealing to arithmetic sequences and properties of such sequences.

2. Arguing by contradiction (excluded middle): To prove that if the opposite angles of a quadrilateral are supplementary, then the quadrilateral can be inscribed in a circle, one can construct a circumcircle through three vertices of a quadrilateral and argue that if the fourth vertex can be in neither the interior nor the exterior of the circle, then it must be on the circumcircle, and therefore the quadrilateral can be inscribed in a circle.

Reasoning when conjecturing and generalizing. In school mathematics, students (and teachers) engage in developing conjectures based on their observations and data they have generated. Given a plausible conjecture—generated by the teacher or generated by students—a teacher must be able to test the conjecture with different domains or sets of objects.

Generalizing is the act of extending the domain to which a set of properties applies from multiple instances of a class or from a subclass to a larger class of mathematical entities, thus identifying a larger set of instances to which the set of properties applies. When generalizing, students may develop a formal argument intended to establish the generalization as being true, and that argument must be evaluated by the teacher. In some instances, the teacher needs to produce an argument that convinces students of a generalization's truth or explains some aspect related to the statement or its domain of applicability.

Teachers also engage in mathematical reasoning in the context of conjecturing and generalizing when they create and use counterinstances of generalizations. The creation of counterinstances requires one to reason about the domain of applicability of a generalization and what results when that domain is constrained or extended. For example, the generalization that multiplication is commutative can be shown to be false when considering matrix multiplication.

Examples of reasoning when conjecturing and generalizing include the following:

1. Analyzing the extent to which properties of exponents generalize from natural number exponents to rational number or real number exponents.
2. Recognizing that a graphical approach to solving polynomial equations is far more generalizable than the usual set of polynomial factoring techniques.
3. Generating counterinstances to challenge the frequent student conjecture that as the area of a rectangle increases, the perimeter increases.
4. Being able to determine the exceptions to the conjecture that two pairs of congruent sides and any pair of congruent angles ensure that two triangles are congruent.

Constraining and extending. Fundamental to mathematical reasoning is constraining, extending, or otherwise altering conditions and forms. Constraints can be removed, altered, or replaced to explore the resulting new mathematics. Mathematical relationships and properties can be tested for extended sets of numbers. As Cuoco, Goldenberg, and Mark (1996) argue, "Mathematicians talk small and think big" (p. 384). Teachers need to move flexibly between related small and big ideas. They use constraining, extending, and altering as ways to refine their creation of valid statements from intuitive notions and observations.

Teachers engage in mathematical reasoning when they consider the effects of constraining or extending the domain, argument, or class of objects for which a mathematical statement is or remains valid while preserving the structure of the mathematical statement. They need to recognize when it is useful to relax or constrain mathematical conditions. The mathematical reasoning involved in constraining and extending enables teachers to create extensions to given problems and questions.

With secondary school mathematics as the bridge between prealgebra mathematics and collegiate mathematics, secondary mathematics teachers often are challenged to explore the consequences of imposing or relaxing constraints. To *constrain* in mathematics means to define the limits of a particular mathematical idea. Constraints can be removed or replaced to explore the resulting new mathematics. When mathematicians tinkered with the constraint of Euclid's fifth postulate, new geometries were formed. When one removes the constraint of the plane in using Euclidean figures, the mathematics being used changes as well. Secondary mathematics teachers regularly encounter situations in which to provide a suitable response, they must tailor a generalization so that it can reasonably be extended to a larger domain of applicability. Teachers with an understanding of the mathematics their students will encounter in further coursework can structure arguments so that they can be extended to a more general case.

Examples of constraining and extending include the following:

1.   When finding the inverse of a function, one must sometimes constrain the domain if one wants the inverse to be a function as well. The inverse of $f(x) = \sin x$ is a function only if the domain of f is restricted.
2.   Extending the concept of absolute value to a modulus definition as the domain is extended from real to complex numbers.
3.   Extending the mathematical object, "triangle," from Euclidean to spherical geometry.
4.   Being cautious of extending the rules of exponents, developed and proved for natural number exponents, to negative, rational, real, or complex exponents.

Mathematical creating

Mathematical creating entails generating new ways to convey a mathematical object, new mathematical objects, and transformations of existing representations of a mathematical entity. The essence of mathematical creating is the production of new mathematical entities through the mathematical activities of representing, defining, and transforming.

Representing (new way to convey). Inherent in mathematical work is the need to represent mathematical entities in ways that reflect given structures, constraints, or properties. The creation of representations is particularly useful in creating and communicating examples, nonexamples, and counterexamples or counterinstances for mathematical objects, generalizations, or relationships. Teachers need to be able to fluently and rapidly construct representations that underscore key features of the represented entity.

Each representation affords different views of the mathematical object, but several different representations can highlight the same feature. Teachers need to be able to assess which features of the mathematical entity each form captures and which features it obscures. They also need to be able to recognize the ways in which a proposed representation may conflict with the features of the mathematical entity. Representing involves choosing or creating a useful form that conveys the crucial aspects of the mathematical entity that are needed for the task at hand.

Some types of mathematical representations are common. Teachers need to be able to create representations of those common forms in ways that reflect conventions. Teachers also need to be able to create less common, and even novel, forms of representation. In this activity, attention to structures, constraints, and properties is critical.

Defining (new object). The mathematical activity of defining is the creation of a new mathematical entity by specifying its properties. Generating a definition requires identifying and articulating a combination of a set of characteristics and the relationships among these characteristics in such a way that the combination can be used to determine whether an object, action, or idea belongs to a class of objects, actions, or ideas.

Teachers of secondary mathematics need to be able to appeal to a definition to resolve mathematical questions, and they need to be able to reason from a definition. They need to create definitions and to assess the definitions that students create or propose.

Modifying/transforming/manipulating (new form). Perhaps the most recognizable form of transforming is symbolic manipulation. Teachers need to see these transformations as purposeful activities undertaken to produce a symbolic form that conveys particular information. Transformations of graphs (e.g., window changes, adjustment of bin sizes) to create more meaningful representations are similarly important. Whether technology-supported or completed by hand, transforming one representation to another representation (of a similar or different form) is fundamental in solving problems.

Integrating strands of mathematical activity

Mathematical modeling provides one example of how the strands of Mathematical Activity intertwine in mathematical work. A popular description of the modeling process starts with a real-world problem that is translated into a formal mathematical system. Mathematical noticing occurs as the modeler specifies the conditions and assumptions that matter in the real-world setting. Devising the model requires mathematical creating informed by mathematical reasoning. After a potential model has been generated, mathematical creating takes over as the model is manipulated until a solution is found. The solution is mapped back to the real world to be tested with the problem through mathematical noticing. If a real-world conclusion does not align with the modeling goal, aspects of the model, such as initial conditions that are assumed, may be constrained, expanded, or altered to form a new model. It is important to note that the issue is one of extent of fit and utility rather than absolute correctness.

Mathematical modeling activities in secondary school might involve authentic modeling tasks that involve the generation of novel models or more restricted modeling work that is done in the service of students learning curricular mathematics, the mathematics that is the focus of classroom lessons (Zbiek & Conner, 2006). A close analysis of mathematical modeling across these contexts suggests that mathematical modeling is a nonlinear process that incorporates the three strands of Mathematical Activity (for an elaboration of modeling activities, see Zbiek & Conner, 2006).

MATHEMATICAL CONTEXT OF TEACHING

Not only should teachers of secondary mathematics be able to know and do mathematics themselves but also their mathematical understanding must prepare them for their particular context, facilitating their students' development of mathematical understanding. In Ryle's (1949) terminology, the context of teaching mathematics requires both knowing how and knowing that. It moves beyond the goal of establishing a substantial and continually growing understanding in mathematics for oneself as a teacher to include the goal of effectively helping one's students develop mathematical understanding. Mathematical understanding in the context of teaching mathematics enables teachers to integrate their knowledge of content and knowledge of processes to increase their students' mathematical understanding.

Probe mathematical ideas

Probing mathematical ideas requires investigating and pulling apart mathematical ideas. Mathematics is dense. One goal in doing mathematics is to compress numerous complex ideas into a few succinct, elegant expressions. Although mathematical efficiency and rigor are essential if one is to engage in complex mathematical thinking, they can also cause confusion, especially for those being newly initiated into the culture of mathematics.

Analyzing mathematical ideas also requires a broad knowledge of mathematical content and associated mathematical activities such as defining, representing, justifying, and connecting. Teachers need mathematical knowledge that will help them pull apart mathematical ideas in ways that allow the ideas to be reassembled as students mature mathematically. They need to recognize and honor the conventions and structures of mathematics and recognize the complexity of elegant mathematical ideas that have been compressed into simple forms.

Examples of analyzing mathematical ideas include the following:

1. Understanding the role of the domain in determining the values for which a function is defined.

2. Recognizing the similarities and differences between multiplying real numbers and multiplying matrices.
3. Exploring the various meanings of division—partitive and quotitive—to recognize that contexts for division involve more than the inverse of multiplication.

Access and understand the mathematical thinking of students

Mathematics teachers should understand how their students are thinking about particular mathematical ideas. A competent teacher uncovers students' mathematical ideas in a way that helps them see the mathematics from a learner's perspective. Teachers can gain some access to students' thinking through the written work they do in class or at home, but much of that information is highly inferential. Through discourse with students about their mathematical ideas, the teacher can learn more about the thinking behind the students' written products. Classroom interactions play a significant role in teachers' understanding of what students know and are learning. It is through a particular kind and quality of discourse that implicit mathematical ideas are exposed and made more explicit.

Students often discuss mathematics using vague explanations or terms that have a colloquial meaning different from their mathematical meaning. A teacher needs to be able to interpret imprecise student explanations, to help students focus on essential mathematical points, and to help them learn conventional terms. Success in such endeavors requires understanding the nuances and implications of students' understanding and recognizing what is right about their thinking as well as features of their thinking that are likely to lead them to unproductive conceptions. Achieving such a balance requires the teacher to have an extensive knowledge of mathematical terminology, formal reasoning processes, and conventions, as well as an understanding of differences between colloquial uses and mathematical uses of terms.

Examples of mathematical understanding needed for accessing students' mathematical thinking include the following:

1. Determining whether a student means *face* or *edge* (or something else) in using the word *side* in a discussion of Platonic solids.
2. Using a collection of useful representations (e.g., graph, table, drawing, or set of examples) that may help a student share mathematical ideas.
3. Designing mathematical tasks that expose students' thinking and maintain a high level of cognitive demand as the tasks are discussed.

Know and use the curriculum

Teachers use the curriculum to help students connect mathematical ideas and progress to a deeper and better grounded mathematics. How mathematical understanding is used to teach mathematics in a specific classroom or with a specific learner or specific group of learners is influenced by the curriculum that helps organize teaching and learning. A teacher's mathematical proficiency and mathematical activities can help to make that curriculum meaningful, connected, relevant, and useful.

Mathematical understanding related to the mathematics curriculum requires a teacher to identify foundational or prerequisite concepts that enhance the learning of a concept. In addition, the teacher needs to understand how the concept being taught can serve as a foundational concept for future learning. The teacher needs to know how the concept fits learning trajectories. The teacher also needs to be aware of common mathematical misconceptions and how those misconceptions may sometimes arise at particular points in this trajectory. Although learning trajectories describe paths for learning mathematics, mathematics teachers need to understand that there is not a single fixed or linear order for learning mathematics but rather multiple ways to approach a mathematical concept and to revisit it. Mathematical concepts and processes evolve in the learner's mind, becoming more complex and sophisticated with each iteration. Mathematical understanding prepares a teacher to enact a curriculum that not only connects mathematical ideas explicitly but also develops a disposition in students so that they expect mathematical ideas to be connected and an intuition so that they see where those connections might be (Cuoco, 2001).

A teacher with mathematical understanding of the context of teaching understands that a curriculum contains not only mathematical entities but also mathematical processes for relating, connecting, and operating on those entities (National Council of Teachers of Mathematics, 1989, 2000). A teacher must have such understanding to set appropriate curricular goals for his or her students (Adler & Davis, 2006).

Examples of knowing and using the curriculum include the following:

1. Understanding the concept of area in a way that includes ideas about measure, descriptions of two-dimensional space, measures of space under a curve, measures of the surface of three-dimensional solids, infinite sums of discrete regions, operations on space and measures of space, foundations of the geometric properties of area, and useful applications involving area.
2. Selecting and teaching functions in a way that helps students build a basic repertoire of function types (Even, 1990).

Assess the mathematical knowledge of learners

Assessing the mathematical knowledge of learners is an integral component of the context of teaching mathematics. During each class, teachers must assess or evaluate students' mathematical understanding. Such assessment is crucial not only for recognizing student error but also in determining the level of students' mathematical understanding for purposes of developing tasks and planning lessons. Assessing students' mathematical knowledge involves much more than assessing a student's ability to follow a procedure. Teachers should possess a mathematical understanding for teaching that helps them identify the essential components of mathematical concepts so they can, in turn, assess a student's ability to use and connect these essential ideas. Determining how students are progressing in class is at the heart of assessing the mathematical knowledge of learners. To determine the mathematical progress of their students, teachers must be attentive to the errors students frequently make.

Examples of assessments that teachers frequently perform include the following:

1. Realizing that a mathematical error may be located in how the student is using and understanding mathematical language, or colloquial language such as "canceling out."
2. Choosing examples for finding solutions of quadratic equations graphically so that the set of examples includes equations with no, one, and two solutions.
3. Using open-ended questions to draw out the source of the student's confusion about the difference between the area of a circle and its circumference.
4. Recognizing the common error of finding the reciprocal or multiplicative inverse of the algebraic expression that defines an $\Re$-to-$\Re$ function when asked to find the inverse of the function.

Reflect on the mathematics in one's practice

Teachers should be able to analyze and reflect on their mathematics teaching practice in a way that enhances their mathematical understanding. There are many ways to reflect on one's practice, and one of the most important is to use a mathematical lens. How did the mathematical complexity of the problem in this lesson change when the students were given a hint? Which of several equivalent definitions is most appropriate when this term is introduced? How was the topic of that test question connected to a topic treated earlier in the course? Thoughtful reflection on problems of practice can be reconsideration of a lesson just taught, or it can be part of the planning for a future lesson. It may occur as the teacher interprets the results of a formal assessment, or it may be prompted by a textbook treatment of a topic.

Teachers often reflect on their teaching as they teach—as they are making split-second decisions. A teacher's decisions about how to proceed after accessing student thinking depend on many factors, including the mathematical goals of the lesson. It is valuable to revisit these quick reflections and decisions when there is time to think about the mathematics one might learn from one's practice.

Examples of reflection on the mathematics in one's practice include the following:

1. Identifying an unconventional notation that students are using and contrasting its properties with those of conventional notation.
2. Analyzing how the topic of a lesson might be presented in a way that shows mathematics as culturally situated.
3. Modifying a mathematical conjecture so that a statement could be proved in other ways.

By focusing on Mathematical Proficiency, Mathematical Activity, and the Mathematical Context of Teaching, the MUST Framework is designed to provide a stereoscopic view of the mathematical understanding a secondary mathematics teacher ought to have. In the remainder of this volume, we attempt to fill out that view with examples of Situations and their uses.

NOTES

1. The MUST Framework as described in this chapter was the product of the work of mathematics education faculty and graduate students at the two institutions (Pennsylvania State University and University of Georgia). In addition to the authors of this chapter, individuals who participated in some way in the development of the framework are Bob Allen, Shawn Broderick, Sarah Donaldson, Kelly Edenfield, Ryan Fox, Brian Gleason, Heather Johnson, Shiv Karunakaran, Hee Jung Kim, Evan McClintock, and Sharon O'Kelley.
2. Although our notion of mathematical noticing may have some features in common with Goodwin's (1994) *professional noticing,* the terms are not equivalent, and one is not derived from the other.
3. Algebraic transformations such as the production of equivalent expressions and equivalent equations are at the core of many school algebra courses.
4. For further discussion of the "student's distributive property," see Skane and Graeber (1993).

REFERENCES

Adler, J., & Davis, Z. (2006). Opening another black box: Researching mathematics for teaching in mathematics teacher education. *Journal for Research in Mathematics Education, 36,* 270–296.

Cuoco, A. (2001). Mathematics for teaching. *Notices of the AMS, 48,* 168–174.

Cuoco, A., Goldenberg, E. P., & Mark, J. (1996). Habits of mind: An organizing principle for mathematics curricula. *Journal of Mathematical Behavior, 15,* 375–402. doi:10.1016/S0732-3123(96)90023-1

Even, R. (1990). Subject matter knowledge for teaching and the case of functions. *Educational Studies in Mathematics, 21,* 521–544. doi:10.1007/BF00315943

Goodwin, C. (1994). Professional vision. *American Anthropologist, New Series, 96*(3), 606–633. doi:10.1525/aa.1994.96.3.02a00100

Hanna, G. (1989). Proofs that prove and proofs that explain. *Proceedings of the International Group for the Psychology of Mathematics Education, 2*, 45–51.

Hersh, R. (1993). Proving is convincing and explaining. *Educational Studies in Mathematics, 24*, 389–399. doi:10.1007/BF01273372

Kilpatrick, J., Swafford, J., & Findell, B. (Eds.). (2001). *Adding it up: Helping children learn mathematics*. Washington, DC: National Academy Press.

National Council of Teachers of Mathematics. (1989). *Curriculum and evaluation standards for school mathematics*. Reston, VA: Author.

National Council of Teachers of Mathematics. (2000). *Principles and standards for school mathematics*. Reston, VA: Author.

Ryle, G. (1949). *The concept of mind*. Chicago, IL: University of Chicago Press.

Shulman, L. (1986). Those who understand: Knowledge growth in teaching. *Educational Researcher, 15*(2), 4–14. doi:10.3102/0013189X015002004

Skane, M. E., & Graeber, A. O. (1993). A conceptual change model implemented with college students: Distributive law misconceptions. In *The Proceedings of the Third International Seminar on Misconceptions and Educational Strategies in Science and Mathematics*. Ithaca, NY: Misconceptions Trust.

Zbiek, R. M., & Conner, A. (2006). Beyond motivation: Exploring mathematical modeling as a context for deepening students' understandings of curricular mathematics. *Educational Studies in Mathematics, 63*, 89–112. doi:10.1007/s10649-005-9002-4

CHAPTER 3

DEVELOPMENT OF PRACTICE-BASED SITUATIONS

Patricia Wilson and Rose Mary Zbiek

Two of the defining characteristics of the Mathematical Understanding for Secondary Teaching (MUST) Framework are its grounding in teaching practice and its focus on mathematics teaching at the secondary level. The strategy we used for remaining faithful to practice at the secondary level was to observe and record interesting mathematical events that took place in mathematics classes taught in Grades 6–12. These events often involved noteworthy student questions, teacher-created tasks, or mathematical conjectures and discussions. We recognized that interesting, relevant discussions took place in school hallways or teachers' workrooms as well as in university classes for students preparing to teach mathematics at the secondary level. Records of these events were turned into formal documents called Situations, which became the data that informed the construction of the MUST Framework. Because it is based on data from events that actually happened, the MUST Framework captures and organizes mathematical understanding that can be useful for secondary teachers in their practice.

Mathematical Understanding for Secondary Teaching: A Framework and Classroom-Based Situations, pages 31–40.

STRUCTURE OF A SITUATION

A Situation is a formal document that contains three main parts: a Prompt, a set of Foci, and a Commentary. Each of the components is discussed in the following sections. The set of Situations our teams developed can be found in chapters 7 through 49.

Prompt

The Prompt is the heart of the Situation and the direct tie of the Situation to practice. A Prompt, as described in Figure 3.1, should stimulate interesting mathematical discussion and may arise from student and teacher insights as well as from difficulties and questions. A Prompt can be viewed from many levels and it often takes multiple readings to appreciate the complexity of the mathematics that would be useful to understand if one is to make the most of the opportunity to bring varied mathematical ideas to bear on events like that captured by the Prompt.

The Prompt in Figure 3.2 is authentic and most algebra teachers can relate to this event, which is the basis of the Quadratic Equations Situation (Situation 20 in chapter 26). It is also good because it can be unpacked to expose important mathematics that generalizes to many events in mathematics classrooms at the secondary level. A quick reading of this Prompt suggests that this event is about understanding an algorithm for solving quadratic equations, but subsequent readings expose the importance of comprehending what it means to solve a quadratic equation and acquiring a variety of ways to think about solutions and processes for arriving at solutions, which opens the door to a variety of Foci.

Set of Foci

The creation of a Prompt is followed by the development of its Foci, each of which conveys mathematical understanding for teaching as part of the Prompt's set of Foci (as noted in Figure 3.3). The goal of a Focus is to unpack the mathematics that may be implicit in the Prompt and make explicit the mathematics that could be useful to a teacher who experiences events related to content that is similar to that described in the Prompt. A group of mathematics educators discussed

---

Prompt. The Prompt sets the stage for the mathematics of the Situation by briefly describing an event from teaching practice. Teaching practice includes preparing, implementing, and reflecting on classroom instruction. The instruction can be either in secondary mathematics or in the preparation and professional development of secondary teachers.

FIGURE 3.1. A working description of a Prompt.

Prompt

In an Algebra 1 class some students began solving a quadratic equation as follows:

Solve for x:

$$x^2 = x + 6$$

$$\sqrt{x^2} = \sqrt{x+6}$$

$$x = \sqrt{x+6}$$

They stopped at this point, not knowing what to do next.

FIGURE 3.2. A Prompt from the Quadratic Equations Situation (Situation 20 in chapter 26).

the Prompt, selected mathematics to form into a Focus that promotes mathematical understanding, and produced an initial collection of Foci. In general, a Focus was designed to communicate with readers who have, or are obtaining, the equivalent of an undergraduate degree in mathematics or mathematics education.

A Prompt can motivate many Foci, but we selected only a few Foci to make a manageable and optimal set for each Prompt. Each Focus should bring something new and different to the set, but it is reasonable that there may be some overlapping ideas among Foci. Selecting a few Foci for a Situation was challenging. The Foci taken collectively were not meant to be representative of the curriculum, but they were selected to illustrate a broad range of mathematical ideas that connect to the content present in the Prompt.

The mathematical topics of the Foci for the Quadratic Equations Situation (Situation 20 in chapter 26) include such things as meanings of *absolute value*, differences between an unknown and a variable, meanings of *solution*, and the

---

Focus. A Focus presents a particular aspect of mathematical understanding for teaching at the secondary level that is relevant to the Prompt. Mathematical understanding for teaching includes concepts, processes, representations, solution methods, interpretations, types of reasoning, properties of mathematical objects, and definitions.

Set of Foci. A set of Foci provides examples of the range and depth of mathematical ideas associated with the Prompt.

FIGURE 3.3. Working descriptions for a Focus and set of Foci.

importance and uniqueness of the zero-product property. For example, in Focus 1, we highlight the primacy of equivalent equations in the process of solving equations and connect the ideas to one meaning of absolute value and to the concept of *solution of an equation*. We point out the zero-product property and its relationship to factoring while considering standard methods of solving quadratic equations by factoring and by employing the quadratic formula in Foci 3 and 4, respectively. In Focus 5, we move to a graphical approach. The set of Foci both explore and expand common methods to solve equations as well as attend to an important property and to the meaning of *solution* and *absolute value*.

The Foci address relevant mathematics but do not address related pedagogical issues. The absence of pedagogical elaborations is not to imply they do not matter but rather to underscore our goal of producing the Situations to develop the MUST Framework. It also honors how events similar to the Prompt may occur in a variety of classes, though the context of each event will be different. The goal of a Focus is not to suggest what a teacher might *do* but rather to explicate *mathematical understanding* that is useful to a teacher. What the teacher decides to do with the mathematical understanding will vary greatly depending on students and school setting.

Each Focus has an introductory, indented statement that emphasizes the main point of the Focus. The indented statements for the Solving Quadratic Equations Situation (Situation 20 in chapter 26) appear in Figure 3.4. A complete set of the Foci for the Situation can be found in chapter 26. The indented statements quickly

---

Mathematical Focus 1

Factoring and using the zero-product property can be used to solve many quadratic equations.

Mathematical Focus 2

All quadratic equations can be solved by completing the square or by using the quadratic formula.

Mathematical Focus 3

A geometric analogy to an area model can be used to represent quadratic equations and their solutions.

Mathematical Focus 4

Solving an equation through algebraic manipulation requires that equivalence is maintained between each form of the equation.

Mathematical Focus 5

Approximate solutions to equations can be found by graphically determining the zeros of the associated function.

FIGURE 3.4. A list of the Mathematical Foci in the Solving Quadratic Equations Situation (Situation 20 in chapter 26).

Mathematical Focus 5

Approximate solutions to equations can be found by graphically determining the zeros of the associated function.

The solutions to an equation in which an expression involving x is equal to 0 (such as $x^2 - 4 = 0$) are comparable to the zeros of a function of x (such as $f(x) = x^2 - 4$). This is because a zero, or x-intercept, of a function is an x-value for which the value of $f(x)$ is 0.

For example, the solutions to $x^2 - 4 = 0$ are $x = 2$ and $x = -2$; and the zeros (the x-intercepts) of the graph of f where $f(x) = x^2 - 4$ are $x = 2$ and $x = -2$. The equation $x^2 - 4 = 0$ and the function f with rule $f(x) = x^2 - 4$ are not the same, as x is an unknown in the equation (represents a specific value), whereas in the function rule, x is a variable (changing quantity). However, the equation and the function are related: The solutions of the equation are the same as the zeros of the function.

Because solutions to equations and zeros of functions are related in this way, graphing a function can be a useful method for solving an equation. However, as noted previously, if the strategy is to find the zeros of the function, the accompanying equation must be equal to 0. In this Situation, this will involve performing equivalence-preserving operations on the equation so that one member is 0:

$$x^2 = x + 6$$
$$x^2 - x - 6 = 0$$

The graph of $f(x) = x^2 - x - 6$ (see Figure 26.1) will indicate, by its zeros, approximate solutions of $x^2 - x - 6 = 0$.

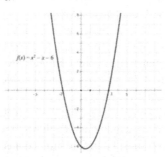

$f(x) = x^2 - x - 6$

Figure 26.1. A graph of $f(x) = x^2 - x - 6$.

A similar method requires graphing the functions f and g with rules $f(x) = x^2$ and $g(x) = x + 6$ (that is, treat each side of the original equation as a function) and determine the points of intersection of the two functions. These are the points at which x^2 and $x + 6$ are equal.

FIGURE 3.5. Focus 5 from the Solving Quadratic Equations Situation (Situation 20 in chapter 26).

convey the essence of the Foci and provide a summary of the set of Foci. For example, the indented statements in Figure 3.4 suggest that the Foci for the Situation, among other things, address several strategies for solving quadratic equations.

Within a set of Foci, a particular Focus can take a mathematical direction that complements the mathematics explicit in the Prompt, such as by exploring a common algorithm, using a different area of mathematics, or employing a different type of representation. For example, Focus 5 is one of five Foci in the Solving Quadratic Equations Situation (Situation 20 in chapter 26). It is an example of how a Focus can address a Prompt with an algebraic or symbolic representation by discussing mathematical understanding of quadratic equations from a graphical point of view. The entire Focus appears in Figure 3.5.

Commentary

A Situation can include a Commentary (as described in Figure 3.6) before the set of Foci or after the set of Foci. The Solving Quadratic Equations Situation (Situation 20 in chapter 26) has two Commentaries. Regardless of its placement, a Commentary provides additional information about the nature of the set of Foci that were chosen. For example, the (first) Commentary shown in Figure 3.7a provides information about the set of Foci and offers a rationale for the selection of Foci. As previously explained, the selection of just a few Foci is difficult and the Commentary offers an opportunity to provide additional information to stimulate the thinking of the reader and to help the reader understand what is important about the selected set as well as what might be missing from the set. For example, the Postcommentary for the Solving Quadratic Equations Situation (Situation 20 in chapter 26) in Figure 3.7b explains how *domain* is central to equation solving and can be used to explain why extraneous roots arise when various procedures are applied. A Commentary also might discuss how the ideas in the Foci could apply or generalize to a mathematical context different from that of the Prompt. The Postcommentary in Figure 3.7b extends extraneous solutions from quadratic and radical settings to logarithmic contexts.

Taken collectively, the Prompt, Foci, and Commentary form a Situation that presents an authentic teaching context and explores mathematics that is useful in teaching secondary mathematics. Each Situation connects multiple mathematical ideas to the context, by virtue of how the Situations were created.

Commentary. The Commentary describes the rationale for or explains the importance of the mathematics in the set of Foci.

FIGURE 3.6. A working description of Commentary.

Commentary

This Situation highlights some issues concerning solving equations (regarding both quadratic equations and equations in general) that are prevalent in school mathematics.

Focus 1 and Focus 2 present two typical methods of solving a quadratic equation: factoring and the quadratic formula. These are included because this Prompt illustrates the importance of having accurate and dependable means by which to attempt to solve quadratic equations. Focus 3 provides a geometric approach for solving $x^2 = x + 6$. Focus 4 provides guidelines for solving any algebraic equation and emphasizes maintaining equivalence. Focus 5 shows the relationship between the solution(s) of an equation and the zero(s) of a function. It contains a graphical approach to solving quadratic equations. The Situation ends with a Postcommentary on the occurrence of extraneous solutions.

(*a*)

Postcommentary

Often, in solving equations, extraneous solutions result. These occur when the original domain is expanded during the course of the solution process.

It was stated previously that $x^2 = x + 6$ and $x = \sqrt{x+6}$ could not be equivalent because they have different solution sets, specifically, the former has two solutions and the latter has only one. To see that $x = \sqrt{x+6}$ has only one solution, consider solving the equation using the principles in Focus 4.

First, however, attention must be given to the domain of x for $x = \sqrt{x+6}$. For $x = \sqrt{x+6}$ to be defined in the real numbers, $x \geq -6$. But if $x = \sqrt{<\text{some expression}>}$, $x \geq 0$. This means that the domain is the intersection of $x \geq -6$ and $x \geq 0$; thus, the domain is $x \geq 0$.

The inverse operation for taking the square root is squaring. Squaring both members of $x = \sqrt{x+6}$ yields the original equation ($x^2 = x + 6$), and the two solutions are 3 and -2. Note that in this original form of the equation, the domain is the set of real numbers; there are no restrictions on the values of x. However, in $x = \sqrt{x+6}$, $x \geq 0$, so the only solution is $x = 3$. Negative 2 is called an *extraneous solution*. It was introduced by expanding the valid domain from the nonnegative real numbers to all real numbers.

(*b*)

FIGURE 3.7. Example of a Commentary (a) and Postcommentary (b), from the Solving Quadratic Equations Situation (Situation 20 in chapter 26). (continues)

Note that different equations will have extraneous solutions as a result of expanding other domains to the real numbers. Consider $\log_3(5x-12)+\log_3 x = 2$. From the first term, $x > \frac{12}{5}$. From the second, $x > 0$. The intersection of these two restrictions is $x > \frac{12}{5}$. Solving this equation yields two solutions because the problem is converted to solving a quadratic equation whose variable has a domain of all real numbers. This domain expansion introduces a possible extraneous solution.

$$\log_3(5x-12)+\log_3 x = 2$$
$$\log_3\big((5x-12)x\big) = 2$$
$$5x^2 - 12x = 9$$
$$5x^2 - 12x - 9 = 0$$
$$(5x+3)(x-3) = 0$$
$$x = -\frac{3}{5} \text{ or } x = 3$$

In fact, the first solution is not in the original domain and is an extraneous solution. The only solution for this logarithmic equation is $x = 3$.

FIGURE 3.7. Continued

PROCESS FOR CREATING SITUATIONS

The process of creating Situations began with collecting events and using them to create Prompts that stimulated good mathematical discussions. Many of the events were witnessed by supervisors of student teachers as they visited schools and were able to observe classes taught by mentor teachers as well as student teachers. University supervisors were included in hallway discussions and planning sessions. They often were able to capture the verbatim conversation. As we proceeded, we realized that there were similar, stimulating events that were remembered from previous interactions with mathematics classes at the secondary level and in university courses preparing teachers to teach mathematics at the secondary level.

A characteristic feature of all the Prompts is that they came from actual events and were not altered to highlight any specific mathematics. Although the Prompts were not altered to convey particular mathematics, we chose Prompts around which to build complete Situations so that the set would represent the strands of secondary school mathematics. We note that the set might emphasize some areas, such as algebra and function. Given the large number of algebra courses needed to meet states' expectations regarding algebra in the secondary school curriculum and the frequency with which student teachers and interns are assigned to algebra classes, it is not surprising that our set of observed events was heavy in examples from algebra and functions.

The next step in creating a Situation was to produce the Foci. A team of mathematics educators, including faculty members and doctoral students, began to pose and create Foci that would illustrate important mathematics. The goal of a Focus was to clearly connect secondary school or college mathematics to the Prompt in a way that would be informative to a teacher faced with the event, and not to provide a response to the Prompt. Each Focus was reviewed, revised, and discussed in multiple sessions.

Most Prompts stimulated the production of a large set of Foci, and then the selection of a diverse, small set of Foci began. Although we were tempted to include 10 to 12 Foci for each Situation, we saw the importance of culling the set of Foci to approximately 4 or 5 critical Foci that helped anyone using a Situation to see key components of mathematics that would be part of useful and connected mathematical understanding for teaching secondary school mathematics. With an eye toward taking different mathematics paths, we blended Foci with related content and then decided which Foci would be most helpful in describing the teacher's mathematical understanding that would make the most of the learning opportunity. Again, there was extensive debate and often further revision of the set of selected Foci.

After the set of Foci was determined, we worked to create an italicized topical sentence for each Focus to capture the essence of mathematical understanding described in the Focus. We used full sentences because our goal was to convey mathematical ideas and not simply to name mathematical topics. A Commentary was then written for the set of Foci, and occasionally we found that it was necessary to include a second Commentary—following the Foci—to extend the discussion.

We called upon mathematicians, statisticians, mathematics educators, supervisors of mathematics, and teachers not involved in their creation to vet the set of Situations. Some Foci were revised again to increase their clarity or to include a point that was not made explicit.

At all stages of development, we attended carefully to the correctness of the mathematics and the relevance of the mathematics to curricula at the secondary level. The entire set of Situations represents a broad, but certainly not complete, description of mathematical understanding that is useful to teachers at the secondary level. The set of Situations was sufficiently broad to allow us to develop the MUST Framework, which is described in chapter 4.

ATTRACTIVENESS OF THE SITUATIONS

The primary function of the Situations was to serve as data from actual mathematics instruction that would identify mathematical understanding that is useful at the secondary level. Creating Situations was part of the process of building the MUST Framework. Because the Situations arise from teaching practice and target important mathematics, it was quite natural for those who were creating Situations to use them in their classes and in professional development sessions. As the Situations were used, the Foci became better articulated and their use also helped the developers to select the critical Foci to be included. It was not expected that those who reviewed the Situations would become vested in using the Situations in instruction. However, the Situations were viewed as particularly useful in preparing mathematics teachers at the secondary level and alternative uses soon emerged. A discussion of a variety of uses of Situations is the topic of chapter 4.

CHAPTER 4

FROM SITUATIONS TO FRAMEWORK

Patricia Wilson and M. Kathleen Heid

The process of creating the Framework for Mathematical Understanding for Secondary Teaching (MUST) based on our Situations has been a challenging task. Any such Framework needs to capture the dynamic nature of teachers' use of mathematics in their teaching rather than simply to list knowledge, skills, and proficiencies for teaching mathematics. The knowledge structure of competent mathematics teachers is much more complex than a specific list of the mathematics that teachers need to know and the mathematical processes that they need to be able to perform, and trying to articulate such a list is likely to be an unproductive venture. Instead, we have designed a framework that showcases important components of mathematical understanding that can be useful in teaching mathematics at the secondary level. The components and the strands that comprise them are not disjoint but rather provide complementary lenses through which to view mathematical understandings that can serve teachers of mathematics at the secondary level.

The framework we have built—with the help of the myriad mathematics educators, mathematicians, and teachers who participated at various points in the process—describes components and strands of mathematical understanding that, we believe, need to be central to mathematics teacher education at both the pre-service and in-service levels, because these components are useful to teachers as

Mathematical Understanding for Secondary Teaching: A Framework
and Classroom-Based Situations, pages 41–55.
Copyright © 2015 by Information Age Publishing

41

they continue to learn mathematics and to facilitate the learning of their students. We have intentionally avoided claiming that teachers need specific mathematics. Rather, we believe that the mathematics described in the components can be useful to teachers of mathematics. A lack of familiarity with a particular piece of mathematics can be overcome by a robust knowledge of other related pieces of mathematics as well as through good classroom argumentation and discussion.

This chapter describes the evolution of the MUST Framework. We hope that the description of the process helps to explain the derivation and the substance of the three major perspectives of Mathematical Proficiency, Mathematical Activity, and Mathematical Context of Teaching, and encourages the reader to consider the multiple, overlapping dimensions of mathematical understanding for secondary teaching.

WORKING FROM THEORY AND PRACTICE

Revisiting goals

As noted in chapter 1, we have attempted to advance previous work, trying to identify the mathematical knowledge that is useful to teachers by focusing on their constantly evolving, dynamic mathematical understanding. Throughout the process, we worked to remain faithful to our goal of basing the framework on actual classroom practice. Ours was a grounded process that acknowledged the importance of situating the framework in the context of mathematics teaching and learning at the secondary level. We were committed to building a framework about mathematical understanding in contrast to frameworks that focus on pedagogical components. Although pedagogical understanding is an extremely important part of the complex puzzle of good mathematical instruction, we believed that including pedagogical issues in a framework for mathematical understanding could detract from the emphasis on the mathematical issues. As a consequence, we began by striving for a framework that focused strictly on mathematics and tried to highlight the mathematics within constructs such as *pedagogical content knowledge* proposed by Shulman (1986). In addition, we wanted the MUST Framework to help to distinguish mathematical understanding for teaching from mathematical understanding for other purposes and professions. We considered mathematical understanding for teaching to be specialized mathematical understanding. Our four foundational goals were (a) to build on previous work by creating a framework that focused on dynamic processes as well as on content, (b) to construct our framework grounded in practice of teaching mathematics at the secondary level, (c) to focus on mathematics rather than pedagogical concerns, and (d) to distinguish core mathematical understanding that is particularly useful for teaching. In the following sections, we describe the process of developing the framework, including the avenues we pursued as well as those we explored and abandoned.

Working from practice

Driven by our goal to build a framework grounded in the dynamic practice of mathematics teaching at the secondary level, we developed descriptions of what we saw as we visited mathematics teachers at work in secondary schools as well as university instructors preparing mathematics teachers for work at the secondary level. We began collecting written prompts from our observations and constructing the Situations as described in chapter 3. Having produced, revised, vetted, and revised again more than 40 Situations, our goal was to use those Situations as data from which to develop a framework for Mathematical Understanding for Secondary Teaching.

Our initial analysis of the Situations used a range of perspectives to describe mathematical knowledge for teaching at the secondary level. At various times, we considered different perspectives such as:

- Content knowledge based on the categories developed in the *Mathematical Education of Teachers* document (Conference Board of the Mathematical Sciences, 2001),
- Connections between mathematical ideas,
- Ways to use technology in doing mathematics,
- Understanding students' mathematical thinking, and
- Understanding applications of mathematics.

In another early approach to analyzing data from the classroom observations, we focused on mathematical ideas that teachers were using or could be using to promote mathematical understanding. We recorded features of mathematics that occurred within instruction such as: vocabulary, representations, conceptions and misconceptions, notation, definitions, and mathematical structure.

From the start, it was evident that there were many ways to build a framework and many ways to view the dynamic construct of mathematical understanding. As we analyzed Situations that arose from visiting classrooms and talking with teachers, salient features emerged. Secondary curricula, especially at the high school level, tend to be structured around a static list of topics to be addressed, and we did not want to be trapped into a list that resembled a scope and sequence of topics. We started with broader mathematical lenses: Mathematical Objects, Big Mathematical Concepts, Mathematical Processes, and Mathematical Structures. A lens of Mathematical Objects would start with the school curriculum but address the larger structure of school mathematics. It would be organized around mathematical objects and their properties, operations, and relationships across objects. For example, a mathematical object that is identified using this lens might be that of function, with properties that include end behavior, operations that include composition, and relationships that include functions with similar rates of change. A lens on Big Mathematical Concepts would center on overarching mathematical concepts such as equivalence, similarity, invariance, linearity, isomorphism, and function. A lens on Mathematical Processes would focus on the actions on mathematical objects involved in doing mathematics (e.g., representing, justifying, and generalizing). A lens on Mathematical Structures would focus on structures involved in reasoning in a closed system and types of mathematical actions such as determining applicable domains for generalizations and identifying consistencies across representations (e.g., inverse as related to a set and an operation on that set, extending the rules for multiplication of integral exponents to rational exponents).

An important part of mathematics instruction is the set of mathematical actions taken by the teacher, and we needed to account for what teachers could do and may do with mathematics in the course of their teaching. Based on the actions that were suggested in the Situations, the following are examples of mathematical actions of teaching that we sought a framework to account for:

1. *Creating a counterexample.* For example, use matrix operations as a counter to the claim that the commutative property holds for any operation called multiplication regardless of the mathematical entities on which it operates. To accomplish this, teachers need to know the properties of the objects and operations involved in the counterexample and variations of those properties.

2. *Creating an example or nonexample.* For example, a teacher may need to create a quadratic polynomial that does not factor over the real numbers in order to illustrate quadratic ($\mathbb{R} \to \mathbb{R}$) functions whose graphs do not intersect the axis for the independent variable (often the x-axis), or use limits or derivatives to generate a symbolic rule for a function whose graph has particular preidentified characteristics.

3. *Fitting a question in a larger setting in order to identify a special case of a broader category of mathematical objects.* For example, a teacher may prove that quadrilaterals with supplementary opposite angles can be circumscribed by a circle and continue to illustrate that not all quadrilaterals can be circumscribed by asking for nonexamples.

4. *Explaining why a process does not generalize when trying to apply the process to a different entity.* For example, use the "students' law of distributivity" (e.g., claiming $\sin(a + b) = \sin(a) + \sin(b)$) to demonstrate that the distributive property of multiplication over addition does not necessarily generalize to other functions and operations.

As we examined and classified actions involved in the mathematical work of teaching mathematics, we recognized their dependence on two phenomena: *mathematical actions* and *mathematical entities*. Any task for teaching mathematics can be characterized by combining actions with entities. Entities can be subdivided into smaller units. We came to think of the actions and entities as the nouns and verbs of the Situations. The lists of actions and entities that we identified in the Situations appears in Figure 4.1.

ACTIONS:

Apply, Choose, Classify, Connect, Create, Demonstrate, Evaluate, Explain, Extend, Identify, Investigate, Order, Prove, Recall, Recognize, Refute, Simplify, Use

ENTITIES:

Mathematical Arguments (Diagrams, Domains, Expressions, Functions, Numbers, Patterns, Shapes)

Representations (Algebraic, Geometric, Numerical, Verbal)

Statements (Conclusions, Conditions, Conjectures, Counterexamples, Definitions, Examples, General arguments, Nonexamples, Questions)

Deductive Arguments (Algorithms, Conventional practices, Deductive arguments, Informal reasoning, Mathematical induction, Procedures, Strategies)

Connections (Outside mathematics, Inside mathematics)

FIGURE 4.1. Actions and entities identified in Situations.

Over the course of our work on the Situations, we generated lists of mathematical actions of teaching mathematics at the secondary level—lists that later formed a database for the development of one of the major components of the framework. One source for entries on those lists was the record of conferences and meetings we held. For our first conference, we convened a group of experts in the area of mathematical knowledge for secondary teaching. They were people who had written curricula or expository pieces for teachers on the topic, and some of the discussion at that conference was directed toward describing mathematical actions of teaching mathematics at the secondary level. On another occasion, members of the research team met in small groups to generate a list of ways in which teachers draw on their mathematical knowledge in the course of teaching mathematics at the secondary level. Notes and transcripts from both of those meetings gave us a list of mathematical actions—what, at first, we called *mathematical tasks of teaching*. We sorted those actions into the categories shown in Figure 4.1. Although the actions we classified were mathematical, the verbs under which we classified them were not specifically mathematical. As a consequence, we established a goal to limit the definition of each verb so that it described mathematical actions. For example, the verb "recognizing" is not limited to a mathematical action, although the actions on our list that involved recognizing were mathematical. Using mathematical actions that involved recognizing as an anchor, the verb, "recognize," was limited to *"Recognizing mathematical properties, constraints, or structure in a given mathematical entity or setting, or across instances of a mathematical entity."* This mathematically delimited definition of the Recognizing category accounted for the following mathematical actions that were included in the list we had generated:

- Recognizing an exhaustive set of cases,
- Recognizing limitations of reasoning from diagrams,
- Recognizing when a set of constraints defines a unique case and when it defines a set of cases,
- Seeing and recognizing structure,
- Having awareness of the importance of mathematical structure (set and operations on a set),
- Identifying special cases, and
- Distinguishing when certain properties hold and when they do not.

The definitions of mathematical actions were useful in helping us make progress toward a framework for Mathematical Understanding for Secondary Teaching that captured the mathematics of classroom-based Situations.

Our creation of mathematically delimited definitions for the actions included a process for validating our definitions. After having created a mathematical definition for an action, we compared it to each of the mathematical tasks of teaching we had classified under that action. When we were having difficulty making sense of what could be meant by a particular description of a mathematical action, we referred to meeting notes and lists from which we had drawn the task. Referring to the context in which the mathematical action of teaching was discussed enabled us to "clean up language" in the description of the action. We felt that doing so could make the document more readable to others and could work to develop our own shared understanding of the actions and the tasks. It also allowed us to maintain consistency with what we perceived to be a reasonable interpretation of the meaning of a particular mathematical task.

Although our list contained most of the verbs in Figure 4.1, we did not include some of the verbs that appeared in the list. We did so for several reasons: Meeting notes and lists did not support a mathematical definition of the verb, the verb seemed too broad to support a specific mathematical definition (e.g., "recall"), or it seemed to make sense to subsume the verb within another verb or fuse it with other verbs (e.g., demonstrate/explain; prove/refute). We also augmented the original list of verbs (with *assess, reason, coordinate,* and *reflect*) as we worked with classifying and reclassifying actions. The final list of mathematical verbs consisted of: *recognize, choose, create, use* representations, *assess* (interpret and adapt) the mathematics of the situation, *evaluate/calculate, extend, connect, demonstrate/explain, prove, investigate, apply, reason, coordinate,* and *reflect.*

Working from policy and theory

At the same time that we were working with the Situations and building the Framework from the ground up, we examined policy documents and theoretical underpinnings that eventually informed our work.

We were influenced by prominent reports including the *Mathematical Education of Teachers* (MET) and *Adding It Up: Helping Children Learn Mathematics.* The MET report (CBMS, 2001) explained that teachers needed a deep knowledge of mathematical content in the specific areas of algebra and number theory; geometry and trigonometry; functions and analysis; data analysis, statistics, and probability; and discrete mathematics and computer science. The MET report was written through the collaboration of mathematicians and mathematics educators. It was grounded in what the combined group knew was necessary for building a firm mathematical foundation as well as in what they understood about learning mathematics. In a similar way, the Mathematics Learning Study Committee wrote *Adding It Up: Helping Children Learn Mathematics* (Kilpatrick, Swafford, & Findell, 2001) about children's mathematical knowledge with advice from mathematicians, teachers, mathematics education researchers, and educational psychologists. The authors of *Adding It Up* described mathematical proficiency in terms of five strands: conceptual understanding, procedural fluency, strategic competence, adaptive reasoning, and productive disposition. Although *Adding It Up* was based on research on learning mathematics in Prekindergarten through Grade 8, it seemed reasonable that the strands and many of the recommendations could be extended to high school mathematics. Moreover, one might also argue that secondary teachers need proficiency that includes not only knowing mathematics, but also being able to use their mathematical knowledge to teach learners of mathematics.

In addition to influential reports, we consulted with mathematicians, teacher educators, educational researchers, and teachers. We held conferences of experts and potential users to look at the evolving framework as well as the Situations. Our first national conference convened researchers who had focused on teachers' mathematical knowledge. For our second national conference, we invited teacher educators, teacher leaders, and mathematicians to study our Situations and Framework and provide feedback based on their research and experiences. Our third national conference was a users conference for teachers, mathematicians, and mathematics educators who would likely put our framework to use in their research and instruction.

We presented our evolving framework at meetings including those sponsored by the National Council of Teachers of Mathematics, the Mathematical Association of America, the American Mathematical Society, the National Council of Supervisors of Mathematics, the Association of Mathematics Teacher Educators, and the International Group for the Psychology of Mathematics Education-North American Chapter. We were continually seeking to improve our ideas and to merge our grounding in practice with what is known about learning mathematics. In the beginning years of our work, we were still struggling with multiple dimensions of mathematical understanding. We were trying to integrate components such as big ideas, structures, strands of proficiency, ways of thinking, analyzing student thinking, and areas of collegiate mathematics. Later in the process, we narrowed our focus to four major perspectives. We were characterizing mathematical understanding by considering perspectives that focused on specific dimensions of understanding. We differentiated this construct from mathematical knowledge for teaching because of our focus on mathematical understanding rather than on the confluence of mathematical knowledge and pedagogical knowledge. The first of these four perspectives was focused on mathematical objects including their properties, representations, operations, and relationships. This perspective was broadly focused on the secondary mathematics curricula. The second of these four perspectives was focused on the previously discussed Big Mathematical Concepts. This perspective helped connect the secondary curricula to the collegiate curricula and remained grounded in practice. The third of the four perspectives focused on the mathematical work of teachers and included actions such as defining mathematical objects, exemplifying, explaining, and justifying. The fourth perspective, also closely related to the work of teaching, focused on the specific processes of doing mathematics such as representing, generalizing, defining, and justifying.[1]

As we attempted to advance previous work, respond to critiques, and remain faithful to classroom practice, the framework evolved. Consequently, there is a variety of framework documents with different names and components. Early versions referred to *Mathematical Knowledge for Teaching at the Secondary Level* and later, *Mathematical Proficiency for Teaching*. The current Mathematical Understanding for Secondary Teaching Framework consists of three perspectives that carry the wisdom of the previous versions. We expect that future work may lead to additional modifications as researchers and educators learn more about the mathematical understanding that is useful to mathematics teachers at the secondary level.

STRUCTURING THE PERSPECTIVES

The current Mathematical Understanding for Secondary Teaching Framework has three major perspectives, and strands describing specific views of mathematical understanding were classified under each perspective. The next section details how we used our data from practice and existing theories to identify the resulting perspectives and strands.

The first perspective: Mathematical proficiency

The first perspective (there is no intention to imply priority by the order of the perspectives) was that of Mathematical Proficiency. Knowledge of mathematical content is present in previous frameworks (e.g., Adler & Davis, 2006; Ball, Thames, & Phelps, 2008; Cuoco, 2001; Cuoco, Goldenberg, & Mark, 1996; Even, 1990) and is emphasized in most teacher preparation programs. Our initial analysis of our Situations showed useful teacher knowledge to include content knowledge and understanding connections, as well as using that knowledge in teaching. The various revisions of the framework all contained references to content knowledge, and proficiency (i.e., having and using content knowledge) became an integral part of the framework. Whether we were focusing on nouns, verbs, entities, or actions, mathematical proficiency was at the core of understanding. The strands of proficiency that were defined by the Mathematics Learning Study Committee and published in *Adding It Up: Helping Children Learn Mathematics* seemed to capture the proficiencies that we had identified. We expanded the set of the five strands of mathematical proficiency to include a sixth strand, Historical and Cultural Knowledge. This sixth strand is particularly important for teachers in exercising their responsibility for situating the study of mathematics historically and culturally. We proceeded to map the Situations to the (now) six strands of mathematical proficiency. We were able to exemplify each of the strands with multiple examples drawn from the Situations, providing us with evidence of a reasonable fit.

The second perspective: Mathematical activity

Although the Mathematical Proficiency strands seemed to capture the general mathematical capabilities needed by teachers, they did not account for the specific mathematical actions that it would be productive for secondary mathematics teachers to carry out in their teaching. The second perspective of the Framework centered on mathematical actions. The category of Mathematical Activity emerged from the previously described list of mathematically delimited actions we had derived from our analysis of the Situations and mathematical actions of teaching generated at our conferences and meetings. The list of mathematically delimited verbs seemed to account for the mathematical actions we had identified as well as for mathematical processes such as representing, generalizing, connecting, or proving, and mathematical strategies such as constraining or conjecturing.

The initial list consisted of the following verbs: *recognize* structure and convention, *connect* within and outside the subject, *represent, constrain* and *extend, generalize, model, exemplify, define,* and *justify*. Although they seemed to capture well the actions we had seen in the Situations we wrote, the list of verbs, along with mathematically delimited definitions and actions captured by those verbs, did not seem broad enough to account for the inclusion of the Situations that might need to be included in the evolution of the framework. In an attempt to

Category of mathematical activity	Mathematical activity
Mathematical noticing	• Recognize mathematical properties, constraints, or structure in a given mathematical entity or setting, or across instances of a mathematical entity. • Consider and select from among known (to the one choosing) mathematical entities or settings based on known (to the one choosing) mathematical criteria. • By recognizing structural similarity, seek and make connections between (features of) representations of the same object.
Mathematical reasoning	• Reason about a mathematical entity in one or more than one way, including, but not limited to: from mathematical definitions, from given conditionals, from and toward abstractions, by continuity, by analogy, and by using structurally equivalent statements. • Given a statement, formulate different levels and types of mathematically and pedagogically viable proofs.

FIGURE 4.2a.   Correspondence between categories/subcategories of Mathematical Activity component and verbs arising from lists of mathematical actions.

broaden our categories, we classified those that we had identified into the four larger categories of mathematical noticing, mathematical reasoning, mathematical creating, and integrating strands of mathematical activity (we later referred to these categories as strands). (See Figures 4.2a and 4.2b for the correspondence between categories/subcategories of the Mathematical Activity component and verbs arising from lists of mathematical actions.) We discovered that each of our previous categories mapped well to one these four übercategories, and that attempting to map the activity categories arising from our Situations work into these larger categories led us to uncover a missing subcategory (Defining), which seemed to fit well into the Mathematical Creating strand. Figure 4.2a and Figure

Category of mathematical activity	Mathematical activity
Mathematical reasoning (continued)	• Take a mathematical action to find out more about the structure, constraints, and/or properties of a mathematical situation or a mathematical object. • Extend domain, argument, or class of objects for which a mathematical statement is/remains valid. • Interpret and/or change certain mathematical conditions/constraints that are relevant to a mathematical activity.
Mathematical creating	• Create, represent, or define a mathematical entity or setting from known (to the one creating) structure, constraints.. or properties. • Modify, transform, or manipulate a representation or definition of a mathematical structure, constraint, property for a given purpose.
Integrating strands of mathematical activity	• Coordinate mathematical knowledge. • Reflect on one's own mathematics doing. • Employ algorithms, definitions, and/or technology in mathematical settings and/or real world quantitative settings.

FIGURE 4.2b.   Correspondence between categories/subcategories of Mathematical Activity component and verbs arising from lists of mathematical actions (continued).

4.2b show the current category system as well as activities from our prior Situations work that map into those categories.

The third perspective: Mathematical context of teaching

One of the original goals of constructing a framework was to characterize the mathematical understanding that was especially important for teaching. Although the first and second perspectives were based on practice at the secondary level and theory related to learning mathematics, they do not identify proficiencies and activities that are unique to teaching. The perspective entitled Mathematical Context of Teaching was embedded implicitly in many of the activities that we identified in the Situations, but it did not appear explicitly. The context of teaching necessitates the mathematical understanding needed to help others gain proficiencies and facility in mathematical activity. We needed a different perspective to highlight the context of teaching implicit in the Situations. Building on the Situations, we could see that mathematical proficiencies and activities were useful in mathematics teaching at the secondary level, but we were still missing the description of the mathematical contexts in which aspects of mathematical understanding are particularly important for mathematics teaching.

Although we relentlessly tried to focus on the mathematics of Mathematical Understanding for Secondary Teaching and not to get diverted into the space of pedagogical content knowledge, we found this technical distinction difficult to implement. The dilemma was how to account for the special understandings that a secondary teacher needed in specific mathematical contexts of teaching without dwelling on the pedagogical. We turned our attention to the mathematical work of teaching, and thought of it as the mathematical understanding that helped teachers access, interpret, and build on the mathematical understanding of students. This perspective on mathematical understanding is not subsumed in pedagogical content knowledge. Teachers need the mathematical knowledge that allows them to analyze mathematical ideas that arise in the classroom. They need mathematical knowledge to ask questions that will give them access to a student's thinking and will help the student reassess his or her understanding. Teachers need to understand the mathematical connections within the curriculum and to understand how changing or supplementing the curriculum influences the development of mathematical ideas. They need mathematical understanding in order to assess the critical components of students' mathematical thinking.

The third perspective of Mathematical Context for Teaching was developed to focus on the mathematical understanding that was useful to implement sound pedagogical strategies effectively in a mathematics classroom. Unlike the others, this perspective focuses sharply on the mathematical understanding that will prepare a teacher to understand and enable someone else's learning of mathematics. Although many mathematical professions require communication of mathematics, they usually do not focus on advancing the mathematical understanding of a student, a core responsibility of teachers in the context of teaching.

The strands of the Mathematical Context of Teaching are clearly grounded in practice. They were created through consideration of conference discussions, teaching experiences, and research literature, rather than through the analysis of the Situations. Activities suggested in many of the strands can be observed in classroom events during which teachers engage students in discussions. To understand the issues underlying the Prompt for each of the Situations, a teacher must be able to analyze the mathematics that was involved. A teacher with sufficient mathematical understanding can access student thinking by raising questions related to the mathematical ideas presented in the Foci of the Situation. Teachers use their mathematical understandings daily to interpret the curriculum and assess students' understanding of core ideas. This ability to apply mathematical understanding in the service of facilitating the learning of others is not typical of the work of other mathematical professions, but is critical to the mathematical understanding for teachers of mathematics. The Mathematical Context perspective was created to capture the contexts in which teachers need to apply Mathematical Proficiencies and Mathematical Activities.

Viability of the situations and framework

The framework is built on incidents from classrooms, but the interpretation of the incidents and the organization of the perspectives of the MUST Framework were the products of our consultation with groups of mathematicians, statisticians, researchers, mathematics educators, supervisors of mathematics, and teachers. It was important to know how the framework resonated with practicing mathematicians, secondary teachers, and other mathematics educators. Although valuable feedback was obtained at each conference and from individual reviewers, it was critical to investigate how well the framework represented ideas of professionals who had not been involved in the development of the framework. Two studies addressed the viability of the Situations and Framework for representing the mathematics that was potentially useful to teachers.

In a study by Conner, Wilson, and Kim (2011), practicing secondary teachers were given the Prompts from classrooms and asked to identify the mathematics that would be useful for teachers who were faced with the opportunities that the Prompt afforded. They were provided with only the Prompt and were not given the Foci or Commentaries. They were asked to list mathematical knowledge and mathematical practices that would be critical for thinking about how to make the most of the learning opportunity. Their lists of important, useful mathematics were remarkably similar to the mathematical knowledge and practices identified by the authors of the Situation Foci and Commentaries. In the same study, preservice teachers were asked to identify mathematical ideas that they should know in order to be able to work with students or events that were described in given Prompts, but they were not provided with the Foci. Again, the preservice teachers provided many of the same mathematical ideas that were explained in the Foci for the given Prompts. This study suggests that the full Situations do focus on mathematical understanding that teachers perceive to be important and useful for the work of teaching, and provides a type of validation for the mathematical understanding described in the MUST Framework.

As the Situations were developed and the perspectives in the framework were created, mathematicians and mathematics educators began to use Situations in their courses and presentations. The Prompts are brief and quickly generated discussion. Students and audiences were captivated by the volume of mathematical ideas that were relevant to any given Prompt. By engaging preservice and in-service teachers in discussion of the Foci in a Situation or in generating new Foci, educators were able to elicit significant mathematical questions as well as expose students' understanding. We see this as different but equally valuable evidence that the Situations are valuable vehicles for identifying components of mathematical understanding. An extensive set of further examples of uses of Situations is discussed in chapter 6.

The framework was also used to guide the research of Hee Jung Kim (2013) as she studied the mathematical understanding of teaching assistants who taught a collegiate course in Trigonometry. Although the framework was designed to artic-ulate mathematical understanding useful for mathematics teachers at the second-ary level, it was helpful in investigating the understanding of instructors at the col-legiate level who were teaching a course that is commonly taught at the secondary level. Kim used the framework to design tasks that revealed understanding about the framework perspectives of Mathematical Proficiency, Mathematical Activity, and Mathematical Work (later, "Context") of Teaching. She created specific tasks that provided opportunities to exhibit understanding of strands within a specific component. In her analysis, she coded the participants' responses according to the mathematical understanding described by the strands. She reported that there was much overlap between framework perspectives. For example, a task designed to elicit Mathematical Proficiency also provided evidence of Mathematical Activity. This is further evidence that the perspectives of the framework represent differ-ent viewpoints and emphases rather than different mathematics. Her study found that the Teaching Assistants had a thorough understanding of complex concepts and processes, but struggled with some fundamental ideas such as the nature of an angle and the definition of the unit circle. Kim credited the MUST Framework with helping her identify not only complex notions but also fundamental ideas.

CONCLUSION

Several underlying premises guided our development of a framework for the Mathematical Understanding of Secondary Teachers.

Central to our work was the commitment that the framework would derive from authentic representations of practice. Each of those representations, embod-ied in the Situations, was drawn from practice-based episodes that some member or members of our team had witnessed.

Our construction of the framework was based on steadfast commitment to rep-resenting and characterizing the mathematics that could be useful to teachers in authentic secondary mathematics settings.

The three perspectives of the framework that arose from the Situations are dif-ferent views of the same phenomenon.

The next chapter describes the process of developing a Situation. It is our hope that the understanding of the process of constructing the Framework from Situations discussed in this chapter will position readers to engage productively in the development of their own Situations.

NOTE

1. This approach, focusing on mathematical processes and the products of those processes was the foundation of research on the personal and classroom mathematics of secondary teachers conducted at The Pennsylvania State University (Zbiek, Heid, & Blume, 2012).

REFERENCES

Adler, J., & Davis, Z. (2006). Opening another black box: Researching mathematics for teaching in mathematics teacher education. *Journal for Research in Mathematics Education, 36,* 270–296.

Ball, D. L., Thames, M. H., & Phelps, G. (2008). Content knowledge for teaching: What makes it special? *Journal of Teacher Education, 59,* 389–407. doi:10.1177/0022487108324554

Conference Board of the Mathematical Sciences. (2001). *The mathematical education of teachers.* Providence RI and Washington DC: American Mathematical Society and Mathematical Association of America.

Conner, A., Wilson, P. S., & Kim, H. J. (2011). Building on mathematical events in the classroom. *ZDM, 43,* 979–992. doi:10.1007/s11858-011-0362-1

Cuoco, A. (2001). Mathematics for teaching. *Notices of the AMS, 48,* 168–174.

Cuoco, A., Goldenberg, E. P., & Mark, J. (1996). Habits of mind: An organizing principle for mathematics curricula. *Journal of Mathematical Behavior, 15,* 375–402. doi:10.1016/S0732-3123(96)90023-1

Even, R. (1990). Subject matter knowledge for teaching and the case of functions. *Educational Studies in Mathematics, 21,* 521–544. doi:10.1007/BF00315943

Kilpatrick, J., Swafford, J., & Findell, B. (Eds.). (2001). *Adding it up: Helping children learn mathematics.* Washington, DC: National Academy Press.

Kim, H. J. (2013). Graduate teaching assistants' mathematical understanding for teaching trigonometry (Unpublished doctoral dissertation). University of Georgia.

Shulman, L. (1986). Those who understand: Knowledge growth in teaching. *Educational Researcher, 15*(2), 4–14. doi:10.3102/0013189X015002004

Zbiek, R. M., Heid, M. K., & Blume, G. W. (July, 2012). *Seeing mathematics through processes and actions: investigating teachers' mathematical knowledge and secondary school classroom opportunities for students.* Paper presented at the quadrennial meeting of the International Congress on Mathematical Education, Seoul, Republic of Korea.

CHAPTER 5

CREATING NEW SITUATIONS AS INQUIRY

Rose Mary Zbiek and Glendon Blume

We developed Situations with the goal of using them to establish the Framework, Mathematical Understandings for Secondary Teaching. Chapter 3 outlines our developmental process. As finished products, Situations have many uses including those outlined in chapter 6. The purpose of this chapter is to explore the generation of novel Situations as an activity for teachers and others.

Situations may develop in different ways. We begin this chapter with consideration of contexts in which development of Situations might be meaningful activity. Then, we turn to sources and means for identifying and articulating Prompts and Foci. The chapter ends with suggestions for facilitating others' work in developing Situations to enrich their mathematical understandings. In this chapter, we use the term *developer* to refer to anyone involved in the development of a Situation.

CONTEXTS FOR SITUATION DEVELOPMENT ACTIVITIES

Ideal contexts for Situation development activities are those in which participants share the goal of developing over time a deeper understanding of the mathematics needed for teaching secondary mathematics. Facilitators might be mathemati-

Mathematical Understanding for Secondary Teaching: A Framework and Classroom-Based Situations, pages 57–63.
Copyright © 2015 by Information Age Publishing

cians or mathematics educators, or some combination thereof. We believe that it is valuable for a participant to engage with Situations as a user prior to engaging in Situation development. Some experience with the format of a Situation, recognition of how the mathematical ideas that underpin the various Foci differ, and an understanding of the "grain size" of Foci can be helpful preparation for developing Situations on one's own.

One context for development of Situations might be a capstone course for undergraduate mathematics or mathematics education majors intending to be middle school or high school mathematics teachers. The specific goals in this setting could be a combination of revisiting secondary school mathematics content and connecting that content to college mathematics. The activity might be framed as a challenge to create one Situation for each strand of the secondary school curriculum and to include content from as many college mathematics courses as possible. In this context, the developers' inquiry through Situation development would entail in-depth examination of K–12 mathematics content and connections between K–12 mathematics topics and related collegiate mathematics.

A second context for Situation development is the engagement of in-service teachers in Situation development as a way to develop their capacity for vertical articulation. With Prompts from a middle school context, teachers could be challenged to develop Foci that connect to elementary school content, to other middle school content, and to high school and college content. The Postcommentary for such Situations might then link the ideas of the Situation to state and local standards and curriculum materials. In this context, the developers' inquiry through Situation development could center on broader understanding of the scope and sequence of important mathematical ideas across K–12 mathematics, the mathematical understandings required by various standards and curriculum materials, and the mathematical connections between K–12 mathematics and collegiate mathematics.

In a third context, groups of student teachers or interns and mentor teachers might engage in developing Situations as a way of preparing mathematically for the beginning of a new unit of study. Such experiences could help teachers to become more familiar with the secondary school content of the unit and to enhance their ability and disposition for connecting their college mathematics to secondary school topics. Mentors would have an opportunity to share the mathematics scope and sequence of their school systems and revisit college mathematics topics. In this context, the developers' inquiry through Situation development might involve the student teacher and mentor in examination and resolution of mathematical issues that arose in the student teacher's preliminary planning.

In a fourth context, college instructors teaching mathematics courses for teachers could use the activity to develop ideas about how to bring secondary school mathematics applications and connections into college mathematics courses or to make such things more explicit. A secondary benefit could be insights into the extent to which their department's course offerings are useful to prospective teachers. In this context, the developers' inquiry through Situation development might be targeted at improving the relevance of collegiate mathematics curricula for prospective middle school and high school teachers.

A fifth context in which Situation development might occur is within the programs of study of mathematics education graduate students. One aspect of preparing mathematics education graduate students as professional development providers or researchers who study professional development is enhancement of their understanding of K–12 mathematics content. Developing Situations offers graduate students an opportunity to think deeply about the content of school mathematics curricula from a teacher's perspective as well as from the perspective of a professional development provider. Groups of graduate students might develop Situations in graduate seminars, or graduate students might develop Situations independently in consultation with faculty members. In this context, the developers' inquiry through Situation development could involve investigating their own understandings of mathematics content and implications from mathematics education research.

Many other contexts for Situation development are possible. Those mentioned here are offered as examples of the range of audiences and goals for which Situation development is appropriate.

TURNING EVENTS INTO PROMPTS

Situations are based on events that are expressed as Prompts. The Prompts contain the mathematical motivation—the key mathematical puzzle or issue that captures attention and needs resolution. Creating a Situation demands first identifying events and then constructing a Prompt that conveys enough of the mathematical motivation but does not cause users to foreclose on a particular Mathematical Focus. Throughout this chapter we use the Division Involving Zero Situation (see Situation 1 in chapter 7) to exemplify particular ideas.

Identifying events

The initial step in developing a Situation is the identification of an event to serve as a Prompt. As noted previously, the mathematical motivation for a Situation is a mathematical question, puzzlement, or issue that captures attention and needs resolution. Events that include students' questions or puzzlement about mathematical issues (e.g., Is $2^{3.5}$ halfway between 2^3 and 2^4?), requests for justification (e.g., Why can't we divide by 0?), or requests for explanations from teacher or peers (e.g., How are parabolas and graphs of quartic functions different?) can be fruitful for Situation development. In addition, for an event to be a "Situation-worthy" event, it needs to be rich enough to admit multiple Foci. Although some student questions might be fruitful for Situation development, other questions might not necessarily be good candidates for Situations. For example, questions such as, "Which ones are the alternate-interior angles?" or "Is the x-coordinate the first one or the second one?" are not necessarily indicators of rich, Situation-worthy events.

Prompts for participant-developed Situations might have (at least) two origins. Given an event supplied by the facilitator (which could arise from an actual classroom event or be a "manufactured event"), each participant or small group of participants could develop their own Prompt. Alternatively, participants could generate Prompts from their individual experiences, or a group of participants might develop Prompts from one or more shared classroom incidents.

In our case, the events were incidents that we directly encountered in our teaching experiences or witnessed in other teaching work. Some of these events are similar to experiences that arise frequently. As an example, the Prompt for Division Involving Zero (see Figure 5.1) comes from a conversation with prospective teachers but the need to consider divisions involving 0 and the possible explanations are similar to those typically posed by middle school or high school mathematics students. Prompts based in real-world experiences could have several other sources, such as vignettes from mathematics education courses or teacher education research studies.

If the goal is to address a predetermined area of mathematics, then it might be more efficient, if not necessary, to manufacture the context for the Prompt. Such

On the first day of class, preservice middle school teachers were asked to evaluate $\frac{2}{0}$, $\frac{0}{0}$, and $\frac{0}{2}$ and to explain their answers. There was some disagreement among their answers for $\frac{0}{0}$ (potentially 0, 1, undefined, and impossible) and quite a bit of disagreement among their explanations:

- Because any number over 0 is undefined;

- Because you cannot divide by 0;

- Because 0 cannot be in the denominator;

- Because 0 divided by anything is 0; and

- Because a number divided by itself is 1.

FIGURE 5.1. Prompt for Situation 1: Division Involving Zero (see chapter 7).

Prompts might arise by examining curriculum materials and considering how students might respond to various tasks.

Writing prompts

Once an event has been identified, the Prompt needs to be written. The Prompt might not be the complete account of a real-world Situation but might convey only part of the entire event. By providing only the relevant details, we found that the focus was on the mathematics rather than other issues, such as classroom management. On the other hand, longer Prompts might help teachers develop the ability to identify mathematical motivations in the complexity of their own classrooms. For example, if the Prompt is based on a question posed by a student in a class, the Prompt might provide the task on which students were working and the student's question but not the teacher's response to the student. The Division Involving Zero Prompt mentions three different cases ($\frac{0}{2}, \frac{2}{0}$, and $\frac{0}{0}$) in the opening lines but ends with student responses related only to $\frac{0}{0}$.

Foci typically are better when they are relatively short. Longer Foci tend to blend two or more major mathematical ideas and thus make it more difficult to delve deeply in reasonable amounts of time. Rather than providing a long dialogue or using student wording for explanations, Division Involving Zero presents student ideas about $\frac{0}{0}$ as bulleted phrases.

DEVELOPING MATHEMATICAL FOCI

The starting point for developing Foci seems to be noting the mathematical motivation in the Prompt. Small-group discussion might be useful in helping all team members to understand the key mathematical issue of the Prompt. The work of developing Mathematical Foci involves both determining the mathematical paths to take and articulating them as Foci.

Identifying mathematical paths

Several different approaches might be useful in initiating thinking about the mathematical paths that possible Foci might take. Perhaps the most natural starting point is having teachers think about what they have done or would do in the Prompt context, assuming that they would not avoid or ignore the mathematical motivation! Initial probing questions might ask how the initial reaction is mathematically connected to the Prompt. Subsequent questions might then pursue whether the initial reaction is mathematically valid or how the reactions might be mathematically adjusted. In discussing Division Involving Zero, attention might be brought first to the fact that the explanations relate to $\frac{0}{0}$. Questions might then ask groups to decide which, if any, of the explanations are mathematically valid.

A different approach to Foci development involves looking at the conceptual understanding that underpins the mathematical motivation. One might look at a major concept in the Prompt and consider its properties. A Focus might be constructed by drawing attention to a property or by using details of the property to reason through the mathematics. A key concept in Division Involving Zero is *division*. Focus 2 arises from using partitive and quotitive meanings of division to analyze ideas in the Prompt.

Foci might also arise by analyzing procedures that are involved in a Prompt. A Focus might evolve by looking at an alternative procedure or by considering the applicability of the procedure in the Prompt setting.

Inspirations might also come from mathematics education research. Knowledge of documented misconceptions and ways to challenge them, for example, might be useful in identifying mathematical ideas for Foci. Understanding that division can be conceptualized in two ways, partitive division and quotitive division as in Focus 2, comes from mathematics education research (Verschaffel, Greer, & DeCorte, 2007).

Ideas for Foci might arise by asking how mathematical entities in the Prompt might be represented in different ways, perhaps through the use of technology. They also could arise by asking how representations in the Prompt might be interpreted in multiple ways. Thinking about the mathematical motivation from the Prompt through a different representation provides a new venue in which to explore mathematical ideas related to the Prompt. In Division Involving Zero, the symbol $\frac{0}{0}$ is interpreted as an indicated division in the Prompt and then as a rate in Focus 4.

Articulating ideas as Foci

The mathematical essence or main point of a Focus should be apparent. It should be captured in one or two sentences. These sentences are important in that they not only provide a tool for determining what needs to be included in a particular Focus but also are useful in distinguishing between different Foci. For example, the indented summary statements associated with Focus 2 and Focus 5 in Division Involving Zero make it clear that both are vertical connections. Moreover, these statements suggest that the former connects to ideas from elementary mathematics and the latter connects to college mathematics.

A second feature of Foci is their length. We chose to keep our Foci concise. Keeping Foci concise has the added benefit of encouraging developers to concentrate on writing mathematical text. In addition, shorter Foci are easier to edit for public sharing and might not require as much reading and preparation time during professional development activities. For example, in Division Involving Zero Focus 3, we omit computational details when explaining how different interpretations of the ratios relate to division involving 0.

GUIDING OTHERS IN THE DEVELOPMENT OF SITUATIONS

Situation development is most likely best done by groups rather than by individuals. Members of groups might best vary in their strengths and experiences in both mathematics and teaching so that a broad range of mathematical content from mathematics of early years through graduate mathematics study is collectively available to the group.

Because it is the critical starting point of a Situation, the Prompt might be provided for initial Situation-development activities. The stimulating event might be one that is shared by a member of the development team or one that is an amalgam of the experiences and stories shared by several members of the team.

Supporting development of Foci often requires the facilitator to help developers to (at least temporarily) squelch their tendencies to respond to the Prompt with a solution rather than an explanation. Similarly, developers might be tempted to explore pedagogical responses rather than mathematical ones.

Brainstorming Foci rather than attempting to write Foci might be a useful starting point. To support this process, developers might need a variety of mathematical resources, such as access to quality web sites and classic books, to check details and to confirm their memories or further develop tentative ideas.

REFERENCES

Verschaffel, L., Greer, B., & DeCorte, E. (2007). Whole number concepts and operations. In F. K. Lester Jr. (Ed.), *Second handbook of research on mathematics teaching and learning* (pp. 557–628). Charlotte, NC: Information Age Publishing.

CHAPTER 6

SUMMARY OF USES OF THE MUST FRAMEWORK AND SITUATIONS

James W. Wilson

Throughout the tenure of this project, we have attended to the *practice* of teaching mathematics. The Prompts in each Situation came from our observations of real mathematics classroom events. Ultimately, we want our work to have impact on that practice by leading to better mathematics understanding for secondary teaching. There is a basic assumption that the improved practice of mathematics teaching is derived from a deep understanding of the mathematics in the curriculum and the connections to the wider body of mathematics. Our Situations are connections to the mathematics identified by examining the Prompts, and the Mathematical Understanding for Secondary Teaching (MUST) Framework represents a synthesis of this work across more than 40 Situations that we produced.

USERS CONFERENCE

As we refined the process of creating the Situations and the MUST Framework we directed the project toward a national conference of potential users. Thus, our third Situations Project conference was a *Users Conference*, held at the University of Georgia in March 2010 with more than 60 participants from across the United

States. We invited participants with a variety of interests in mathematics teacher education, postsecondary mathematics teaching, mathematics professional development, secondary school mathematics teaching, research in mathematics education, and mathematics curriculum development. This chapter will attempt to synthesize the ideas from the Users Conference. Although participants are identified with their proposals in much of the chapter, sometimes the input of particular participants is indicated without any reference document other than "conference participant." The Appendix contains a list of the participants at all three conferences. We also have made some use of Situations Project materials in our research and in our Mathematics Education undergraduate and graduate courses and in some mathematics courses at the University of Georgia and The Pennsylvania State University. Some of these were reported at the Users Conference, and some are included in this chapter as accounts of our experiences and ideas for use.

It became clear early on that "uses" meant not only uses of the products—the Framework and the Situations—but also uses of the processes. The conference was organized so that individuals and groups produced proposals for using either the materials or something from our approach. Many at the conference proposed direct uses of the materials with prospective teachers. Others concentrated on use of particular sets of the materials as a stimulus for mathematics professional development with in-service teachers. Several proposals addressed using the materials in postsecondary mathematics courses. These uses included not only the Situations as mathematics explorations but also use of the Framework as a guide for analysis and discussion. Almost every proposal included some element of research or assessment. Finally, many proposals cited uses of our materials and processes as a step to create materials (e.g., videos) for other work in mathematics education.

Some proposed uses were more oriented to incorporating the process of generating Situations from classroom Prompts into course or workshop syllabi. Certainly, the experience of producing the Situations and the Framework had been a stimulating journey for many graduate students and faculty members who worked on the project. These processes can enable analysis and discussion about the various connections among mathematics topics.

Although most participants offered ways to use the existing Situations and Framework, the Users Conference also was a venue for discussing the value of our practice-based approach in contrast to an approach that is developed from a collegiate mathematics perspective. Instead of focusing on uses of the Situations and Framework, some participants who took the collegiate mathematics perspective focused on offering suggestions about alternative ways to present some mathematics topics.

Many of the conference participants had some prior experience with the Situations Project, either as participants in one of the previous conferences for mathematicians and mathematics teacher educators or in using Situations or the Framework from the project within their own teaching or professional development activities. Therefore, their contributions to the conference were based not only on proposed uses but also on their own experience with the ideas and materials. On the other hand, the faculty and graduate students who worked on this project were active participants in mathematics teacher education, both in-service and preservice. There were many opportunities to informally try out ideas for the use of materials as they evolved. This chapter includes summaries of some of those experiences. The conference also included participants for whom the materials were new. These participants embraced some ideas and encouraged expansion, further development, and clarity in some areas.

The following sections include discussion on proposed uses of the Situations Project materials in several interrelated categories:

- Research in Mathematics Education;
- Professional Development—Preservice Teacher Preparation;
- Professional Development—In-service Teacher Learning;
- Teaching Mathematics Content Courses
- Graduate Mathematics Education; and
- Assessment.

The descriptions tend to be brief summaries of the potential uses and they require elaboration and adaptation to particular contexts. Proposals at the conference tended to be two or more pages rather than a paragraph as presented here.

USES IN RESEARCH IN MATHEMATICS EDUCATION

The Situations Project is a long-term project of scholarly inquiry. It is a research enterprise and taken collectively, the MUST Framework and the Situations represent innovations for exploring mathematical understanding for secondary teaching. All of this work begins with observations of practice and emphasizes the exploration of mathematics that may be relevant to those observations of practice.

The Users Conference generated several suggestions for research based on the MUST Framework and the Situations.

Study of results of professional development

When the Situations are used as the basis for professional development with any elementary school, middle school, or high school teachers, corresponding research questions such as the following might be addressed:

1. What mathematics do teachers learn from this professional development?
2. What impact does the professional development have on teachers' planning and preparation?
3. What impact does the professional development have on teachers' instructional practice?

4. What impact does the professional development have on student achievement?

Almost every proposal at the Users Conference had comments about potential research or assessment questions.

Study of the mathematical disposition of subgroups of master's degree students

In master's-level programs focused on mathematics education, many students hold undergraduate degrees in mathematics. This includes students from outside mathematics education who want to teach high school mathematics and are acquiring initial certification, current high school teachers who seek a graduate degree and want to continue to teach high school mathematics, mathematics students who want to teach mathematics at the community college level, and mathematics students who do not plan to teach. Looking at how these four groups might respond to the mathematics in one or more Situations could shed light on the mathematical dispositions of each group. Understanding those dispositions could lead to improved programs most appropriate to the groups.

Study of mentoring practices with student teachers

The mentoring of student teachers by an experienced critic teacher is essential, yet very little is known about structuring the discourse between student teachers and the critic teachers to engage mathematical thinking. The MUST Framework is proposed as a research tool to characterize the types of feedback that mentor teachers give to their student teachers or to develop awareness of mathematical understanding to be observed and discussed in mentoring settings.

Study of Situations for generating mathematical discussions

As mentioned in chapter 4, a study by Conner, Wilson, and Kim (2011) reported on the use of a small set of Situations with preservice and in-service mathematics teachers in a professional-development context. Their research found that the Situations led these prospective and in-service teachers to in-depth discussion and understanding of the mathematics relevant to the Prompts. Students reported that the process often extended the range of their mathematical knowledge and understanding, sometimes recalling material they had not considered before or sometimes bringing them to explore new ideas. The material provoked increased levels of mathematical understanding.

Empirical validation studies

The MUST Framework represents an extensive set of hypotheses about the nature of mathematics understanding and its connection to practice. Several conference participants called for validation studies of the MUST Framework with experienced mathematics teachers. Data are needed to verify or clarify the ideas of *mathematical proficiency*, *mathematical activities*, and *mathematical context of teaching*.

Further development of the MUST Framework

The MUST Framework is the key conceptual product of the Situations Project. In many ways, the particular Situations are not as central to our work and are not necessarily unique. Many additional Situations might be possible, and the particular discussions of Mathematical Foci that have been produced are not unique. The MUST Framework was produced out of the Situations based on mathematical issues found in classrooms—thus, it is a type of "reverse engineering" result.

Research on preservice and induction-year(s) teacher education

Klerlein proposed a research project at the conference to use the MUST Framework to identify the development of mathematical understanding across a preparation program and the induction years for prospective middle grades teachers. This program included a new organization that was not developed using the MUST Framework. Rather, the MUST Framework was to be a lens for examining the paths these teachers take in becoming mathematics teachers. The program incorporated several problem-based preservice mathematics courses, an induction year, and pedagogy that included mathematical "play."

Investigating the MUST Framework and Situations in action: An investigation of mathematics teaching

Hatfield proposed this use with dual purpose—first as an implementation of a professional development in a school or as a capstone course in teacher preparation, and second as a research project tied to that implementation. The proposal was designed with the following stages:

Stage 1. Help practicing secondary mathematics teachers to use the MUST Framework as a conceptual backdrop for studying a variety of Situations with the aim of strengthening participants' pedagogical content knowledge.

Stage 2. Engage the participants in their identification of new Situations, directly connected to what they are teaching. They would write and refine Focal analyses.

Stage 3. Each would conduct one or more lessons in which they would enact their usages of one or more approaches based on the mathematical ideas in the given, or constructed, Situation/Foci discussions.

Stage 4. Using video snippets, they would come together to engage in collaborative analyses of lesson events and student thinking. These could lead to further refinements or extensions of their lessons.

Stage 5. After the professional-development project has ended, follow up with participants to ascertain the extent to which they are continuing to use MUST Framework/Situations in support of their teaching and their professional work.

USES IN PROFESSIONAL DEVELOPMENT—PRESERVICE TEACHER PREPARATION

Preservice secondary mathematics teacher education involves extensive study of mathematics content and mathematics education (i.e., mathematics teaching and learning). The Situations and Framework may find direct use as course materials for study and discussion. Quite often, however, a powerful strategy is to move from direct use of the materials for discussion to a mode of generating Situations.

Use of Situations in an undergraduate mathematics teacher education course

At The Pennsylvania State University, Blume has used draft versions of the Situations in an elective course, Understanding Mathematics in Classroom Situations. He chose 23 draft Situations selected for content not directly focused on algebra. These included trigonometry, geometry, number concepts, and statistics. The class met for 3 hours per session and usually examined one to three Situations per class. Typical use was to pose some version of a Prompt and then engage the students to react to the Prompt with attention to the mathematics ideas underlying the Prompt rather than on pedagogical issues. This was followed by discussion of the students' reactions to the Prompt, developing or elaborating on the mathematical concepts underlying the Situation. Next, there was opportunity for further development of some topics that arose from the discussion or development of ideas from Foci that were not addressed in the small groups and ensuing discussion. Homework assignments included explorations that constituted extensions of mathematical ideas encountered in class discussion, proofs of results that were conjectured to be true but not proved during class, and identification of additional mathematical ideas related to the Situation (e.g., a mathematical idea that might constitute the basis for an additional Focus for that Situation). Among the challenges faced were finding adequate time for following up on the mathematics topics that arose, making decisions about which mathematics content to pursue in depth, maintaining a focus on content rather than pedagogy, and difficulty in crafting the course into a coordinated sequence of topics.

Developing a Situation into an assessment instrument

Situation 1: Division Involving Zero was used by Seaman in a Logic of Arithmetic class at the University of Iowa. Seaman adapted the Situation for an in-class writing assignment. Students were asked to decide what $\frac{0}{0}$ meant and how to address the student who thought $\frac{0}{0}$ should be 1. The Mathematical Foci of Situation 1 provided a framework for examining student responses and follow-up to the lecture that had emphasized partitive and quotitive division. Seaman's use provided a formative assessment for deciding further emphases to be incorporated in the lectures.

Using Situations in an undergraduate mathematics methods course

DuCloux and HyeonMi Lee also used Situations to facilitate assessment and feedback within a secondary mathematics methods course. The Prompts of Situation 35: Sin 32° (see chapter 41) and Situation 11: Powers (see chapter 17) were provided to the students in an online discussion. Students were asked (a) How would you respond to the student? and (b) What mathematical knowledge would a teacher need to know tPoro provide a conceptually sound response? This assessment was near the end of the teacher education sequence just prior to student teaching. DuCloux and Lee were also examining research questions on mathematical proficiencies of preservice secondary mathematics teachers and on misconceptions that might be revealed. Such online uses require special attention to technical issues, establishing an online response/discussion format, and issues with lack of depth or superficial and imprecise explanations. Others at the conference had proposals for the online use of Situations with considerable warning about the special issues, both technical and motivational.

Use of the MUST Framework as a metacognitive tool

Portnoy discussed using the Framework as a metacognitive tool in a course at the University of New Hampshire, Analysis for Secondary Teachers. It is a course designed to deepen prospective secondary teachers' understanding of secondary school mathematics. Students are given context-based mathematical problems to investigate. Portnoy explained that these challenging problems require a deeper understanding of the secondary mathematics than future students in the secondary school will be expected to develop. Prospective student teachers are asked to find solutions to the problems and to find strategies that might be useful in explaining the solutions to secondary students. They are asked to keep a problem-solving journal that details their investigation and solution(s). The proposed use of the Framework as a metacognitive tool is to have the students code their journals according to the categories (Mathematical Proficiency, Mathematical Activities, and Mathematical Context of Teaching) and subcategories of the MUST Framework.

Use of the Prompts from Situations in a first secondary mathematics methods course

Jaqua proposed using discussion and analysis of the Prompts in the Situations to review the high school mathematics curriculum. Jaqua's model for a first methods course proposes that the Situations would be a basis for prospective secondary teachers to explore their mathematics, discover what additional mathematics they need to know, solve problems and review curriculum at a deeper level, and determine connections to diverse mathematics content. Jacqua proposed giving the students the Prompt to a Situation and then requiring each to post online a discussion of the mathematics relevant to the Prompt. Students would read and respond substantially to the posts of at least two classmates. Face-to-face discussion of the submitted responses would be done in the next class meeting.

Use in undergraduate mathematics methods and content courses

We have received and granted requests for mathematics educators to try out and use the Situations Project materials in their courses. These have been liberally granted. We have not, however, systematically followed up and received feedback from these requests. The uses have included summer courses for in-service teachers, preservice methods and content courses for middle school teachers, mathematics content courses for elementary teachers, geometry courses for prospective secondary teachers, and others.

Secondary methods course for the big picture

McCrory, Winsor, TeCroney, and Edenfield proposed to use the MUST Framework to help students to see the "big picture," including the complexity of knowledge and proficiency they are developing for teaching. They described their goal of using the Situations and Framework to help preservice mathematics teachers develop the inclination and ability to deconstruct and connect mathematics ideas. The emphases were on using the Situations to (a) provide a context for understanding ideas in the MUST Framework, (b) develop experience analyzing and deconstructing mathematical ideas (especially in conjunction with their work in Calculus 1), (c) develop a disposition to keep exploring mathematical ideas even when a single explanation is apparent, (d) develop and appreciate the benefits of collaborative work in mathematics, and (e) learn to plan or prepare a lesson by attending to the mathematical ideas related to the lesson. Situation 6: Absolute Value Equations and Inequalities (see chapter 12) was used illustrate their proposal.

USES IN PROFESSIONAL DEVELOPMENT— IN-SERVICE TEACHER LEARNING

Discussion of in-service teacher professional development activities covered a wide range of activities, from the 1-day professional development to activities that spanned an academic year. Some were targeted to particular school contexts, and others were more general.

Facilitators' guides

At the Users Conference in March 2010, Briars proposed development of Facilitators' Guides for those who conduct professional development workshops. A facilitators' guide would be built around the use of a particular Situation, giving background information and specific suggestions for professional-development-session activities. The National Council of Supervisors of Mathematics (NCSM) followed up on this idea and proposed that a small number of Facilitators' Guides be produced. Other participants of the Users Conference supported the ideas of this proposal.

In the spring of 2012, NCSM personnel and faculty members from the Situations Project developed a prototype Facilitators' Guide for use with Situation 1: Division Involving Zero (see chapter 7). This prototype (Blume, Briars, Heid, Mitchell, Schrock, Viktora, Wilson, & Zbiek, 2013) was presented at the 2013 NCSM Annual Meeting in Denver. The discussions of the prototype were very positive and the decision was made by NCSM to work with the Situations Project to produce a total of six Facilitators' Guides for the following Situations:

Situation 1: Division Involving Zero (chapter 7)

Situation 2: Product of Two Negative Numbers (chapter 8)

Situation 21: Graphing Quadratic Functions (chapter 27)

Situation 34: Circumscribing Polygons (chapter 40)

Situation 35: Calculation of Sine (chapter 41)

Situation 38: Mean and Median (chapter 44)

Each of the guides contains specific information and resources to assist professional development facilitators to conduct short-term professional development workshops around the content of one specific Situation. Each guide contains suggestions for time schedule, activities, and assessment. To assist facilitators to plan and conduct effective professional development workshops, specific comments are included on the mathematical background required for each Focus in the Situation. Each Facilitators' Guide is indexed to a set of pertinent Common Core standards (National Governors Association Center for Best Practices & Council of Chief State School Officers, 2010). Suggestions for reflection and assessment are also included.

In another proposal, Boone argued that well-designed facilitators' guides and/ or video demonstrations/models of the Situational Prompts (and/or various Foci) should be created to support both facilitators and teachers. Professional development opportunities will be necessary to help teachers think about the mathematics inherent in the posed Situations (as well as other Situations that occur in a teacher's classroom). Classroom teachers need to become more adept in determining the mathematics within a posed Situation, thinking deeply and comprehensively about the mathematics, and creating and implementing lessons that develop students' mathematical thinking and understanding.

The importance of teacher professional development and the training of professional development facilitators should not be underestimated. Others at the Users Conference presented comments for assisting professional development facilitators with using the MUST Framework and Situations as a means for mathematics explorations. Many of these ideas are implemented in the NCSM Facilitators' Guides.

Professional learning activities for in-service mathematics teachers

Viktora and Jakucyn discussed using the MUST Framework and a Situation in a professional development day. They felt that such use as the basis of activities for a professional development day might be especially valuable for teachers who are experienced teachers. One of the benefits of using these materials in this way for in-service teachers is to help them make connections across secondary mathematics and to give them a more complete understanding of the overall curriculum. These materials can be used as a springboard to encourage further teacher reflection and collaboration.

One approach would be to begin by presenting the Prompt of an appropriate Situation and encourage the teachers to think about the mathematics involved not only in the specific Situation but also in a larger context. Thinking more generally about the mathematics can be challenging, and requires time for the participants to be prepared for discussion. The Foci of the Situation can be introduced and discussed, one at a time, followed by a conversation about how this activity can be helpful. The MUST Framework should be introduced, giving teachers time to read the document. (Given the constraints teachers face, it is probably not realistic to assume that teachers will have read the document ahead of time.) As part of the effort to encourage understanding of the MUST Framework, it is important to discuss how the Situations were used to develop the MUST Framework. Later in the day, several other Situations can be used. For instance, it might be valuable to group participants by course committees to analyze Situations appropriate to their courses. The day should end with an appropriate culminating activity that, hopefully, will encourage teachers to use these materials as a springboard to improving the mathematical quality of their lessons.

Viktora and Jakucyn also proposed that the same goals might be addressed in a capstone mathematics course organized with the MUST Framework and a set of the Situations.

Planning a summer workshop and follow-up sessions for secondary mathematics teachers

Gober proposed use of the MUST Framework as a guide in planning the activities for a summer workshop and follow-up sessions for secondary mathematics teachers. In her proposal, the MUST Framework helps a teacher to consider and plan learning experiences related to the different aspects of mathematical understanding for teaching: Mathematical Proficiency, Mathematical Activity, and Mathematical Context of Teaching. In this proposal, the MUST Framework could also be shared with the workshop participants as a guide for thinking about their own practice. As part of the workshop, teachers would examine some of the Situations that relate to the mathematics they are exploring and delve deeper into the pertinent mathematics (Mathematical Proficiency and Activity). During the school year, teachers would meet regularly to plan units for the curricula and discuss previously taught units. As teachers plan and share their reflections on lessons that have been taught, contexts for the development of new Situations may arise. Teachers have opportunities to work together to generate new Situations related to their own practice and/or apply knowledge and understanding of the Situations and related mathematics discussed in the summer workshop (Mathematical Context of Teaching). The MUST Framework can also be used to guide the design and implementation of workshop and follow-up activities. The workshops can be tied to the interpretation and implementation of the state mathematics standards.

Developing mathematics teaching proficiency with middle school teachers

Lappan, Borko, and Graysay emphasized the use of the MUST Framework to develop metacognitive awareness of middle school teachers as part of school-based, ongoing inservice to develop proficiency in mathematics teaching. The Situations Project materials, both Framework and Situations, would be used as a template for working with tasks focusing on mathematical areas of the curriculum. One example of a Situation to use with middle-school inservice is Situation 24: Temperature Conversion (see chapter 30), around which the professional development leaders would develop activities for teacher participation and discussion. In examining the Foci of the Situation, questions such as the following would be explored:

- How do your ideas compare to the set of ideas in the Foci?
- Where is the overlap? What ideas in this set are not present in ours? What ideas in our set are not present in theirs?
- What did you learn from this activity?
- What new mathematical ways of thinking did this activity stimulate for you?

- What new insights did this activity give you about the mathematics that would support your students' thinking?
- What new insights did this activity give you about your students' learning?
- What new insights did this activity give you about your teaching?

For homework, each teacher would identify a Prompt from his or her classes that was mathematically problematic for the teacher or the students. Teachers would describe the setting and write the Prompt, Commentary, and Foci, using the Situations Template as a guide. The next inservice session would be built around sharing and discussing these teacher-produced Situations. The cycle of homework and preparation of draft Situations for discussion would continue. The final professional development session would focus on sharing final reflections, discussing and addressing questions to assess the value of the program, such as:

- How have the MUST Framework and the development of Mathematical Foci been useful to my growth as a teacher?
- As a community, how can we use the MUST Framework and Mathematical Foci to continue enhancing our knowledge, dispositions, and practice?

Biweekly professional development group of 8 to 10 mathematics teachers

The context would be an urban high school 2-hour after-school session, two sessions per month. Benson suggested the use of Prompts from the Situations as mathematics stimuli to engage and focus mathematics teachers on a task of common interest. He was undecided whether to share the MUST Framework with teachers but would use it as a guide. Likewise, the initial session would be to get the group of mathematics teachers to suggest Mathematical Foci for the Prompt. Eventually, a goal would be to have teachers create their own Situations based on the discussions. Whether to use the project Situation Foci would be decided on a case-by-case basis.

Biweekly or monthly electronic dialogue for mathematics teachers

Harrington proposed mathematics teacher dialogues built from selected Situations but suggested the use of electronic discussion and a focus on the three broad areas from the MUST Framework: Mathematical Proficiency, Mathematical Activity, and Mathematical Context of Teaching. The structure would have a teacher leader or participant observe an event in a classroom that was fertile enough for expanded conversation and post a description to the electronic discussion board. Facilitating questions would follow as needed to generate discussion about mathematics that might be helpful.

Use of Situations and Framework with in-service elementary teachers in professional development settings

Although our project collected Prompts primarily from secondary settings, several participants at the Users Conference envisioned the use of selected Situations and the process of examining Prompts and examining or creating Foci with elementary teachers. Seaman proposed a summer workshop setting for elementary teachers in which selected Situations would be chosen that address particular issues (e.g., fractions) that assessments have indicated to be a need for elementary teachers. The summer workshops would have a format of 5 consecutive 8-hour days. The MUST Framework would be used to align discussions and solutions with Mathematical Proficiency, Mathematical Activity, and Mathematical Context of Teaching.

The workshop would probably be built around a small number of Situations, perhaps adapted so that the content was within the grasp of elementary teachers. A selected Situation would be the initial focus for thinking about the mathematics in the Prompt. The sequence might follow a pattern of discussion of the problem, work on the problem, and then follow-up discussion. A postworkshop assignment requiring 2 days of work and consisting of a mathematics inquiry or exploration based on ideas generated in the workshop would be completed with electronic reporting of that postworkshop work. Additional Situations from the project could be the source of the postworkshop work.

In-service teacher professional development

Harrington proposed using the MUST Framework for inservice professional development and using the Situations as specific contexts for discussion of mathematical proficiency, mathematical activity, and mathematical context of teaching. His experience suggests that this needs to be a full-day teacher inservice. The format of the inservice, using small-group and large-group work sessions, begins with an overview and discussion of the MUST Framework, followed by discussion of a Situation or multiple Situations. To avoid having teachers view this workshop as a make-it-take-it for their next lesson, he would select Situations and direct discussion toward more general proficiency.

Looking across Situations through the MUST Framework

The use proposed by Lannin and Kastberg selects a set of Situations that are related (for example, ones addressing the use of *variable* or *domain*). Foci across Situations would be examined and discussed by teachers as a means of relating them to the MUST Framework. The goal would be to have teachers explore mathematics ideas that bridge several Situations and to examine the roles that these more global ideas play in the mathematics curriculum. The ultimate goal is to ask, "How does the analysis of Situations impact practice?"

Using the MUST Framework and Situations in a lesson study model of inservice

This proposal came from Winking, using the Gwinnett County's High School Mathematics Institute. The Institute meets for 1 week in the summer and teachers explore tasks to be used in classrooms for the Georgia Performance Standards (GPS). The summer week is followed by meetings twice per month throughout the following year. Teachers are given a potential GPS task and, using the MUST Framework as a guide, they are to use the strategies of lesson study to develop a teaching episode. The identification of an appropriate range of Foci is essential to this process. Several examples were suggested from a web site at Phoenix High School.

Lesson study with secondary mathematics teachers

Hix proposed a synergy between the MUST Framework and Situations and the inservice process of lesson study. One challenge of implementing lesson study is that teachers in the United States often do not understand how to attend to the mathematics of a lesson. Attending to the mathematics of a lesson includes the development of mathematics necessary for the teacher to carry out the task of teaching. This mathematics goes beyond the mathematics of the lesson, and it is the facet that often eludes teachers. Considering the Situations and multiple Foci, the teachers may be able to more directly and deeply develop this mathematics. The teachers would begin their lesson study during the summer, studying the MUST Framework and using the Situations to begin their lesson study activities.

Personal reference documents in the MUST Framework and Situations

Jakucyn and Viktora proposed that the MUST Framework and the total set of Situations might be viewed as a personal reference guide for mathematics teachers to use in personal planning for teaching episodes. Once a Situation is selected relevant to the teacher's episode planning, the Foci will help to provide a better idea of the depth and breadth of the mathematics. This will enable the teacher to use the ideas of mathematical proficiency, mathematical activity, and the mathematical context of teaching. Part of the goal is to encourage teacher reflection and these materials provide a range of support for that reflection.

An extension of this reference guide or library concept is to make the collection one that grows as new Situations might be developed locally or shared with others.

Master's course in curriculum design for secondary mathematics teachers

Burke observed his experiences with master's degree students who often do curriculum redesign for the purpose of action research or a capstone project. He proposed using the MUST Framework and Situations for a decision-making course on guiding the redesign of particular curriculum units. The course would present students with examples of Situations, each of which have five Foci—one elucidating each of the Mathematical Context of Teaching components of the MUST Framework. Students would be asked to contribute more options within each Focus that went beyond the examples included in the Focus. This would give students the opportunity to illustrate or recognize how specific aspects of Mathematical Proficiency and Mathematical Activity can emerge as critical features of the mathematics their own students might do. Collectively, the Situations should give multiple focused expositions of the type of proficiency that would be helpful in a variety of topics and contexts for curriculum decision-making.

With this grounding in the MUST Framework and Situations as pointers to the multifaceted nature of understanding and doing mathematics, the students would design their own Situations both using Prompts provided to them and in Situations for which they would provide the Prompts. Once students have gained insight into the rich possibilities revealed by the MUST Framework and have addressed other issues related to curriculum design (students with special needs, cultural relevance, methods for engaging students, use of technology and other instruments, pedagogical decision-making, etc.), they would be asked to redesign a unit of mathematics in their curriculum. They would be asked to teach the unit and use daily journals to note student learning, wondering, and questioning. They would be asked to produce two or three pivotal Situations that capture the highlights of the mathematical thinking that was achieved in the unit. They would be asked to pay particular attention to the Foci on assessing and accessing student thinking in the write-up of these Situations while still reflecting on and producing Foci related to the other components of MUST Framework. These Situations would be submitted as part of their overall evaluation and analysis of the unit and its effect on student learning.

Using the MUST Framework to guide teachers in their analysis of current curricular materials

This proposal from Sheehy takes the use of the MUST Framework and Situations into the schools where teachers deal daily with the implementations of mandated standards. In this case, the curriculum standards to be implemented were the Georgia Performance Standards, but the ideas here would apply to Common Core State Standards (National Governors Association Center for Best Practices & Council of Chief State School Officers, 2010) as well. The setting was envisioned as a small group of teachers from the same school meeting to select materials and plan lessons. The role of the Situations might be that of a curriculum-materials resource as they are coordinated with standards and resources the teachers have available.

The MUST Framework would be used to guide discussions teachers might have in identifying the need to use a particular task with students. Then, that task would be used for developing sets of Foci related to the task, rather than to a Prompt. The MUST Framework would help teachers attend to mathematical proficiency, mathematical activities, and the mathematical context of teaching. By generating a list of Mathematical Foci related to each task, teachers will be able to reflect on their own mathematical knowledge as well as that of their colleagues. Teachers take the time to study mathematics together, discuss multiple means of representation, identify any misconceptions or gaps in their own understanding of a given topic, offer predictions about the concepts that will be difficult for students to grasp, and connect this planning to their teaching practice.

Building from Mathematical Foci toward Statistical Foci

In this proposal, there is a plea to use and extend the MUST Framework for teachers to think specifically about how the lenses of Mathematical Proficiency, Mathematical Activity, and Mathematical Context of Teaching can be applied to the practice of teaching statistics. It would include having teachers augment the MUST Framework to involve more statistics ideas. Situation 38: Mean and Median (see chapter 44) exemplifies a Situation that could help to direct teacher discussions of things both mathematical and statistical. There is a dual purpose here: One is for teachers to know the mathematical structures that are useful in understanding statistical concepts, and the other is for developing better understanding of the statistics concepts and ways of thinking that should be used when engaged in data analysis and probability tasks.

Use of Situations and the MUST Framework with in-service elementary teachers in professional development settings

Even though most of the materials from the Situations Project are oriented to the middle school or secondary school level, many participants saw adaptations that were appropriate for use with the teaching of elementary-level mathematics. Seaman developed this proposal, with several options, for use in professional development activities in mathematics summer-workshop settings for elementary teachers. Workshops have generally consisted of 5 consecutive 8-hour days. In this proposal, emphasis would be placed on getting teachers to discuss and understand Mathematical Proficiency, Mathematical Activity, and the Mathematics Context of Teaching from the MUST Framework. The Prompts would form the launch point for student inquiry and engagement. This would be followed by engaging teachers in thinking about and discussing multiple/alternative approaches to responding to the Prompt, such as are suggested in the Mathematical Foci. This would expose teachers to a variety of problems arising in school mathematics contexts that generate sophisticated mathematics questions and require precise definitions and arguments to resolve. Hopefully, this would foster a feeling of mathematical empowerment in teachers that they are not alone in being unfamiliar with certain mathematics they may have encountered in learning and teaching settings. Additionally they will see that simple-sounding mathematics questions may require sophisticated mathematics to resolve. Even for teachers who may not have the mathematics background to generate mathematical responses to the Prompts, the hope is to encourage them to contribute as much as they can and then use the Mathematical Foci to help fill in gaps in solutions or explanations.

USES IN TEACHING MATHEMATICS CONTENT COURSES

Many of the proposed uses of the Situations Project materials dealt with instruction in mathematics courses. Most of these were upper-division courses for teachers.

Use of the MUST Framework as an organizational guide for content courses for middle school or high school

Charlene Beckmann discussed ways to use the Situations Project Materials in content courses for middle school or high school students. The MUST Framework was viewed as a guide for engaging students to think deeply about the mathematics in which they engaged. Not only does the Framework provide a guide for the instructor in organizing courses but, as Situations are examined, it also provides a structure for students to identify elements of mathematical proficiency, mathematical activities, and the mathematical context of teaching. These uses envisioned adapting the Situations for various courses but also creating Situations to enhance the coherence of each course.

Using the MUST Framework in teaching advanced graduate mathematics courses

As a mathematician, Smith envisioned using the MUST Framework in teaching a range of advanced graduate mathematics courses. The Framework succeeded in offering a clear and convincing argument that mathematical teaching and learning must involve three aspects: internalizing procedures and facts, creative use of knowledge in problem solving, and facilitating learning through understanding how we think about mathematics, (i.e., knowing, doing, and teaching). Smith viewed the MUST Framework as useful to an individual teacher as a guide to planning a course, as a guide for conducting an individual class, and a guide for assessing the success of the class. Smith assigned the MUST Framework for students to read and absorb. The use envisioned here would give the students a Prompt and challenge them to develop their own path, or a collective path, to a set of Foci. Smith's ideas included teaching college courses for mathematics majors, mathematics courses for teacher candidates, and high school courses for honors mathematics students.

The MUST Framework as a guide for teaching an abstract algebra course for teachers

Cofer proposed uses of the Situations Project materials in an abstract algebra course for prospective teachers. The MUST Framework would serve as a background document for the instructor to use in organizing and assessing course activities. Situations would be selected and adapted for relevant abstract algebra topics. The mathematical goal would be to scaffold students' understanding of advanced and theoretical algebra concepts by connecting the key ideas to familiar mathematics relevant to the school curriculum.

Using the MUST Framework and the Situations to organize capstone courses for teacher preparation

Capstone courses are a key component of mathematics teacher preparation at many institutions. The use of Situations in such courses was proposed by several Users Conference participants. Hollylynne Lee proposed using the MUST Framework and engaging students in explicit discussions of Mathematical Proficiency, Mathematical Activity, and the Mathematical Context of Teaching as various topics were explored from an advanced perspective. This capstone course would coordinate the use of various Situations with the presentation of material in a textbook that might be used.

Transition to higher mathematics course: A course emphasizing mathematics proof and problem solving

Benson proposed using the processes from the Situations Project in the development of a course for secondary teachers in a transition to higher mathematics course or for a course on mathematics proof and problem solving. A set of Situations would be developed for this course using the Situations Project model. In other words, the syllabus would be built around a set of Situation-like episodes developed explicitly for each course.

Organizing a calculus course

McCrory and Winsor proposed using Situations within a calculus course to show different representations of key concepts. They would select or create specific Situations but not explicitly use the MUST Framework.

Algebra for elementary and middle school mathematics teachers: Functions

Senk proposed incorporating the use of the MUST Framework and selected Situations into two existing courses at Michigan State University. These were upper-division mathematics courses for preservice elementary and middle school mathematics teachers majoring in mathematics: Algebra for Elementary and Middle School Mathematics Teachers and Functions and Calculus for Elementary and Middle School Mathematics Teachers. The MUST Framework would be used to focus the students on the mathematical activities, mathematical proficiency and the mathematical context of teaching. These courses are to broaden and deepen the mathematics understanding of future teachers, and attention to the MUST Framework is a way of doing that. Typical use of the Situations would present the Prompt followed by individual or group exploration and discussion followed by class discussions of the mathematics relevant to the Prompt. Examination of the set of Foci in the Situation might then be pursued. Also, homework could be developed from some Situations. Homework might involve assigning an entire Situation to study, it might involve presenting the Prompt and having the students produce potential Foci, or it might involve having the students develop a particular Focus.

A writing in mathematics course

Bona described a course, Writing in Mathematics, required of all mathematics majors at the University of Illinois at Chicago. The Situations provide a rich source of ideas for mathematics essay topics.

Material for exploring misconceptions about $f(x + y) = f(x) + f(y)$

Royster proposed developing a Situation-like episode to investigate the common misconceptions about $f(x + y) = f(x) + f(y)$. Misconceptions include the assertion that it is never true and assertions that it is true for cases for which it is not true. The developed material would lead students to explorations of conditions under which the equation is true for all arguments. The material would be appropriate for a capstone course.

USES IN GRADUATE MATHEMATICS EDUCATION

Many of the proposals from the Users Conference could be adapted to elements of graduate study in mathematics education. In this section, however, some work at the University of Georgia is discussed. The first of these is a 3-semester-hour course, Mathematics Connections, for which the syllabus was built around use of the Situations Project materials. The second is a brief account, reported at the Users Conference, of using the development of a Situation as a class project in a master's degree seminar. The third is an account of an applied project for a Specialist in Education degree student who was a high school mathematics teacher who used Prompts drawn from his classes. In addition, doctoral students have used the MUST Framework to help build rationale for research questions.

A course on mathematical connections

The Mathematics Connections course at the University of Georgia is a mathematics teaching field course for graduate students—in-service teachers working toward the Master of Education degree or candidates pursuing initial certification at the master's level in the Master of Arts in Teaching degree. All students would have mathematics background equivalent to an undergraduate major in mathematics.

James Wilson developed a syllabus and taught this course built around four components:

1. The Common Core State Standards in Mathematics (CCSSM) (National Governors Association Center for Best Practices & Council of Chief State School Officers, 2010);
2. The CCSSM Standards for Mathematical Practice;
3. The MUST Framework and Situations
   a. Calculation of Sine (Situation 35, chapter 41)
   b. Inverse Trigonometric Functions (Situation 16, chapter 22)
   c. Locus of a Point on a Moving Segment (Situation 27, chapter 33)
   d. Perfect-Square Trinomials (Situation 23, chapter 29)
   e. Simultaneous Equations (Situation 18, chapter 24)
   f. Area of Plane Figures (Situation 30, chapter 36)
   g. Exponent Rules (Situation 10, chapter 16)
   h. Adding Square Roots (Situation 14, chapter 20)
   i. Product of Two Negative Numbers (Situation 2, chapter 8)
   j. Summing the Natural Numbers (Situation 4, chapter 10)
   k. Powers (Situation 11, chapter 17)
   l. Circumscribing Polygons (Situation 34, chapter 40)
   m. Division Involving Zero and Situation 1 Facilitators' Guide From NCSM (Situation 1, chapter 7); and
4. The creation of new Situations.

The course objectives were:

1. To learn about and develop an understanding of the CCSSM grade-level standards;
2. To become aware of and implement the CCSSM Standards for Mathematical Practice;

3.   To explore Mathematical Understanding for Secondary Teaching (MUST) via the components of Mathematical Proficiency, Mathematical Activity, and the Mathematical Context of Teaching;
4.   To explore MUST principles via examination of Situations prompted from classroom episodes;
5.   To create MUST Situations from possible classroom episodes;
6.   To engage in independent investigations of mathematics topics using the MUST Framework; and
7.   To examine correspondence among the MUST Framework, CCSSM standards, and the NCTM *Principles and Standards for School Mathematics* (2000).

The first few meetings of the course were devoted to discussions of the MUST Framework, the Situations Project generally, the CCSSM Standards for Mathematical Content, and the CCSSM Standards for Mathematical Practice. These meetings accomplished a general orientation to processes and products of the Situations Project.

Throughout the course, the typical format of the class sessions was developed around discussions. In using the Situations from the project, the students preferred to explore the Prompt, collectively or individually generate some ideas for Mathematical Foci, engage in discussion, and then examine the Foci from the project. Usually the first examination of a Prompt from a project Situation was in a homework assignment. All materials were available on the class website.

After discussion of about six of the Situations, we moved to creating drafts of new Situations. This was a critical phase of the course that required direction of the instructor and willingness to discuss the process as well as potential products. Students were given a template for the Situation format with Heading, Prompt, Commentary, Foci, and Postcommentary. They were encouraged to generate Prompts from their own experience, but they were also given a list of Prompts from the project that had not been developed into final Situations.

The creation of Situations involved the basic ideas described in Chapters 3 and 5. The completion, discussion, review, and revision of Situations in final form were significant activities for the course. After examining and discussing the first five Situations in the preceding list, the course included a group project to respond to the following Prompt:

An Algebra II class has been examining the product of two linear expressions:

$$(ax + b)(cx + d).$$

Well into the class, a student asks, "What would happen if we DIVIDED one linear expression by another?"

Two groups worked independently on developing a set of topics that would eventually be developed into Foci. There were discussion sessions within and across the groups. The writing of the Foci demanded much more than the students anticipated, but they came to believe that the writing and revision process contributed greatly to their understanding of the topics. The two teams came up with slightly different final products—different in terms of the Foci and different in the discussion of the Foci. Regardless, they all agreed that the process of creating a completed Situation, using the template to standardize the process, contributed to a more thorough understanding of the topic. In each of the two groups' products, essential points of asymptotes, undefined values, division of polynomials, and roots were addressed.

During the second half of the course, each student was expected to complete three Situations in final form. Students presented their drafts and engaged the class in discussions. Every student-generated Situation went through multiple discussions and revisions. All drafts and final write-ups were posted on the course web page. The topics for the final Situation write-ups were:

Quadrilaterals;

Pythagorean theorem;

What is π?;

Limit and sum of a series;

Solving logarithmic equations;

Parentheses and brackets;

180 degrees in a Euclidean triangle;

Adding fractions;

Increasing or decreasing functions;

Infinity + infinity;

Zero vs. nothing;

Slopes of perpendicular lines;

Complex roots;

Line of best fit;

Exponential bases;

Translating functions;

False identity;

Multiplying binomials;

Exponential rules;

Irrational lengths;

Numbers raised to the zero power;

Congruent triangles vs. similar triangles;

Zero slope; undefined slope; and

Simultaneous equations.

This use of the Situations materials and the Situations Project strategies fit nicely for this particular course with a small cadre of well-prepared graduate students. The free-flowing and creative student discussions introduced some unanticipated topics and required some additional preparation by the instructor. From the perspective of the instructor, these course activities demanded more depth and breadth of mathematical knowledge than had other approaches to this course.

Looking back
The following are some observations about use of the Situations Project processes and materials in this course.

1. The course demanded that the instructor respond to a wide range of mathematics content knowledge, make suggestions to students, and work to redirect some of the issues that grew out of the discussions.
2. Students had a strong tendency to connect course content to their knowledge of existing school curricula. Suggestions from the instructor or from the discussions were often needed to broaden the perspective.
3. Connecting discussions to the CCSSM content standards was hampered by the views these students held of the CCSSM standards. Basically, the CCSSM content standards were thought to represent all that was to be included in the school mathematics curriculum rather than a minimum level. If a Situations topic was not explicitly mentioned in the Common Core materials, there was resistance to considering it.
4. We found, via discussions, much agreement between the CCSSM Standards for Mathematical Practice and the MUST Framework of Mathematical Proficiency, Mathematical Activity, and the Mathematical Context of Teaching.
5. Students expressed a desire to revise the Situations from the original Situations Project. This may have been an artifact of our organization of the course, in that they reviewed the Prompt, and then generated their own list of topics that might be considered for Foci. When they examined the Project-produced Foci, they often found topics that they thought to be

important that were not present or were stated in a different way. On the other hand, they usually found additional topics in the Project-produced Foci that they had not uncovered. Regardless, this process was helpful for getting students to think critically about the mathematics content represented in the Prompt.

Master's seminar on mathematical knowledge for teaching

The goal was to understand the process of specifying mathematical knowledge for teaching. The seminar chose to pursue this goal by looking at a particular mathematical entity (a real-valued function that is the quotient of two first-degree polynomials) and using a much earlier draft of the MUST Framework as background for examining this topic. No Situations were used directly. Students in the seminar had studied and discussed several of the Situations. We felt the process of developing a Situation write-up for a new topic would be instructive in the seminar format.

This was a master's-level graduate seminar in which all participants were either experienced teachers or master's degree candidates who had completed a mathematics teaching internship or student teaching in mathematics at the secondary level. Three of them were graduate students with teaching experience outside the United States. Class met once per week.

The following Prompt was proposed:

A high school algebra class was studying multiplication of binomials (as well as factoring and roots) of the form $ax + b$ and $cx + d$. A student asks, "What if we divide instead of multiply?"

The students organized into small groups and worked on developing the Situation write-up over a period of 3 weeks. There were extensive revisions.

A significant activity, individually as well as in the group, was to produce a graph of an example and play with changing the parameters a, b, c, or d. This helped the seminar participants to identify special cases and general patterns. For instance, using a graphing calculator, students produced the graph of $y = \dfrac{x+4}{x+2}$ (see Figure 6.1). In other words, the parameters were $a = 1$, $b = 4$, $c = 1$, and $d = 2$.

Then, by substituting n for one of the parameters and animating the graph, a family of graphs could be examined. This eventually led to algebraic characterizations of various observations about the curves, asymptotes, domain, roots, and the like.

The word *play* was chosen intentionally. In no case did these students want to argue that the graphs were proving anything.

By the third week, the class had produced the following Commentary and Foci:

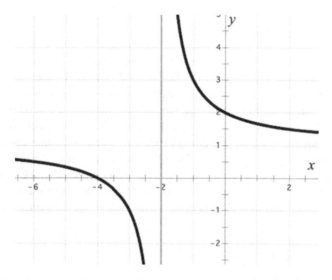

FIGURE 6.1. Graph of $y = \dfrac{x+4}{x+2}$.

Commentary

In our Foci, we are assuming that we have a new function $f(x) = (ax + b) / (cx + d)$. We prove that this is a function within Mathematical Focus 3 (asymptotes). With this new function, we find that there are restrictions on the domain and range, unlike the function $f(x) = (ax + b)(cx + d)$. In this Situation, we do not consider complex numbers. We see that the graph of this function is always a hyperbola, with the exception of the degenerative (or undefined) cases. To completely discuss hyperbolas, we found it important to discuss how to represent the roots and intercepts, graphically and symbolically. It is also necessary to show how the coefficients a, b, c, and d affect the graph. In the last two Foci, we explain the limits and the inverse of the function.

Mathematical Foci

Mathematical Focus 1: Roots and intercepts

Mathematical Focus 2: How the parameters a, b, c, and d affect the graph

Mathematical Focus 3: Asymptotes

Mathematical Focus 4: Limits

Mathematical Focus 5: Inverse of $f(x)$

Mathematical Focus 6: The function is always a hyperbola, with the exception of degenerative or undefined cases.

As the instructor, the author found several recurring issues in this course. These included (a) struggle with precision of language; (b) constant pressure to change focus to pedagogy rather than mathematical knowledge; (c) lack of knowledge of, or interest in, applications or historical perspective; (d) pressure to be guided by (even satisfied with) what is in some particular syllabus or textbook; (e) redundancy; (f) difficulty with being explicit about their perceived connections to the MUST Framework; and (g) unwillingness to relate this work to the mathematics of more general functions that are the ratio of two polynomials, one or both of which may be of degree greater than 1. On the other hand, there was strong evidence that the seminar participants engaged in challenging reflection about this task.

Specialist in education applied project

The Specialist in Education degree at the University of Georgia is a degree program for 1 year of study beyond the master's degree. In mathematics education, a very significant component of the degree is an applied project that is developed with some connection to the student's teaching assignment. One such project at the University of Georgia was developed by a teacher at the local high school who held an undergraduate degree in mathematics from the University of Tennessee and the Master of Education degree from the University of Georgia. The applied project developed from interest in our Situations Project but not as a graduate student working on the project. Using the Project guidelines for Situation development from Prompts observed from classroom episodes, the teacher developed 10 Prompts from his classes. Then he developed the following 10 Situations building on those Prompts:

PN1. Dividing an Inequality by a Negative Number

PN2. Slopes of Perpendicular Lines

PN3. Absolute Value in the Complex Plane

PN4. Simplifying Rational Expressions

PN5. Finding the Mean of a Set

PN6. Modular Arithmetic

PN7. Complex Numbers

PN8. Functional Representation

PN9. Constructing a Tangent

PN10. Area of Intersection of Two Circles

This was a very challenging applied project. This teacher did not have the benefit of a team of people working with him to share ideas for Foci. Each Situation required several revisions. On the other hand, three of his Situations were eventually adopted by the Situations Project.

USES IN ASSESSMENT

Assessment activities making use of the MUST Framework and the Situations were proposed at the Users Conference. Portnoy suggested creating an instrument to assess the MUST level of novice teachers upon graduating from various teacher education programs. For example, at the University of New Hampshire there are two paths to certification, an undergraduate path and a postgraduate path. The first requires the candidate to spend 1 semester (usually the final semester of the program) student teaching. The second places graduates in yearlong internships at cluster sites (similar to professional development schools). A question that arises is whether one could evaluate and compare these two programs using measures based on the MUST Framework.

At a different level, assessment activities are built into each of the NCSM facilitators' guides. These assess the specific goals of a particular Situation and the professional development associated with it.

Understanding how to assess statistical thinking and reasoning

The MUST Framework and a Situation were proposed by Franklin to help high school teachers come to better understand how to assess statistical thinking and reasoning. Rather than tests of procedural fluency, the proposed assessment used the MUST Framework with a statistics Situation and a set of guided questions to evaluate the goals embedded in the Situation. This particular proposal was geared either to in-service high school teachers in professional development workshops or to students in preservice mathematics education courses. The statistics example was Situation 38: Mean and Median (chapter 44). Some of the guided questions suggested were:

1. Is this student's approach the one you would have used?
2. If not, what approach would you present?
3. Discuss the strengths and weaknesses of the student's write-up. How does the student's approach compare to the other approach(es) suggested within your group?
4. What type of thinking and/or reasoning was used by the student (mathematical or statistical)? Explain.
5. Of the approaches under consideration (student and group), how do these approaches connect with the Framework (MP, MA, and MCT)?

6.   What are the intended mathematical and/or statistical goals that you see being assessed with the Prompt?
7.   How do these goals connect with the Framework?
8.   What goals are missing from the Prompt that you consider important to assess?
9.   How would you improve the Prompt to meet these missing goals?
10. Develop a scoring rubric for the new and improved Prompt. Write a model solution for the Prompt. Also, note other possible acceptable solutions. Clearly outline what you would consider a complete solution, a substantial solution, a developing solution, and a minimal solution from the student.
11. After developing the rubric, revisit the Prompt. Do you believe the rubric allows you to adequately assess the intended goals?
12. How might you change your teaching to better meet these goals?

Use of the Situation structure in a mathematics instruction examination

James Wilson implemented an assessment in a University of Georgia graduate mathematics education class based somewhat on Situation 2: Product of Two Negative Numbers (see chapter 8). Situation 2 had not been explicitly studied by this class. In the assessment, a modified version of the Situation 2 Prompt was given and then the students were asked to provide two responses. The first was to list as many mathematics perspectives as they could determine. This was similar to making a topic list for a range of Foci. The second was to describe an instructional sequence they would use to help ninth-grade students understand why the product of two negative numbers is positive. It was assumed that ninth graders would "know the rule."

The results were surprising. The lists of mathematics topics varied greatly from one to as many as eight. Only 3 of the 10 teachers listed some version of using the properties of number systems (e.g., inverses, distribution, identities) as suggested in CCSSM (National Governors Association Center for Best Practices & Council of Chief State School Officers, 2010). On the second part, however, the proposed instruction from 7 of the 10 experienced teachers attempted some version of using multiplication as repeated addition as a means to develop understanding. None of them saw any unjustified statements in the process. The Common Core objective of justifying that $(-1)(-1) = 1$ on the basis of the properties of number systems was cited by 7 of the students but none of them implemented that approach in their proposed instruction.

The assessment was successful, not in deciding progress, but rather in redirecting the class discussions to some useful and in-depth explorations. This led to several examinations of the use of the number-system properties, geometric interpretations, and use of applications and models.

CONCLUDING REMARKS

The uses described in this chapter reflect the participants' enthusiastic response to using Situations and the MUST Framework in research, professional development, mathematics content courses, and assessment. The participants argued for uses with graduate and undergraduate students, and some extended the use from working with secondary teachers to include elementary teachers. The uses of the Situations and MUST Framework are shared in this chapter with the expectation that these ideas can be refined and adapted to the reader's work. We urge the reader to build on the current work and try new uses while working to improve mathematical understanding for secondary teaching.

REFERENCES

Blume, G. W., Briars, D., Heid, M. K., Mitchell, S., Schrock, C.,Viktora, S., Wilson, J., & Zbiek, R. M. (2013). *Sample facilitators' guide. Division involving zero.* University Park, PA and Athens, GA: Mid-Atlantic Center for Mathematics Teaching and Learning, Center for Proficiency in Teaching Mathematics, and National Council of Supervisors of Mathematics.

Conner, A., Wilson, P. S., & Kim, H. J. (2011). Building on mathematical events in the classroom. *ZDM Mathematics Education, 43*, 979–992. doi:10.1007/s11858-011-0362-1

National Council of Teachers of Mathematics. (2000). *Principles and standards for school mathematics.* Reston, VA: Author.

National Governors Association Center for Best Practices & Council of Chief State School Officers. (2010). *Common core state standards for mathematics.* Washington, DC: Authors. Retrieved from http://www.corestandards.org/assets/CCSSI_Math%20 Standards.pdf

CHAPTER 7

DIVISION INVOLVING ZERO

Situation 1 From the MACMTL–CPTM Situations Project

Bradford Findell, Evan McClintock, Glendon Blume,
Ryan Fox, Rose Mary Zbiek, and Brian Gleason

PROMPT

On the first day of class, preservice middle school teachers were asked to evaluate $\frac{2}{0}, \frac{0}{0}$, and $\frac{0}{2}$ and to explain their answers. There was some disagreement among their answers for $\frac{0}{0}$ (potentially 0, 1, undefined, and impossible) and quite a bit of disagreement among their explanations:

- Because any number over 0 is undefined;
- Because you cannot divide by 0;
- Because 0 cannot be in the denominator;
- Because 0 divided by anything is 0; and
- Because a number divided by itself is 1.

Mathematical Understanding for Secondary Teaching: A Framework and Classroom-Based Situations, pages 95–101.
Copyright © 2015 by Information Age Publishing

COMMENTARY

The mathematical issue centers on the possible values that result when 0 is the dividend, the divisor, or both the dividend and the divisor in a quotient. The value of such a quotient would be 0, undefined, or indeterminate, respectively. The Foci use multiple contexts within and beyond mathematics to represent and illustrate these three possibilities. Connections are made to ratios, factor pairs, Cartesian product, area of rectangles, and the real projective line.

MATHEMATICAL FOCI

Mathematical Focus 1

An expression involving real number division can be viewed as real number multiplication, so an equation can be written that uses a variable to represent the number given by the quotient. The number of solutions for equations that are equivalent to that equation indicates whether the expression has one value, is undefined, or is indeterminate.

One can think of a rational number as being the solution to an equation. If division expressions involving 0 also represent rational numbers, equations involving these expressions should have consistent results. To find the solution of the equation $\frac{0}{2} = x$, consider the equivalent statement $2x = 0$, which yields the unique solution $x = 0$. To see the impossibility of a numerical value for a rational number with a 0 in the denominator, consider the equation $\frac{0}{0} = x$, and its potentially equivalent equation, $0x = 0$. Because any value of x is a solution to this equation, there are infinitely many solutions; hence, there is no unique solution, and so the expression $\frac{0}{0}$ is indeterminate. Using the same thinking, if $\frac{2}{0} = x$, then $0x = 2$. No real number x is a solution to this equation, so the expression $\frac{2}{0}$ is undefined.

Mathematical Focus 2

One can find the value of whole number division expressions by finding either the number of objects in a group (a partitive view of division) or the number of groups (a quotitive view of division).

In partitive division, a given total number of objects is divided equally among a number of groups. A nonzero example would be $\frac{12}{3}$, in which 12 objects are shared equally among 3 groups and the question concerns how many objects would be in one group. Similarly, $\frac{0}{2}$ can be thought of as 0 objects in 2 groups, which means 0 objects in each group. Additionally, the expression $\frac{0}{0}$ is a model for dividing 0 objects among 0 groups. In other words, if 0 objects are shared by 0 groups, how many objects are in 1 group? There is not enough information to answer this question, so the expression $\frac{0}{0}$ is indeterminate. If the number of objects in a group is 3, or 7.2, or any size at all, 0 groups would have 0 objects. Similarly, $\frac{2}{0}$ is a model for the question: If 2 objects are shared by 0 groups, how many objects are in 1 group? In this case the number of objects in the group is undefined, because there are 0 groups.

Using a quotitive view of division, the expression $\frac{12}{3}$ is interpreted as a model of splitting 12 objects into groups of 3 and asking how many groups can be made. So $\frac{0}{2}$ can be thought of as splitting 0 objects into groups of 2, which means 0 groups of size 2. The expression $\frac{0}{0}$ models the splitting of 0 objects into groups of size 0, and asks how many groups can be made. Because there could be any number of groups, there are an infinite number of solutions, so the expression is indeterminate. Lastly, the expression $\frac{2}{0}$ models the splitting of 2 objects into groups of 0 and asking how many groups can be made. Regardless of how many groups of 0 are removed, no objects are removed. Therefore, the number of groups is undefined.[1]

Mathematical Focus 3

The mathematical meaning of $\frac{a}{b}$ (for real numbers a *and* b *and sometimes, but not always, with* b $\neq 0$*) arises in several different mathematical settings, including slope of a line, direct proportion, Cartesian product, factor pairs, and area of rectangles. The meaning of $\frac{a}{b}$ for real numbers* a *and* b *should be consistent within any one mathematical setting.*

There are mathematical situations in which ratios are necessary, and a quotient can be reinterpreted as a ratio. For example, the slope of a line between two points in the Cartesian plane can be defined as the ratio of the change in the y-direction to the change in the x-direction, or as the rise divided by run. In the case of two coincident points, the change in the y-direction and the change in the x-direction are both 0, which means that the rise divided by run is $\frac{0}{0}$. There are an infinite number of lines through two coincident points, and so the slope is indeterminate. In the case of two points lying on the same vertical line whose y-coordinates differ by a, the change in the y-direction is a, and the change in the x-direction is 0. It might be tempting to claim that because the slope of a vertical line is undefined, $\frac{a}{0}$ is undefined. However, this claim is exactly what needs to be shown.

The model for direct proportion, $y = kx$, represents a family of lines through the origin. For y and nonzero x as the coordinates of points on a line given by $y = kx$, the ratio $\frac{y}{x}$ equals k, which is constant. If this ratio held for the coordinates of the origin, it would be $\frac{0}{0} = k$. However, no one value of k would make sense as the value of $\frac{0}{0}$ because the origin is on every line represented by an equation of the form y = kx. Thinking about the equation $y = kx$ in terms of number relationships also leads to the conclusion that the value of $\frac{0}{0}$ cannot be determined: If $y = kx$ and $x = 0$, then $y = 0$ and k can be any real number, just as in Focus 1. It is important to note that in the case in which $x = 0$ and $y \neq 0$, such as $\frac{2}{0}$, it is difficult to explain via direct proportion; if $y = kx$, then $x = 0$ and $y \neq 0$ is an impossible circumstance.

A different mathematical context for looking at division involving 0 is the Cartesian product. A nonzero example is this: If 12 outfits can be made using 3 pairs of pants and some number of shirts, how many shirts are there? There must be 4 shirts, as this would give 12 pants–shirt combinations. Similarly, if 0 outfits can be made using 2 pairs of pants and some number of shirts, there must be 0 shirts. If 0 outfits can be made using 0 pairs of pants and some number of shirts, the number of possibilities for the number of shirts is infinite. Lastly, how many shirts are there if there are 2 outfits and 0 pairs of pants? No possible number of shirts can be used to make 2 outfits if there are 0 pairs of pants.

In the context of factor pairs, a division expression with an integral quotient represents an unknown integer factor of the dividend. For $\frac{12}{3}$, 3 and the quotient are a factor pair for 12. In this expression, 12 can be written as the product of 3 and the quotient: $12 = 3 \times 4$. For $\frac{0}{2}$, 2 and the quotient are a factor pair for 0. Therefore, the quotient must be 0, because $0 \times 2 = 0$. For $\frac{0}{0}$, 0 is part of an infinite number of factor pairs for 0 and so the expression is indeterminate. For $\frac{2}{0}$, 0 is not part of any factor pair for 2, thus the expression is undefined.

One side length of a rectangle is the quotient of the area of the rectangle and its other side length. Suppose that rectangles can have side lengths of 0. If a rectangle has area 12 and height 3, what is its width? The width would be 4. If a rectangle has area 0 and length 2, its width is 0, suggesting that 0 divided by 2 is 0. If a rectangle has area 0 and height 0, what is its width? Any width is possible, suggesting that 0 divided by 0 is indeterminate. If a rectangle has area 2 and height 0, what is its width? It is impossible for a rectangle to have area 2 and height 0, suggesting that 2 divided by 0 is undefined.[2]

Mathematical Focus 4

Contextual applications of division or of rates or ratios involving 0 illustrate when division by 0 yields an undefined or indeterminate form and when division of 0 by a nonzero real number yields 0.

If Angela makes 3 free throws in 12 attempts, what is her rate of success? If Angela makes 3 free throws in 12 attempts, her rate is $\frac{1}{4}$. If Angela makes 0 free throws in 2 attempts, her rate is 0. If Angela makes 0 free throws in 0 attempts, her rate could be any of an infinite number of rates. On the other hand, because it is not possible for Angela to make 2 free throws in 0 attempts, it is not possible to determine her rate.

Determining the speed of an object over a given period of time is another rate context. If one travels 12 miles in 3 hours, how fast is one traveling? Traveling 12 miles in 3 hours yields a rate of 4 miles per hour. If one travels 0 miles in 2 hours, one is traveling 0 miles per hour. If one travels 0 miles in 0 hours, how fast is one traveling? An infinite number of speeds are possible. If one travels 1 mile in 0 hours, how fast is one traveling? This situation is impossible because traveling for 0 hours means one is not traveling at all. [Note that there is a sense of infinite speed here, so it might be tempting to define $\frac{1}{0}$ as infinity. However, this leads to further complications, as noted in Focus 5.]

Additionally, the idea of rate is prevalent when calculating unit price, such as when purchasing multiple quantities of an item in a store. If $12 buys 3 pounds of tomatoes, what is the cost of 1 pound? If $0 buys 2 pounds of tomatoes, then 1 pound can be bought for $0. If $0 buys 0 pounds of tomatoes, there are an infinite number of possible costs for 1 pound. If $2 buys 0 pounds of tomatoes, it is not possible to determine the number of dollars needed to buy 1 pound.[3]

Mathematical Focus 5

Slopes of lines in two-dimensional Cartesian space map to real projective one-space in such a way that confirms that the value of $\frac{a}{b}$ when b = 0 is undefined if a ≠ 0 and indeterminate if a = 0.

In the Cartesian plane, consider the set of lines through the origin, and consider each line (without the origin) to be an equivalence class of points in the plane. Except when $x = 0$, the ratio of the coordinates of a point gives the slope of a line—the line that is the equivalence class containing that point. The origin must be excluded because it would be in all equivalence classes, which suggests that $\frac{0}{0}$ would be the slope of any line through the origin [see Focus 3]. Note that the slope of a line through the origin is equal to the y-coordinate of the intersection of that line and the line $x = 1$. The slope then establishes a natural one-to-one correspondence between the equivalence classes (except for the equivalence class that is the vertical line, because it does not intersect the line $x = 1$) and the real numbers. Thus, the real numbers give all possible slopes, except a slope for the vertical line.

When $x = 0$, all points in the equivalence class lie on a vertical line, the y-axis. (Again the origin must be excluded from this equivalence class.) As positively sloped lines approach vertical, their slopes approach ∞, suggesting the slope of the vertical line to be ∞. As negatively sloped lines approach vertical, their slopes approach $-\infty$, suggesting the slope should instead be $-\infty$. However, there is only one vertical line through the origin, so it cannot have two different slopes. To resolve this ambiguity, one might decide that ∞ and $-\infty$ are the same "number" because they should represent the same slope. This set of all possible slopes then consists of all real numbers and one more number, which might be called ∞. Imagine beginning with the extended real line, $\mathbb{R} \cup \{\infty, -\infty\}$, and "gluing together" the points ∞ and $-\infty$ so that they are the same point. This is the real projective one-space, $\mathbb{R} \cup \{\infty\}$.[4]

POSTCOMMENTARY

For situations involving division with 0, there are three types of forms: 0, undefined, and indeterminate. The indeterminate form has particular importance in calculus. Given a function, f, that would be continuous everywhere except that $f(a)$ is indeterminate, a functional value can sometimes be selected to make a related function that is continuous everywhere. For all of its domain values except a, the new function would have the same values as the given function. For example, in the case of the function defined by $f(x) = \dfrac{\sin x}{x}$ where x is in radians, the function f is continuous for all real numbers except 0, because the functional value at $x = 0$ is the indeterminate form $\dfrac{0}{0}$. The piecewise-defined function, $f(x) = \begin{cases} \dfrac{\sin x}{x}, & x \neq 0 \\ 1, & x = 0 \end{cases}$ is continuous for all real numbers. In this case, the fact that the limit of interest was 1 was used: $\lim\limits_{x \to 0} \dfrac{\sin x}{x} = 1$ where x is in radians. However, in other cases, limits related to $\dfrac{0}{0}$ do not have to be 1, or even an integer. For example, $\lim\limits_{x \to 0} \dfrac{2\sin x}{3x} = \dfrac{2}{3}$. These examples illustrate that, depending on the function, an indeterminate form can sometimes be replaced by a limiting value.

NOTES

1. These types of arguments are ones many students give, and they are important from the point of view of learners being able to attach to this situation meanings with which they already are familiar. Although they are based on real-world ideas (ideas that are informed by starting with a nonzero number of items and breaking those into a nonzero number of groups), the language of dividing collections of real physical objects into groups with fewer objects per group breaks down or becomes meaningless when there are 0 groups. What this indicates is that the metaphor used for attaching everyday meaning to division, namely dividing collections of physical objects into groups with fewer objects per group, is no longer a usable metaphor. The breakdown of this metaphor is that it is not a precise mathematical argument that explains what is problematic about the mathematics used in the missing-factor definition of quotient.

2. See Note 1 for a comment regarding the limitations of these real-world analogies.

3. See Note 1 for a comment regarding the limitations of these real-world analogies.

4. "Gluing together" the ends of the real line creates the same entity as the one-dimensional unit "sphere," namely, the unit circle. That may be easier to picture than real projective one-space and matches the intuition of taking the two supposed "ends" of the real number line, ∞ and $-\infty$, and gathering them up together to form the top "point" (north pole) of the unit circle.

CHAPTER 8

PRODUCT OF TWO NEGATIVE NUMBERS

Situation 2 From the MACMTL–CPTM Situations Project

**Ryan Fox, Sarah Donaldson, M. Kathleen Heid,
Glendon Blume, and James Wilson**

PROMPT

A question commonly asked by students in middle school and secondary mathematics classes is "Why is it that when you multiply two negative numbers together, you get a positive number answer?"

COMMENTARY

Students are able to visualize the addition and subtraction of integers, but multiplication of integers, particularly signed numbers, seems to be more abstract. Representing multiplication of quantities less than 0 is difficult. The Foci make this abstract concept more concrete by providing multiple ways to think about the multiplication of negative numbers, some of which only suggest that the product

*Mathematical Understanding for Secondary Teaching: A Framework
and Classroom-Based Situations*, pages 103–115.
Copyright © 2015 by Information Age Publishing
All rights of reproduction in any form reserved.

of two negative numbers should be positive and some of which establish definitively that the product of two negative numbers is positive via a general proof. Focus 1 applies a repeated addition model to multiplication of negative numbers, but that model is shown to have limitations. Real-world applications often are used to suggest that the product when multiplying two negative numbers should be positive; Focus 2 offers one such application involving an employee's pay. Focus 3 employs a visual approach using scalar properties of vectors to suggest that the product should be positive. In Focus 4, the distributive property and other properties of the real numbers are used to show, for specific cases and in general, that the product of two negative numbers should be a positive number. Focus 5 develops an intuitive, pattern-finding approach to suggest that the product should be positive. A geometric argument based on similar triangles appears in Focus 6. Focus 7 offers an analysis based on some concepts from abstract algebra.

MATHEMATICAL FOCI

Mathematical Focus 1

> *Repeated addition suggests that the product of a negative integer and any negative number is a positive number.*

Repeated addition is one way in which whole number multiplication typically is introduced. The product 3×5 can be thought of as adding 5 three times (starting from 0): $0 + 5 + 5 + 5$, or 15. Similarly, multiplication involving a negative integer, for example, 3×-5, can be thought of as repeated addition: Starting from 0, there are three addends, each of which is -5: $0 + (-5) + (-5) + (-5)$. Addition of negative integers leads to the result $3 \times -5 = -15$. Due to the commutative property of multiplication, the product -5×3 also equals -15. However, the repeated addition model for the product -3×-5 is not straightforward because it is difficult to interpret the product as -3 addends, each of which is -5. In Table 8.1, in moving from one row to the row below it (decreasing the first factor in the product by 1), an addend of -5 is being taken away from (subtracted from) the sum. The pattern suggests that the entry after 0 would be computed by subtracting an addend of -5 from 0, which is the same as adding the opposite of -5, meaning that 5 is the result, and suggesting that the product of two negative integers is a positive integer.

If one factor is negative but not an integer, the repeated addition model can be applied. For example, if the product in Table 8.1 had been -3×-5.1 (or even -5.1×-3, which is equal to -3×-5.1 because real number multiplication is commutative), each addend would have been -5.1, but the number of addends and the signs of the products would remain the same.

The repeated addition model for multiplication of negative numbers can suggest that the product of a negative integer and another negative number is positive; however, the model is not helpful when both factors are negative and neither is an integer. For example, if one multiplies -3.7×-5.1, one would need to interpret

TABLE 8.1   The Sum That Results When -5 Is Used a Decreasing Number of Times as an Addend

Product	Number of addends	Sum
3 × -5	3 addends, each of which is -5	(-5) + (-5) + (-5), or -15
2 × -5	2 addends, each of which is -5	[(-5) + (-5) + (-5)] – (-5), or -10
1 × -5	1 addend, which is -5	[(-5) + (-5)] – (-5), or -5
0 × -5	0 addends, each of which is -5	[(-5)] – (-5), or 0
-1 × -5	One fewer addend, or -1 addends, each of which is -5	[0] – (-5), or 0 + (+5), which is 5
-2 × -5	-2 addends, each of which is -5	[0 – (-5)] – (-5), or 0 + (+5) + (+5), which is 10
-3 × -5	-3 addends, each of which is -5	[0 – (-5) – (-5)] – (-5), or 0 + (+5) + (+5) + (+5), which is 15

what is meant by -3.7 addends, each of which is -5.1. Also, if both factors were negative irrational numbers, the repeated addition model would require one to interpret what was meant by an irrational number of addends. Although the repeated addition model for multiplication by a negative number is illuminating for negative integer multiplication and suggests—but does not prove—that the product of any negative integer and any other negative number is positive, it should not be interpreted to imply that the product of any two negative numbers is positive.

Mathematical Focus 2

Real-world instances that involve adding or subtracting positive or negative amounts can be used to suggest that the product of two negative numbers is a positive number.

Debts, debits, or deductions, as well as savings, credits, or deposits, can be used to illustrate that the product of two negative numbers is a positive number. Removal or deduction of an amount from an employee's paycheck is negative, whereas addition of an amount to an employee's paycheck is positive. Suppose that an employer deducts $120 per month from an employee's paycheck for health insurance. After 6 months, the deduction (debit) is 6 × (-$120), or -$720. Multiplication models the total amount.

If the employer were to offer the employee a special benefit of removing those deductions (debits), that removal of a negative amount would be a negative action on a negative value. Just as implementing the six deductions can be modeled by 6 × (-$120), removal of six -$120 deductions can be modeled by -6 × (-$120).

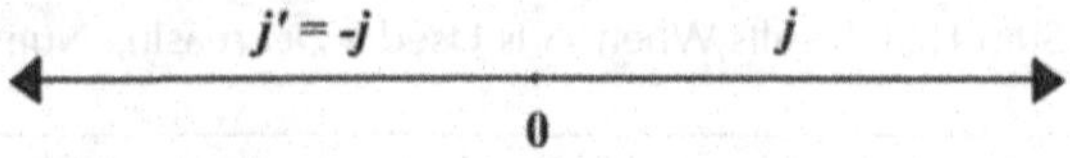

$$j' = -j \qquad\qquad j$$

0

FIGURE 8.1. The vectors j and –j.

Removal of those deductions would result in a gain in pay, over 6 months, of $720 for the employee, suggesting that (-6) × (-$120) = $720. Thus, a negative number of negative transactions—a negative number times a negative number—results in a positive number of dollars in pay.

Mathematical Focus 3

Products of negative numbers can be represented as the composition of two reflections.

Vectors on a number line can offer insight into the product of two negative numbers. When a vector on the number line is multiplied by the scalar -1, the vector is reflected about the origin. Consider the vector j in Figure 8.1. Multiplication of j by the scalar -1 yields the vector j', or $-j$. A second reflection, resulting from multiplication of this new vector, j' or $-j$, by the scalar -1, would yield its reflection about the origin, the vector $-j'$ or $-(-j)$ as illustrated in Figure 8.2.

The product of the two reflections is the original vector, j, that is

$$(-1)[(-1)j] = 1j.$$

But by the associative property of scalar multiplication,

$$(-1)[(-1)j] = [(-1)(-1)]j,$$

suggesting that for real numbers, (-1)(-1) = 1. Extending this to any two negative scalars suggests that the product of those two scalars, or the product of two negative numbers, is positive.[1]

Mathematical Focus 4

The distributive property of multiplication over addition can be used to illustrate and justify that the product of two negative numbers is a positive number.

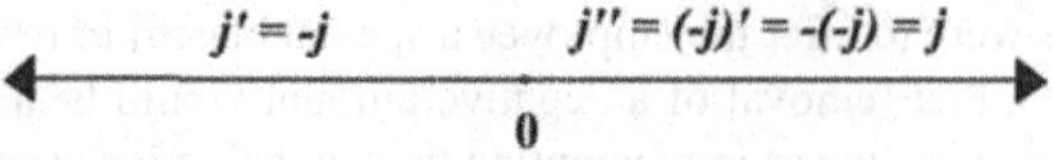

$$j' = -j \qquad\qquad j'' = (-j)' = -(-j) = j$$

0

FIGURE 8.2. The vectors –j and –(–j).

The distributive property of multiplication over addition, $a(b + c) = ab + ac$, enables one to write the product of a number and a sum as the sum of two products, which facilitates writing the product $(a + b)(c + d)$ as $ac + bc + ad + bd$. Starting from the product $7 \times 3 = 21$, one can write

$$(11 - 4)(5 - 2) = 21.$$

Writing subtraction as addition of the opposite yields

$$[11 + (-4)][5 + (-2)] = 21.$$

Applying the distributive property successively yields

$$[11 \times 5] + [(-4) \times 5] + [11 \times (-2)] + [(-4) \times (-2)] = 21.$$

Given that one has previously established that the product of a negative number and a positive number is a negative number,

$$[55] + [-20] + [-22] + [(-4) \times (-2)] = 21, \text{ or}$$

$$13 + [(-4) \times (-2)] = 21$$

Solving for the term $[(-4) \times (-2)]$ produces

$$[(-4) \times (-2)] = 8,$$

providing an example of the product of two negative numbers being a positive number. An infinite number of such examples is possible, each involving use of the distributive property after rewriting each factor as a difference. In each case, one can use factors of known products (in this case, 7 and 3 as factors of 21) to find the value of the product of two negative numbers.

Other illustrations of the use of properties of the real numbers are possible.[2] For example, the distributive property allows one to determine the value of the expression $2 \times 3 + -2) \times 3 + (-2) \times (-3)$ in two different ways, and doing so can help to establish that $(-2) \times (-3) = 6$.

$2 \times 3 + (-2) \times 3 + (-2) \times (-3)$	$= 2 \times 3 + (-2) \times [3 + (-3)]$ factoring out -2),
	$= 2 \times 3 + (-2) \times [0]$ because $a + (-a) = 0$,
	$= 2 \times 3 + 0$ because $a \times 0 = 0$,
	$= 2 \times 3$ because $a + 0 = a$.

Taking the same expression and factoring out (3) from the first two terms:

$2 \times 3 + (\text{-}2) \times 3 + (\text{-}2) \times (\text{-}3)$

$= [2 + (\text{-}2)] \times 3 + (\text{-}2) \times (\text{-}3)$	factoring out 3
$= [0] \times 3 + (\text{-}2) \times (\text{-}3)$	because $a + (-a) = 0$,
$= 0 + (\text{-}2) \times (\text{-}3)$	because $0 \times a = 0$,
$= (\text{-}2) \times (\text{-}3)$	because $0 + a = 0$.

So, in the one case the expression equals 2×3, and in the other it equals $(\text{-}2) \times (\text{-}3)$. The symmetric and transitive properties of equality then guarantee that $(\text{-}2) \times (\text{-}3)$ is the same number as 2×3, illustrating that the product of these two negative numbers, -2 and -3, is the positive number, 6. However, this argument can be generalized to establish that the product of any two negative numbers is a positive number.

For two positive real numbers, a and b, consider rewriting the expression $ab + (-a)(b) + (-a)(-b)$ in two different ways.

$ab + (-a)(b) + (-a)(-b)$	$= ab + (-a)[b + (-b)]$	factoring out $-a$,
	$= ab + (-a)(0)$	because $x + (-x) = 0$,
	$= ab + 0$	because $-x \times 0 = 0$,
	$= ab$	because $x + 0 = x$.

And,

$ab + (-a)(b) + (-a)(-b)$	$= [a + (-a)](b) + (-a)(-b)$	factoring out b,
	$= (0)b + (-a)(-b)$	because $x + (-x) = 0$,
	$= 0 + (-a)(-b)$	because $0 \times x = 0$,
	$= (-a)(-b)$	because $0 + x = x$.

By the symmetric and transitive properties of equality, $(-a)(-b) = ab$. If a and b are positive real numbers as assumed initially, $(-a)(-b)$ is the product of two negative numbers because the additive inverse of each of those positive numbers is a negative number. Also, ab is positive because it is the product of two positive numbers. Therefore, the product of any two negative numbers is a positive number.

Mathematical Focus 5

Investigating patterns in real-valued functions yields insight into the product of two negative numbers.

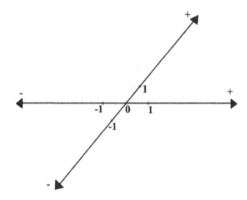

FIGURE 8.3. Two intersecting lines with unit distances marked in both positive and negative directions.

A table of values for a function of the form $f(x) = ax$, where $x \geq 0$ (see Table 8.2 for an example using $a = -5$) offers a way to develop a pattern involving the product of two numbers. The pattern established in Table 8.2 is that as the value of x decreases by one, the value of negative 5 times x increases by 5. As the value of x continues to decrease from positive numbers, through 0, to negative numbers (see Table 8.3), the value of $f(x)$ increases first from negative numbers to 0. The increase of 5 in $f(x)$ with each decrease of 1 in x can be established by examining the rate of change of $f(x)$. Because $f(x)$ is a linear function, its rate of change (slope of its graph) is a constant, -5, or $\frac{-5}{1}$. This means that the value of $f(x)$ decreases by 5 for each increase of 1 in x. If this pattern in the function values continues to hold, then $f(x)$ will be positive when x is negative. For $-5 \times -2 = 10$, the product of two negative numbers is a positive number. The pattern suggests that this will be true for any product of -5 and a negative number.[3]

TABLE 8.2 Values of $f(x) = -5x$ for Positive and Zero Values of x

x	$f(x) = -5x$	$f(x)$
4	$f(x) = -5 \times 4$	-20
3	$f(x) = -5 \times 3$	-15
2	$f(x) = -5 \times 2$	-10
1	$f(x) = -5 \times 1$	-5
0	$f(x) = -5 \times 0$	0

TABLE 8.3   Values of $f(x) = -5x$ for Positive, Zero, and Negative Values of x

x	$f(x) = -5x$	$f(x)$
1	$f(x) = -5 \times 1$	-5
0	$f(x) = -5 \times 0$	0
-1	$f(x) = -5 \times -1$	5
-2	$f(x) = -5 \times -2$	10

Mathematical Focus 6

A geometric model based on similar triangles can suggest that the product of two negative numbers is a positive number.

Suppose that two lines intersect at a common point labeled as 0 for each line. From the definition of a negative number as the additive inverse or opposite of a positive number, label positive and negative directions on each line. Using the multiplicative identity, 1, indicate a unit in each direction on each line (see Figure 8.3). Any point on either line has an orientation, positive or negative, determined by its location.

Positive number times a positive number

Begin by considering a positive value, *a*, on the horizontal line and a positive value, *b*, on the slanted line. Next construct a line segment from 1 on the horizontal line to *b* on the slanted line. Construct a parallel line segment from *a* on the horizontal line to its intersection with the slanted line. From the geometry of similar triangles, the location of the intersection is the product *ab* (see Figure

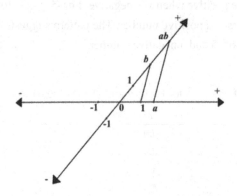

FIGURE 8.4.   The product *ab* is positive when both *a* and *b* are positive.

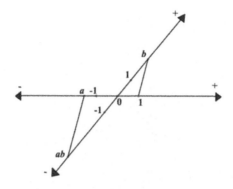

FIGURE 8.5. The product *ab* is negative when *a* is negative and *b* is positive.

8.4). This is also the classic construction of the fourth proportion, $1:b = a:x$, which leads to $x = ab$.

Because the triangles are similar, the location of the intersection represents the product of *a* and *b*. This case suggests that the product of two positive numbers is positive.

Negative number times a positive number

Consider the preceding construction but beginning with *a* located in the negative direction and *b* located in the positive direction (see Figure 8.5). The construction proceeds in the same way, with *a* on the horizontal, *b* on the slant line, and construction of the segments from the 1 location to the *b* location and then a parallel segment from the *a* location to the intersection point. Again, the intersection point will be the product *ab*, but now this product is located in the negative direction. This suggests that the product of a negative number and a positive number is a negative number. Again this is a direct consequence of the similar triangles.

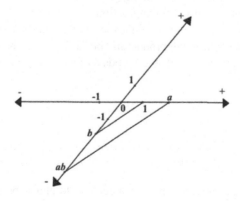

FIGURE 8.6. The product *ab* is negative when *a* is positive and *b* is negative.

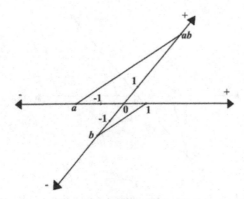

FIGURE 8.7. The product *ab* is positive when both *a* and *b* are negative numbers.

Positive number times a negative number

Consider the previous construction but beginning with *a* located in the positive direction and *b* located in the negative direction (see Figure 8.6). The construction is done in the same way, with *a* on the horizontal, *b* on the slanted line, and construction of the segments from the 1 location to the *b* location and then a parallel segment from the *a* location to the intersection point. Again, the intersection point will be the product *ab*, and again, this product is located in the negative direction. This suggests that the product of a positive number and a negative number is a negative number.

Negative number times a negative number

Again, consider a similar construction but beginning with both *a* located in the negative direction and *b* located in the negative direction (see Figure 8.7). The construction proceeds in the same way, with *a* on the horizontal, *b* on the slanted line, and construction of the segments from the 1 location to the *b* location and then a parallel segment from the *a* location to the intersection point. Again, the intersection point will be the product *ab*, but this product is located in the positive direction. This suggests that the product of a negative number and a negative number is a positive number.

Mathematical Focus 7

> The product of two negative numbers can be shown to be positive by using properties of the real number system, including the identity and inverse properties.

The set of real numbers forms an ordered field under the operations of addition and multiplication as they are typically defined. In a field, additive identities and additive inverses exist. In addition, the distributive property of multiplication over addition holds. These properties can be used to show that the product of two negative numbers is positive. Negative real numbers are the additive inverses (opposites) of positive real numbers. Zero acts as the additive identity in this field. Choose any positive element from the set of real numbers and call it *a*. The following sequences of statements illustrate the use of the various properties[4] to show that the product of two negative numbers is positive.

$$a + (-a) = 0$$

$$(-1) \times (a + (-a)) = (-1) \times 0$$

$$(-1)(a) + (-1)(-a) = 0$$

$$-a + (-1)(-a) = 0$$

$$a + -a + (-1)(-a) = a + 0$$

$$(-1)(-a) = a$$

This establishes that the product of -1 and the additive inverse of any negative real number is that real number. This result is used within the following sequence of statements to show that $(-1) \cdot (-1) = 1$.

Next, let *a* and *b* be any two negative real numbers. Therefore, $a = -c = (-1)c$ and $b = -d = (-1)d$ for some positive real numbers *c* and *d*. Then

$ab = [(-1)c] \, [(-1)d]$

$\qquad = [[(-1)c](-1)]d$ by the associative property of multiplication

$\qquad = [(-1)(c(-1))]d$ by the associative property of multiplication

$\qquad = [(-1)[(-1)c]]d$ by the commutative property of multiplication

$\qquad = [[(-1)(-1)]c]d$ by the associative property of multiplication

$\qquad = (1c)d \qquad\qquad$ using $(-1)\,(-1) = 1$

$\qquad = cd \qquad\qquad\quad$ by the multiplicative identity.

Thus, the product of two negative real numbers is equal to the product of two positive real numbers, which previously has been established to be positive. So the product of two negative real numbers is positive.

POSTCOMMENTARY

Having multiple ways to think about the multiplication of negative numbers is often helpful in teaching. There are many intuitively compelling reasons to have the product of two negative numbers be positive (e.g., extending the pattern generated by multiplying positive whole numbers by negatives, the result when vectors are reflected over 0 on a number line), but in all cases those amount to giving motivation for what is taken as the usual statement about the product of two negative numbers. Conceptual models and patterns such as those described in Focus 3 and Focus 5 (see also Crowley & Dunn, 1985) can be invoked to provide support for believing the result, but do not prove it.

Nonmathematical analogies[5] often are used to motivate the rule that the product of two negative numbers is a positive number. For example, one might draw an analogy between filming the filling of a glass of water and multiplication. One can film filling (positive) or emptying (negative) a glass of water, and one can run the film forward (positive) or backward (negative). Running the film backward as the glass is emptying will show it filling (suggesting that a negative times a negative equals a positive). Although such analogies are seductive because they "work"—they simply remind the user of the result (two negatives yield a positive)—they do not capture the mathematical essence of some model of multiplication (e.g., repeated accumulation of equal-sized groups). This movie analogy merely suggests that there is another situation in which two negative actions lead to a positive result. Other nonmathematical analogies include a person's action of moving into or out of a city and that person's categorization as a good person or a bad person. When two negatives occur, namely, a bad person moves out of the city, it is good for the city (a positive result). Again, this only suggests that another situation exists in which two negatives lead to a positive result.

Extension of operations and properties from one number system to an expanded number system (e.g., from integers to rational numbers or to real numbers) requires identification of a collection of assumptions (axioms) that hold for the expanded number system. Although it is tempting to apply the number properties established for one number system to another number system, it is necessary to establish the reasonableness of those properties for each new number system.

NOTES

1. Note that this Focus assumes that the behavior of vectors on the number line under multiplication by a negative scalar resembles the behavior of real numbers under multiplication by a negative number.

2. An illustration on page 31 of Hefendehl-Hebeker (1991) that the product of two negative numbers is a positive number includes the use of the additive inverse property, multiplication property of equality, and the symmetric and transitive properties of equality.

3. Note that this pattern does not suggest that the product of any two negative numbers should be positive, only that the product of -5 and a negative number should be positive. One might create tables of values for functions of the form $f(x) = kx$ for a variety of negative values of k and use the pattern across those tables as the basis for an empirical argument

(but not a proof) that suggests that the product of any two negative numbers is positive.

4. Note that this sequence of statements relies on several properties having been established prior to their use here. For example, the zero-product property ($a \times 0 = 0$) is used to replace $(-1) \times 0$ by 0.

5. One can call these analogies *nonmathematical* because they do not constitute the basic elements of a mathematical argument. There are no clearly identified starting objects, or primitives, accompanied by axioms from which the mathematics community determines allowable conclusions that can be drawn or logical arguments that can be formed about the primitives based on those axioms.

REFERENCES

Crowley, M. L., & Dunn, K. A. (1985). On multiplying negative numbers. *Mathematics Teacher, 78*, 252–256.

Hefendehl-Hebeker, L. (1991). Negative numbers: Obstacles in their evolution from intuitive to intellectual constructs. *For the Learning of Mathematics, 11*(1), 26–32.

CHAPTER 9

CROSS MULTIPLICATION

Situation 3 From the MACMTL–CPTM Situations Project

Rose Mary Zbiek, M. Kathleen Heid, Brian Gleason, and Shawn Broderick

PROMPT

An Algebra 1 class was working through an example, for which the teacher had written the following on the whiteboard:

$$\sqrt{\frac{12}{81}} = \sqrt{\frac{3 \cdot 4}{3 \cdot 27}} = \sqrt{\frac{4}{27}} =$$

A student asked, "When we're doing this kind of problem, will it always be possible to cross multiply?"

COMMENTARY

When learning new content, students often attempt to apply procedures with which they are comfortable even though those procedures may not be mathematically appropriate for the new content. This appears to be the case in the Prompt,

Mathematical Understanding for Secondary Teaching: A Framework and Classroom-Based Situations, pages 117–119.
Copyright © 2015 by Information Age Publishing

117

in which a student imposes a process used to solve proportions on this problem involving the simplification of a radical involving a fraction. The first Focus explains the student's apparent thinking and the associated misapplication of the process known as *cross multiplication*. The second Focus gives the conditions that led to the apparent cross multiplication and explains when this condition will arise.

MATHEMATICAL FOCI

Mathematical Focus 1

Cross multiplication is a process often taught to students learning how to solve proportions. The mnemonic often becomes associated with solving problems involving fractions.

Cross multiply is the name given to a process often used to solve for an unknown in a proportion. In the proportion $\frac{a}{b} = \frac{c}{d}$, one "multiplies across" the equals sign in a diagonal fashion $\frac{a}{b} \times \frac{c}{d}$ to obtain the equation $a \cdot d = b \cdot c$. It appears that the student, while thinking about the problem on the board, has recognized that $3 \cdot 27 = 81$ (using the 3 from the numerator) and that $3 \cdot 4 = 12$ (using the 3 from the denominator) and has confused the process of simplifying fractions with the process of cross multiplication. Presumably, the student is thinking of something along the lines of $\sqrt{\frac{12}{81}} = \sqrt{\frac{3 \cdot 4}{3 \cdot 27}}$.

On the other hand, the student could have been thinking about the use of cross multiplication to check the equivalence of fractions. For example, the student may have been thinking about whether one could "check" the equality $\sqrt{\frac{12}{81}} = \sqrt{\frac{3 \cdot 4}{3 \cdot 27}} = \sqrt{\frac{4}{27}}$ by checking whether $12 \cdot 27 = 81 \cdot 4$ (with or without the square root signs).

Mathematical Focus 2

Simplifying fractions requires finding factors (greater than 1) that are common to both the numerator and the denominator.

The student's question might be more appropriately rephrased to the following: Will it always be possible to find a common integral factor in the numerator and denominator of a fraction? The answer to this is yes if the fraction is not in simplified form and one is interested only in integral factors greater than 1. This may lead to situations similar to the previous one, in which it appears as if some form of cross multiplication is being used, although it is not.

$$\sqrt{\frac{12}{81}} = \sqrt{\frac{3 \cdot 4}{3 \cdot 27}}$$

$$\sqrt{\frac{100}{225}} = \sqrt{\frac{25 \cdot 4}{25 \cdot 9}}$$

The following counterexample shows that, when there is no common factor greater than 1, the misapplied cross multiplication does not produce the original fraction.

$$\sqrt{\frac{14}{15}} = \sqrt{\frac{2 \cdot 7}{3 \cdot 5}} \neq \sqrt{\frac{3 \cdot 7}{2 \cdot 5}} = \sqrt{\frac{21}{10}}$$

CHAPTER 10

SUMMING THE NATURAL NUMBERS

Situation 4 From the MACMTL–CPTM Situations Project

Shari Reed, AnnaMarie Conner, Ryan Fox, Shiv Karunakaran,
M. Kathleen Heid, Evan McClintock, Heather Johnson,
Kelly Edenfield, Jeremy Kilpatrick, and Eric Gold

PROMPT

In a mathematical modeling course for prospective secondary mathematics teachers, discussion focused on finding an explicit formula for a sequence expressed recursively. During the discussion, students expressed a need to sum the natural numbers from 1 to n. After several attempts to remember the formula, a student hypothesized that the formula contained n and $n + 1$. Another student said he thought that it was $\frac{n(n+1)}{2}$ but was not sure. During the ensuing discussion, a third student asked, "How do we know that $\frac{n(n+1)}{2}$ is the sum of the integers from 1 to n? And won't that formula sometimes give a fraction?"

Mathematical Understanding for Secondary Teaching: A Framework
and Classroom-Based Situations, pages 121–133.

121

COMMENTARY

The set of integers from 1 to n is an ordered sequence of natural numbers. Symmetry and the ordered nature of the sequence allow for rearrangements that facilitate finding a sum; under any rearrangement of the elements of a discrete set, the cardinality remains the same, as does the sum of the elements. A formula for the sum of the natural numbers from 1 to any number can be developed from, and verified for, specific instances, and the formula can be proved using a variety of methods, including mathematical induction.

MATHEMATICAL FOCI

Mathematical Focus 1

There are two possibilities for the natural number n: *It can be even, or it can be odd. Because of this, when* n *is a natural number,* $\frac{n(n+1)}{2}$ *is also a natural number.*

If n is odd, then $n + 1$ is even and therefore divisible by 2. Hence $\frac{n(n+1)}{2}$ is a natural number. If n is even, then it is divisible by 2. And hence $\frac{n(n+1)}{2}$ is a natural number.

Proofs of the preceding two results—the sum of the first n natural numbers when n is an odd natural number and the sum when n is an even natural number—are included in the Appendix for Focus 1. These proofs are based on symbolic manipulation of the formula for the sum of the first n natural numbers.

Mathematical Focus 2

Specific examples suggest a general formula for the sum of the first n *natural numbers. Strategic choices for pairwise grouping of numbers is critical to the development of the general formula.*

Case 1: n is even

The sum to be found is as follows:

$$1 + 2 + 3 + \ldots + (n-2) + (n-1) + n$$

If n is even, then pairing the first and last terms, then the second and second-to-last terms, and continuing inward through the sequence yields $\frac{n}{2}$ pairs, with each pair summing to $n + 1$. Therefore, the sum of the sequence is $(n+1)\frac{n}{2}$.

Case 2: n is odd

Consider the sum of the first n natural numbers when n is odd (see Appendix for Focus 2 for the specific example of $n = 9$).

$$1 + 2 + 3 + \ldots + (n - 2) + (n - 1) + n$$

As before, the sums of pairs are $1 + n$, $2 + (n - 1)$, $3 + (n - 2)$, and so on. This time, there are $\frac{n-1}{2}$ pairs, each pair with a sum of $n + 1$, and one term, the middle term $\frac{n+1}{2}$, is not paired.[1] Therefore, the sum from 1 to n is

$$\left(\frac{n-1}{2}\right)(n+1) + \left(\frac{n+1}{2}\right) = (n+1)\left(\left(\frac{n-1}{2}\right) + \frac{1}{2}\right) = (n+1)\left(\frac{n-1+1}{2}\right) = \frac{(n+1)n}{2} .$$

Mathematical Focus 3

Because the first n *natural numbers form an arithmetic sequence, properties of such sequences can be used to find their sum.*

The terms of a finite arithmetic sequence have a kind of symmetry,[2] and the difference between consecutive terms is constant. The commutative and associative properties of addition of a finite collection of real numbers allow the terms to be regrouped so their sum can be calculated more efficiently.

Let S be the sum of the first n natural numbers.[3] Then

$$S = 1 + 2 + 3 + \ldots + (n - 1) + n.$$

When the terms for S are written twice, once in increasing order and once in decreasing order, and corresponding terms are added, each sum is $n + 1$. By the commutative and associative properties of addition and properties of equality,

$$
\begin{aligned}
S &= 1 + 2 + 3 + \ldots + n \\
S &= n + (n-1) + (n-2) + \ldots + 1 \\
S+S &= (1+n) + (2+n-1) + (3+n-2) + \ldots + (n+1) \\
2S &= (n+1) + (n+1) + (n+1) + \ldots + (n+1)
\end{aligned}
$$

$$2S = n(n+1)$$

$$S = \frac{n(n+1)}{2}$$

Therefore, the sum of the first n natural numbers is $\frac{n(n+1)}{2}$.

A manipulation of the general arithmetic sequence in developing the formula for the sum of the first n natural numbers can be found in the Appendix for Focus 3.

Mathematical Focus 4

Geometric arrays provide opportunities to derive the formula for the sum of the first n *natural numbers.*

Model 1: Rearranging a triangular array of dots

One representation of the first n natural numbers is a triangular array of dots, having n rows, in which the number of dots in the first row is 1, and the number of dots in each successive row increases by 1, so that in the nth row there are n dots (see Figure 10.A2 in the Appendix for Focus 4 for a specific example when $n = 4$; see Figure 10.1a for a general example). When n is even, there are $\frac{n}{2} \times (n + 1)$ dots in the rectangular array. Hence, the sum of the first n natural numbers when n is even is $\frac{n(n+1)}{2}$.

Now consider the sum of the first n natural numbers when n is odd. If n is odd, there are $\frac{n+1}{2}$ rows and n columns in the rectangular array, as shown in Figure 10.2. There are $\frac{n+1}{2} \times n$ dots in the rectangular array. Hence, the sum of the first n natural numbers for n odd is $\frac{n(n+1)}{2}$.

Model 2: Duplicating a triangular array of dots

As previously stated, one representation of the first n natural numbers is a triangular array of dots, having n rows, in which the number of dots in the first row is 1, and the number of dots in each successive row increases by 1, so that in the nth row there are n dots.

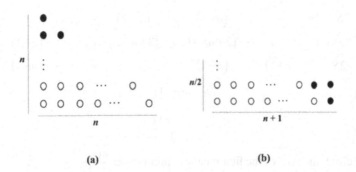

(a)   (b)

FIGURE 10.1. Array (a) representing the first n natural numbers when n is even and a rearrangement (b) of that array in which row 1 is appended to row n, row 2 is appended to row $(n - 1)$, and so on.

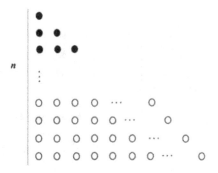

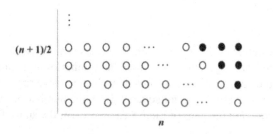

FIGURE 10.2. Array representing the first *n* natural numbers when *n* is odd and a rearrangement of that array in which row 1 is appended to row (*n* – 1), row 2 is appended to row (*n* – 2), and so on.

Copy the original triangular array, rotate it 180° clockwise, and join it with the original to form a rectangular array of dots. The rectangular array contains *n* rows and *n* + 1 columns, as shown in Figure 10.3. The number of dots in the rectangular array is *n*(*n* + 1). The rectangular array is composed of two triangular arrays, each containing a number of dots equal to the sum of the first *n* natural numbers. Because the rectangular array comprises two triangular arrays, the sum of the first *n* natural numbers is half the number of dots in the combined (rectangular) array, or $\dfrac{n(n+1)}{2}$.

Mathematical Focus 5

Decomposition and recomposition of plane geometric figures preserve area. Geometric figures provide opportunities to derive the formula for the sum of the first n *natural numbers and to develop other relationships involving that sum.*

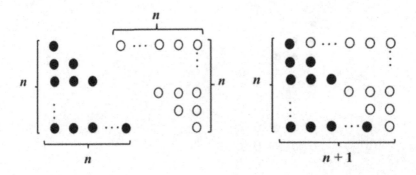

FIGURE 10.3. Duplication of an n-row triangular array of dots resulting in an $n \times (n + 1)$ rectangular array of dots.

Another representation of the first n natural numbers is a staircase: a triangular array of unit squares, having n rows, in which the number of unit squares in the first row is 1, and the number of unit squares in each successive row increases by 1, so that in the nth row there are n unit squares. Both the number of unit squares in the triangular array and the sum of their areas represent the sum of the first n natural numbers.

Model 1: Completing a rectangular array

Another unit-square-area model uses the idea of completing a rectangular array by adding additional unit squares to the rows in a staircase. To complete a rectangular array of unit squares from an n-step staircase of unit squares, append $n - 1$ unit squares to the first (top) row in the staircase, $n - 2$ to the second row in the staircase, and so on, with $n - r$ unit squares being appended to the rth row in the staircase. The resulting rectangular array has n rows of unit squares and n columns of unit squares—a total of n^2 unit squares—and the sum of the areas of those squares is n^2. Figure 10.4 illustrates the construction of the staircase for $n = 4$.

If S represents the sum of the first n integers (and consequently the number of squares in the n-step staircase of unit squares), then $S = 1 + 2 + 3 + ... + n$, and the sum of the areas of the unit squares in the rectangular array is $S + (n - 1) + (n - 2) + ... + 2 + 1$. Because the sum of the areas of the unit squares is n^2, $S + (n-1) + (n-2) + ... + 2 + 1 = n^2$. Adding n to both members of this equation yields

$$S + (n - 1) + (n - 2) + ... + 2 + 1 + n = n^2 + n.$$

Substitution yields $2S = n(n + 1)$. Hence, $S = \dfrac{n(n+1)}{2}$. See Appendix to Focus 5 for a model using a slightly different representation.

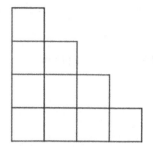 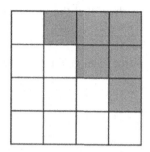

FIGURE 10.4. Completion of a 4 × 4 rectangular array of unit squares from a four-step staircase of unit squares.

Model 2: Combining two triangular numbers into a square array

It can be shown that the sum of a given triangular number and its preceding triangular number equals the square of the index number of the given triangular number. Symbolically, the expression is as follows:

$$T_n = \frac{n(n+1)}{2}$$

$$\Rightarrow 2T_n = n(n+1)$$

$$\Rightarrow T_n + T_n = n(n+1)$$

$$\Rightarrow T_n + T_{n-1} + n = n^2 + n$$

$$\Rightarrow T_n + T_{n-1} = n^2$$

OR

$$T_n + T_{n-1} = \frac{n(n+1)}{2} + \frac{(n-1)n}{2}$$

$$\Rightarrow T_n + T_{n-1} = \frac{n[(n+1)+(n-1)]}{2}$$

$$\Rightarrow T_n + T_{n-1} = \frac{n(2n)}{2}$$

$$\Rightarrow T_n + T_{n-1} = n^2$$

Mathematical Focus 6

The principle of mathematical induction offers a more mathematically rigorous method for verifying a formula for the sum of the terms in a sequence than does a technique that relies on pictures, which may be imprecise.

For the sum of the first n natural numbers, S_n, the principle of mathematical induction can be used to prove that the formula $S_n = \frac{n(n+1)}{2}$ holds for all natural numbers. The proof does not derive the formula but proves that it holds.

The sum of the first n natural numbers can be expressed as $\sum_{i=1}^{n} i$. The inductive hypothesis is that $S_n = \frac{n(n+1)}{2}$. When $n = 1$, $S_1 = 1 = \frac{1(1+1)}{2}$, which establishes the base case. Assume that for natural number m, $S_m = \sum_{i=1}^{m} i = \frac{m(m+1)}{2}$. It is now necessary to show that the inductive hypothesis holds for $m + 1$; that is, that

$$\sum_{i=1}^{m+1} i = \frac{(m+1)(m+2)}{2}.$$

The sum from 1 to $m + 1$ can be expressed as $m + 1$ added to the sum from 1 to m:

$$\sum_{i=1}^{m+1} i = \left(\sum_{i=1}^{m} i\right) + m+1 = \frac{m(m+1)}{2} + m+1 = \frac{m(m+1)}{2} + \frac{2(m+1)}{2} = \frac{(m+1)(m+2)}{2}.$$

By the principle of mathematical induction, the sum of the first n natural numbers is $\frac{n(n+1)}{2}$.

POSTCOMMENTARY 1

The sum of the first n natural numbers is a specific case of the sum of the first n k^{th} powers of natural numbers.

Although the formula for the sum of the first n natural numbers is itself a generalization, it can be generalized further. The formula is the case of $k = 1$ for the sum of the first n k^{th} powers of natural numbers.

Polya (1981, ch. 3) describes how Pascal solved the problem of finding the sum of the kth powers of the first n natural numbers. This iterative method depends on expanding $(n + 1)^{k+1}$ consecutively for $k = 1, 2$, and so on.

As a model of Pascal's method, the following illustrates how to find the sum of the first n natural numbers. Let S_n be the sum of the first n natural numbers. To find S_n, expand $(n + 1)^2$ and list its values for several values of n to see the pattern.

$$(n + 1)^2 = n^2 + 2n + 1$$

$$\Rightarrow (n + 1)^2 - n^2 = 2n + 1$$

$$2^2 - 1^2 = 2 \cdot 1 + 1 \tag{1}$$

$$3^2 - 2^2 = 2 \cdot 2 + 1 \tag{2}$$

$$4^2 - 3^2 = 2 \cdot 3 + 1 \tag{3}$$

and so on, until one obtains

$$(n + 1)^2 - n^2 = 2n + 1 \tag{n}$$

Adding the left members of equations (1) through (n) and adding the right members of equations (1) through (n) and equating the sums yields $(n + 1)^2 - 1 = 2(S_n) + n$. Then

$$S_n = \frac{(n+1)^2 - n - 1}{2} = \frac{(n+1)^2 - 1(n+1)}{2} = \frac{(n+1-1)(n+1)}{2} = \frac{n(n+1)}{2}.$$

Using this method, one can find the sum of the first n squares of natural numbers by expanding $(n + 1)^3$ and listing its values for several values of n to see the pattern. Summing the set of equations from (1) through (n) yields the sum of the first n squares as

$$\frac{n(n+1)(2n+1)}{6}.$$

The method can be extended to find the sum of the first n cubes, the sum of the first n fourth powers, and so on, as long as the sums of all the previous powers have been found.

POSTCOMMENTARY 2

In Focus 4, triangular arrays of dots were used to represent the sum of the first n natural numbers. In other words, the sum of the first n natural numbers is a *triangular* number. A triangular number is a special case of a *figurate* number, that is, a number that can be represented by a regular geometrical arrangement of equally spaced points (Gillette, Gillette, Hobson, & Weisstein, n.d.; Weisstein, n.d.).

APPENDICES

Appendix for Mathematical Focus 1

The argument can be expressed symbolically:
For n odd, let $n = 2m + 1$, with m a natural number, then

$$\frac{n(n+1)}{2} = \frac{(2m+1)(2m+1+1)}{2} = \frac{2(2m+1)(m+1)}{2} = (2m+1)(m+1).$$

For n even, let $n = 2k$, with k a natural number, then

$$\frac{n(n+1)}{2} = \frac{(2k)(2k+1)}{2} = (k)(2k+1).$$

So, for $n \in \mathbb{N}$, whether n is odd or even, $\frac{n(n+1)}{2}$ is a natural number.

Appendix for Mathematical Focus 2

In the array in Figure 10.A1, the first row pairs 1 open dot with 9 solid dots, the second row pairs 2 open dots with 8 solid dots, the third row pairs 3 open dots with 7 solid dots, and the fourth row pairs 4 open dots with 6 solid dots. Because

5 is not paired with another number, there are only 5 open dots in the fifth row. Hence, the array contains four rows of 10 dots each and one row of 5 dots, giving a total of 45 dots in the array. Again, the sum of the first 9 natural numbers is $4 \times 10 + 5 = 45$.

Appendix for Mathematical Focus 3

Stated another way, the sum of the first n natural numbers is equal to the arithmetic mean of 1 and n, $\dfrac{(n+1)}{2}$, multiplied by n, the number of natural numbers being summed. This approach was, according to legend, used by Gauss when he was asked as a schoolboy to find the sum of the first hundred natural numbers.

More generally, a finite arithmetic sequence with n terms can be written as $(a_1), (a_1 + d), (a_1 + 2d), \ldots, (a_1 + (n-1)d)$, where each term differs from the previous term by a constant difference, d. The sum of the terms of the sequence, (a_1), $(a_1 + d), (a_1 + 2d), \ldots, (a_1 + (n-1)d)$, can be derived in the following way:

$$\sum_{k=1}^{n} a_k = \sum_{k=1}^{n} \left[a_1 + (k-1)d \right]$$

$$= na_1 + d \cdot \sum_{k=1}^{n} (k-1)$$

$$= na_1 + d \cdot \sum_{k=1}^{n-1} (k)$$

$$= na_1 + d \cdot \frac{(n)(n-1)}{2}$$

$$= \frac{n}{2} \left[2a_1 + d(n-1) \right]$$

$$= \frac{n}{2} \left[a_1 + a_1 + d(n-1) \right]$$

$$= \frac{n}{2} \left[a_1 + a_n \right]$$

In other words, the sum of the terms of a finite arithmetic sequence is the arithmetic mean of the first and last terms times the number of terms in the sequence.

The first n natural numbers—that is, $1, 2, 3, \ldots, n-1, n$—form an arithmetic sequence with a constant difference of 1 and $a_1 = 1$ and $a_n = n$. Substituting these values into the preceding equation yields the desired formula:

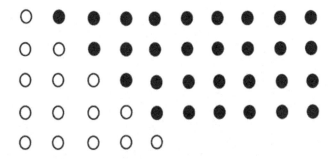

FIGURE 10.A1.   Dot array illustrating the sum of the first nine natural numbers.

$$\sum_{k=1}^{n} a_k = \frac{n(a_1 + a_n)}{2} = \frac{n(1+n)}{2} = \frac{n(n+1)}{2}, \text{ or } S_n = \frac{n(n+1)}{2}.$$

Appendix for Mathematical Focus 4

Mathematical Focus 4 considered the sum of the first n natural numbers when n is even. It showed that one way to derive the formula for the sum of the first n natural numbers is to rearrange the triangular array of dots into a rectangular array and that, when n is even, there will be $\frac{n}{2}$ rows and $n+1$ columns in the rectangular array. Figure 10.A2 illustrates the specific case for $n = 4$.

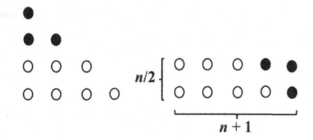

FIGURE 10.A2.   Rearrangement of a four-row triangular array of dots into a rectangular array of dots.

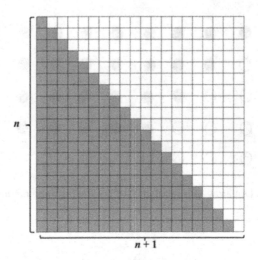

FIGURE 10.A3. Illustration for $n = 19$ of a rearrangement of an n-row triangular staircase and a duplicate copy of it into an $n \times (n + 1)$ rectangular array.

Appendix for Mathematical Focus 5

Model 3: Duplicating a staircase

Copy the original triangular array of unit squares, rotate it 180° clockwise, and join it with the original to form a rectangular array of unit squares (see Figure 10.A3). The dimensions of the rectangular array of unit squares are n by $(n + 1)$, and the sum of the areas of the unit squares in the rectangular array is $n(n + 1)$. The sum of the areas of half the unit squares in the rectangular array equals the sum of the areas of the unit squares in one triangular array, so the sum of the areas of the unit squares in one triangular array is $\dfrac{n(n+1)}{2}$. Because the sum of the areas of the unit squares in the triangular array is the sum of the first n natural numbers, the sum of the first n natural numbers is given by $\dfrac{n(n+1)}{2}$.

NOTES

1. This lack of pairing of the middle term is illustrated in the Appendix for Focus 2.
2. The sum of the kth term and the $n - (k - 1)^{\text{th}}$ term is constant.
3. The sum of the terms in a sequence is often referred to as a *series*.

REFERENCES

Gillette, M., Gillette, R., Hobson, N., & Weisstein, E. (n.d.). *Triangular number—from Wolfram MathWorld*. Retrieved from http://mathworld.wolfram.com/Triangular-Number.html

Polya, G. (1964/1981). *Mathematical discovery: On understanding, learning, and teaching problem solving*. New York, NY: John Wiley & Sons.

Weisstein, E. (n.d.). *Figurate number—from Wolfram MathWorld*. Retrieved from http://mathworld.wolfram.com/FigurateNumber.html

CHAPTER 11

MODULAR ARITHMETIC

Situation 5 From the MACMTL–CPTM Situations Project

**Pawel Nazarewicz, Glendon Blume,
Heather Johnson, Svetlana Konnova, and Jeanne Shimizu**

PROMPT

A group of high school Mathematics Club members was examining the concept of modular arithmetic. They were working in mod 5, and as they were becoming familiar with mod 5, a student asked whether it is possible to write fractions in mod 5. She wondered about the meaning of expressions such as $\frac{3}{4} \bmod 5$.

COMMENTARY

The symbol $\frac{3}{4}$, and more generally, $\frac{a}{b}$ such that a and b are integers and $b \neq 0$, can be interpreted in a variety of ways: as a single rational number (commonly called *fraction*), as a ratio of two numbers, or as a quotient of two numbers. However, these interpretations may cause confusion when dealing with operations within integer rings $\mathbb{Z}_n$ (where n is a positive integer). Thus, it is important to

*Mathematical Understanding for Secondary Teaching: A Framework
and Classroom-Based Situations*, pages 135–145.
Copyright © 2015 by Information Age Publishing

move beyond the previously mentioned common interpretations of the symbol $\frac{a}{b}$ and regard it as only a symbol.

When doing modular arithmetic, it does not make sense to refer to $\frac{a}{b}$ (where b is not a factor of a) as a fraction, because the congruence relation mod m (for m a positive integer) is defined only for integers. This is discussed in Focus 1. However, one can refer to $\frac{a}{b} \bmod m$ as an expression that has meaningful interpretations. The expression $\frac{a}{b} \bmod m$ can be interpreted to represent a times the multiplicative inverse of b in mod m (see Focus 3). As described in Focus 2, if m is prime, then $\frac{a}{b} \bmod m$ will have meaning because each nonzero element of $\mathbb{Z}_m$ has a multiplicative inverse ($\mathbb{Z}_m$ is a finite field). However, if m is not prime, a nonzero element, b, may or may not have a multiplicative inverse, and as a consequence, $\frac{a}{b} \bmod m$ may or may not be interpretable as a times the multiplicative

inverse of b. In Mathematical Focus 4, $\frac{p}{q}$ is interpreted as the solution, x, to the congruence statement $qx \equiv p \bmod m$, where p, q, and m are integers, m is prime, and q is not congruent to 0 mod m. Mathematical Focus 5 addresses the idea of congruence classes for numbers mod m and the conditions necessary for the expressions $\frac{a}{b} \bmod m$ and $\frac{c}{d} \bmod m$ to be in the same congruence class.

Given an expression of the form $\frac{a}{b} \bmod m$, one can ask how to find a value that it can represent in mod m. Mathematical Focus 6 presents a type of Greedy Algorithm that can be used, in general, to find such a value.

MATHEMATICAL FOCI

Mathematical Focus 1

The congruence relation mod m *(for* m *a positive integer) is defined only for integers.*

By definition, if a, b, and m are integers with $m > 0$, then "a is said to be congruent to b modulo m, if $m|(a - b)$" (Strayer, 1994, p. 38). In other words, integer a is congruent to integer b modulo m if positive integer m is a factor of $(a - b)$.

The statement "*a is congruent to b modulo m*" is written $a \equiv b \bmod m$, where b is called the *residue* or the *remainder* and m is called the *modulus*. Commonly used residues for mod m are nonnegative integers less than m. For example,

$30 \equiv 2 \bmod 4$ because 4 is a factor of $(30 - 2)$. (Note: If $a - b$ is not integrally divisible by m, then it is said that "a is not congruent to b modulo m.") Thus, by definition, if $\frac{3}{4}$ is interpreted to represent a number (e.g., a point on the number line halfway between $\frac{1}{2}$ and 1), then $\frac{3}{4} \bmod 5$ does not make sense because $-$ is not an integer.[1]

Mathematical Focus 2

Modular arithmetic occurs in a mathematical system of elements, operations on those elements, and properties that hold for those operations with those elements. For prime m, $\mathbb{Z}_m$, *the integers modulo* m, *form a mathematical system that is a finite field. If* m *is not prime,* $\mathbb{Z}_m$ *is only a ring.*

A *field* is a set, F, of elements together with two binary operations, addition (denoted as +) and multiplication (denoted as *), that satisfies the field axioms listed in Table 11.1.

$\mathbb{Z}_m$, *the integers modulo* m *(for* m *prime)*, consists of a set of integers $\{0, 1, 2, ..., (m - 1)\}$ together with the operations of integer addition and integer multiplication. $\mathbb{Z}_m$ (m prime) forms a mathematical system that is a *finite field*, because $\mathbb{Z}_m$ has a finite number of elements and satisfies all the field axioms (see Niven & Zuckerman, 1966, p. 65, for a proof that $\mathbb{Z}_m$ is a field if and only if m is prime).

TABLE 11.1   Field Axioms

Axiom	Addition	Multiplication
Closure[2]	Set F is closed under addition: a and b in F implies $a + b$ is in F.	Set F is closed under multiplication: a and b in F implies $a * b$ is in F.
Associativity	For all a, b, and c in F, $(a + b) + c = a + (b + c)$.	For all a, b, and c in F, $(a * b) * c = a * (b * c)$.
Commutativity	For all a, b in F, $a + b = b + a$.	For all a, b in F, $a * b = b * a$.
Existence of identities	There is an element 0 in F such that for all a in F, $a + 0 = a$.	There is an element 1 in F such that for all a in F, $a * 1 = a$.
Existence of inverses	For all a in F, there is an element $-a$ in F such that $a + (-a) = 0$.	For all $a \neq 0$ in F, there is an element a^{-1} (or $\frac{1}{a}$) in F such that $a * (a^{-1}) = 1$ or $a * (\frac{1}{a}) = 1$.
Distributivity of multiplication over addition	For all a, b, and c in F, $a * (b + c) = a * b + a * c$.	

Each element of $\mathbb{Z}_m$ can be interpreted as a representative of an equivalence class created by the congruence relation $a \equiv b$ modulo m. For example, in $\mathbb{Z}_m$ there are 5 elements, typically denoted by the standard class representatives 0, 1, 2, 3, and 4.

$$[0] = \{..., -10, -5, 0, 5, 10, ...\} = \{5n, n \in \mathbb{Z}\}$$

$$[1] = \{..., -9, -4, 1, 6, 11, ...\} = \{5n+1, n \in \mathbb{Z}\}$$

$$[2] = \{..., -8, -3, 2, 7, 12, ...\} = \{5n+2, n \in \mathbb{Z}\}$$

$$[3] = \{..., -7, -2, 3, 8, 13, ...\} = \{5n+3, n \in \mathbb{Z}\}$$

$$[4] = \{..., -6, -1, 4, 9, 14, ...\} = \{5n+4, n \in \mathbb{Z}\}$$

Because $\mathbb{Z}_5$ is a finite field and thus closed, if $\frac{3}{4} \bmod 5$ has meaning, then it must be some element of one of the equivalence classes for which one of the elements of $\mathbb{Z}_5$ is a representative.

Mathematical Focus 3

A meaning can exist[3] for $\frac{p}{q} \bmod m$ by considering $\frac{p}{q} \bmod m$ to represent the product of p and the multiplicative inverse of q in mod m, where p and q are integers, and q is not congruent to 0 mod m.

A multiplicative inverse (if it exists) is an element, a^{-1}, such that $a^{-1} \cdot a = a \cdot a^{-1} = 1$. When working in the rational numbers, the number $\frac{1}{a}$ is the multiplicative inverse of a ($a \neq 0$), because $\frac{1}{a} \cdot a = a \cdot \frac{1}{a} = 1$. When working in a modular system with a prime modulus, each nonzero element in the set will have a multiplicative inverse (as noted previously, Niven & Zuckerman, 1966, p. 65, gives a proof that $\mathbb{Z}_m$ is a field if and only if m is prime).

To answer the question, "What is the meaning of $\frac{3}{4} \bmod 5$?" one can interpret $\frac{3}{4} \bmod 5$ to represent the product of 3 and $\frac{1}{4}$, such that the symbol "$\frac{1}{4}$" is interpreted as the multiplicative inverse of 4 in mod 5. Note that the product of a nonzero number and its multiplicative inverse is one.[4]

For example, to find the multiplicative inverse of 4, consider each of the non-zero congruence classes mod 5,

$$[1] = \{\ldots, -9, -4, 1, 6, 11, \ldots\} = \{5n+1, n \in \mathbb{Z}\}$$

$$[2] = \{\ldots, -8, -3, 2, 7, 12, \ldots\} = \{5n+2, n \in \mathbb{Z}\}$$

$$[3] = \{\ldots, -7, -2, 3, 8, 13, \ldots\} = \{5n+3, n \in \mathbb{Z}\}$$

$$[4] = \{\ldots, -6, -1, 4, 9, 14, \ldots\} = \{5n+4, n \in \mathbb{Z}\},$$

multiply the general expression for a representative of the class by 4, and determine whether the resulting product is congruent to 1 mod 5. This is equivalent to asking the question, "Is 5 a factor of the number that is one less than the resulting product?"

For [1], $(4)(5n + 1) = 20n + 4$. Given that 5 is not a factor of $(20n + 4) - 1$, $(4)(5n + 1)$ is not congruent to 1.

For [2], $(4)(5n + 2) = 20n + 8$. Given that 5 is not a factor of $(20n + 8) - 1$, $(4)(5n + 1)$ is not congruent to 1.

For [3], $(4)(5n + 3) = 20n + 12$. Given that 5 is not a factor of $(20n + 12) - 1$, $(4)(5n + 1)$ is not congruent to 1.

For [4], $(4)(5n + 4) = 20n + 16$. Given that 5 is a factor of $(20n + 16) - 1$, $(4)(5n + 1)$ is congruent to 1. Therefore, the multiplicative inverse of 4 is 4.

Given that $3(4) = 12$ and $12 \equiv 2 \bmod 5$, if one interprets the expression $\frac{3}{4} \bmod 5$ to represent (the product of 3 and the multiplicative inverse of 4) mod 5, then the expression $\frac{3}{4} \bmod 5$ represents 2 mod 5.

Mathematical Focus 4

A meaning can exist for $\frac{p}{q} \bmod m$ by considering $\frac{p}{q} \bmod m$ to represent x such that $qx \equiv p \bmod m$, where p, q, and m are integers, and q is not congruent to 0 mod m.

Based on what it means to be congruent modulo m, the congruence classes mod 5 are:

$$[0] = \{..., -10, -5, 0, 5, 10, ...\} = \{5n, n \in \mathbb{Z}\}$$

$$[1] = \{..., -9, -4, 1, 6, 11, ...\} = \{5n+1, n \in \mathbb{Z}\}$$

$$[2] = \{..., -8, -3, 2, 7, 12, ...\} = \{5n+2, n \in \mathbb{Z}\}$$

$$[3] = \{..., -7, -2, 3, 8, 13, ...\} = \{5n+3, n \in \mathbb{Z}\}$$

$$[4] = \{..., -6, -1, 4, 9, 14, ...\} = \{5n+4, n \in \mathbb{Z}\}$$

To answer the question, "What is the meaning of $\frac{3}{4} \bmod 5$?" one can interpret $\frac{3}{4}$ to represent "x such that $4x \equiv 3 \bmod 5$." Examining the set of values congruent to 3 mod 5 for multiples of 4, without loss of generality, choose the smallest positive multiple of 4, namely 8, and solve the resulting congruence statement for x.

$$4x \equiv 8 \bmod 5$$

$$x \equiv 2 \bmod 5$$

Therefore, if $\frac{3}{4} \bmod 5$ is interpreted to represent x such that $4x \equiv 3 \bmod 5$, then x, and thus $\frac{3}{4} \bmod 5$, represents 2 mod 5.

Mathematical Focus 5

A necessary and sufficient condition for $\frac{p}{q} \bmod m$ and $\frac{r}{s} \bmod m$ to be in the same congruence class is $ps \equiv (qr) \bmod m$.

The following is a statement of what is to be proved.

For integers p, q, r, s, and m ($q, s \not\equiv 0 \bmod m$ and m prime), $\dfrac{p}{q} \bmod m$ and $\dfrac{r}{s} \bmod m$ are in the same congruence class if and only if $ps \equiv (qr) \bmod m$, or

$$\frac{p}{q} \equiv \frac{r}{s} \bmod m \Leftrightarrow ps \equiv (qr) \bmod m \cdot$$

The proof that follows uses the interpretation that $\dfrac{x}{y} \bmod m$ represents (x multiplied by the multiplicative inverse of y) mod m for integers x, y, m ($y \not\equiv 0 \bmod m$ and $m > 0$), and the symbol, y^{-1}, will be used to represent the multiplicative inverse of y.

<u>To prove</u>: For integers p, q, r, s, and m, where q and s are not congruent to 0 mod m and where m is prime:

$$\frac{p}{q} \equiv \frac{r}{s} \bmod m \Leftrightarrow ps \equiv (qr) \bmod m .$$

<u>Proof:</u> Suppose, for integers p, q, r, s, and m, where q and s are not congruent to 0 mod m and where m is prime, $\dfrac{p}{q} \equiv \dfrac{r}{s} \bmod m$.

$$\frac{p}{q} \equiv \frac{r}{s} \bmod m \Leftrightarrow p\left(q^{-1}\right) \equiv \left\{r\left(s^{-1}\right)\right\} \bmod m .$$

Multiplying each side of the congruence by s,

$$\Rightarrow p\left(q^{-1}\right)s \equiv \left\{r\left(s^{-1}\right)s\right\} \bmod m .$$

Applying the associative property for multiplication, because the product of a number and its multiplicative inverse is 1, and because 1 is the identity element,

$$\Rightarrow p\left(q^{-1}\right)s \equiv r \bmod m .$$

Applying the associative and commutative properties for multiplication,

$$\Rightarrow ps\left(q^{-1}\right) \equiv r \bmod m .$$

Multiplying each side of the congruence by q,

$$\Rightarrow ps\left(q^{-1}\right)q \equiv (rq) \bmod m .$$

So, applying the associative property for multiplication, because the product of a number and its multiplicative inverse is 1, and because 1 is the identity element,

$$\Rightarrow ps \equiv (qr) \bmod m .$$

Therefore, if $\dfrac{p}{q} \equiv \dfrac{r}{s} \bmod m$, then $ps \equiv (qr) \bmod m$.

Suppose, for integers p, q, r, s, and m, where q and s are not congruent to 0 mod m and where m is prime, $ps \equiv (qr) \bmod m$. One can prove that $\dfrac{p}{q} \equiv \dfrac{r}{s} \bmod$.

Multiplying each member of the congruence $ps \equiv (qr) \bmod m$ by s^{-1}, the multiplicative inverse of s, applying the associative and commutative properties of multiplication, and using the properties of inverses and identities yields

$$p \equiv (qrs^{-1}) \bmod m.$$

Thus,

$$p \equiv \left\{ q\left(\dfrac{r}{s} \right) \right\} \bmod m$$
.

Multiplying each member of the congruence by the multiplicative inverse of q, q^{-1}, applying the associative and commutative properties of multiplication, and using the properties of inverses and identities results in

$$pq^{-1} \equiv \left\{ qq^{-1}\left(\dfrac{r}{s} \right) \right\} \bmod m.$$

Because the product of a number and its multiplicative inverse is 1, and because 1 is the multiplicative identity,

$$pq^{-1} \equiv \dfrac{r}{s} \bmod m, \text{ or } \dfrac{p}{q} \equiv \dfrac{r}{s} \bmod m.$$

Therefore, if $ps \equiv (qr) \bmod m$, then $\dfrac{p}{q} \equiv \dfrac{r}{s} \bmod m$.

So, $\dfrac{p}{q} \bmod m$ and $\dfrac{r}{s} \bmod m$ are in the same congruence class if and only if $ps \equiv (qr) \bmod m$.

Applying this theorem to $\dfrac{3}{4} \bmod 5$ leads to several conclusions:

i. $\dfrac{3}{4} \bmod 5$ is in the same congruence class as $\dfrac{p}{q} \bmod 5$ if and only if $3q \equiv 4p \bmod 5$.

ii. $\dfrac{3}{4} \bmod 5$ and $\dfrac{6}{8} \bmod 5$ are in the same congruence class.

The products (3)(8) and (6)(4) are both congruent to 4 mod 5. This result is not surprising, given that $\dfrac{3}{4}$ and $\dfrac{6}{8}$ are equivalent fractions in the real number system.

iii. $\frac{3}{4}\bmod 5$ and $\frac{3k}{4k}\bmod 5$ $(k \neq 0)$, are in the same congruence class. The products $(3)(4k)$ and $(4)(3k)$ are both congruent to $(12k)\bmod 5$.

iv. $\frac{3}{4}\bmod 5$ and $\frac{6}{13}\bmod 5$ are in the same congruence class. The products (3) (13) and $(6)(4)$ are both congruent to 4 mod 5. This may be counterintuitive because $\frac{3}{4}$ and $\frac{6}{13}$ are not equivalent fractions in the real number system.

v. $\frac{3}{4}\bmod 5$ and $\frac{3+5k}{4+5k}\bmod 5$ $(k$ an integer) are in the same congruence class. The products $(3)(4 + 5k)$ and $(4)(3 + 5k)$ are congruent to 12 mod 5, and therefore, 2 mod 5.

vi. $\frac{3}{4}\bmod 5$ and $\frac{3+5j}{4+5k}\bmod 5$ $(j$ and k integers) are in the same congruence class. The products $(3)(4 + 5k)$ and $(4)(3 + 5j)$ are both congruent to 12 mod 5, which is congruent to 2 mod 5.

Mathematical Focus 6

The value of $\frac{p}{q}\bmod m$ (q and m are relatively prime, m prime) can be found using a type of Greedy Algorithm.

To find a value for $\frac{p}{q}\bmod m$, where q and m are relatively prime and m is prime, one can use an algorithm described in Weisstein (n.d.). That algorithm is similar to a Greedy Algorithm—an algorithm used to recursively construct a set of objects from the smallest possible constituent parts.

Let $q_0 \equiv q \bmod m$ and find $p_0 = \left\lceil \dfrac{m}{q_0} \right\rceil$, where $1 \leq q_0 \leq m-1$ and where $f(x) = \lceil x \rceil$ is the ceiling function that gives the least integer greater than or equal to x.

Next, compute $q_1 \equiv (q_0 \cdot p_0) \bmod m$, where $1 \leq q_1 \leq m - 1$. From $i = 1$, iterate $p_i = \left\lceil \dfrac{m}{q_i} \right\rceil$ and $q_{i+1} \equiv (q_i \cdot p_i) \bmod m$, where $1 \leq q_{i+1} \leq m - 1$ until $q_n = 1$. Then

$\dfrac{p}{q} \equiv \left(p \cdot \prod_{i=0}^{n-1} p_i \right) \bmod m$. (This method always works for m prime, and it may or may not work if m is not prime.)

Applying this method to find $\frac{3}{4} \bmod 5$, $p = 3$, $q = 4$, and $m = 5$. So, $q_0 \equiv 4 \bmod 5$, and $p_0 = \left\lceil \dfrac{m}{q_0} \right\rceil = \left\lceil \dfrac{5}{4} \right\rceil = 2$. Then, $q_1 \equiv q_0 \, p_0 \equiv 4 \cdot 2 \equiv 3 \bmod 5$, and $p_1 = \left\lceil \dfrac{m}{q_1} \right\rceil = \left\lceil \dfrac{5}{3} \right\rceil = 2$. Finally, $q_2 \equiv q_1 \cdot p_1 \equiv 3 \cdot 2 \equiv 1 \bmod 5$, making $n = 2$. So, $\frac{3}{4} \equiv p \cdot \prod_{i=0}^{2-1} p_i \equiv p \cdot p_0 \cdot p_1 \equiv 3 \cdot 2 \cdot 2 \equiv 2 \bmod 5$. Therefore, $\frac{3}{4} \equiv 2 \bmod 5$.

POSTCOMMENTARY

A meaning for $\frac{3}{4} \bmod 5$ exists because 5 is prime and therefore every nonzero element in $\mathbb{Z}_m$ has a multiplicative inverse. However, for $\mathbb{Z}_m$ with m composite, the multiplicative inverse of a nonzero element may not exist.

Suppose one wished to find the value represented by $\frac{3}{4} \bmod 6$. Consider the multiplication table for mod 6:

•	0	1	2	3	4	5
0	0	0	0	0	0	0
1	0	1	2	3	4	5
2	0	2	4	0	2	4
3	0	3	0	3	0	3
4	0	4	2	0	4	2
5	0	5	4	3	2	1

The products *1 times 1* and *5 times 5* both equal 1. Therefore, 1 is its own multiplicative inverse, and 5 is its own multiplicative inverse. Also, no other product of two values equals 1. Therefore, multiplicative inverses for 0, 2, 3, and 4 do not exist. Because (the multiplicative inverse of 4) mod 6 does not exist, $\frac{3}{4} \bmod 6$, as defined to be the product, mod 6, of 3 and the multiplicative inverse of 4 does not exist.

NOTES

1.   Even though a single expression such as $\frac{3}{4} \bmod 5$ might not make sense in an obvious way, one could still assess the truth or falsehood of an equation such as $\frac{37}{12} = \frac{3}{4} \bmod 5$ by determining whether $\left(\dfrac{37}{12} - \dfrac{3}{4} \right)$ is a whole number multiple of 5 (that is, whether there is a positive whole

number k such that $\left(\dfrac{37}{12}-\dfrac{3}{4}\right)=k\cdot 5$). There is no such k, so the equation is false. But an equation such as $\dfrac{189}{12}=\dfrac{3}{4}\bmod 5$ is correct.

2. The definition of a field typically assumes that the binary operations (addition and multiplication) map F × F to F, so closure of F under both operations is implicit.

3. Although when m is prime each element of $\mathbb{Z}_m$ has a multiplicative inverse, when m is composite some elements of $\mathbb{Z}_m$ do not have a multiplicative inverse. In $\mathbb{Z}_8$, for example, 3 has an inverse (i.e., 3), but 2 does not have a multiplicative inverse.

4. This result extends to "equivalent fractions" in that one can use in place of a multiplicative inverse any expression that is equivalent to that multiplicative inverse. One can show that $\left(\dfrac{p}{q}\right)\bmod m$ is equivalent to $\left(\dfrac{p\cdot r}{q\cdot r}\right)\bmod m$ for all $p, q, r, p\cdot r, q\cdot r < m$, because $\dfrac{1}{q\cdot r}=\dfrac{1}{q}\cdot\dfrac{1}{r}$. Note that, if $0 < q, r, q\cdot r < m$ and if m is prime, the inverses of $q, r,$ and $q\cdot r$ exist. However, if $r \geq m$, the equivalence is not guaranteed. For example,
$$\frac{3}{4}\bmod 5 \equiv 3\times\frac{1}{4}\bmod 5 \equiv 3\times 4\bmod 5 \equiv 2\bmod 5$$
and $\dfrac{3\times 7}{4\times 7}\bmod 5 \equiv \dfrac{21}{28}\bmod 5 \equiv 21\times\dfrac{1}{28}\bmod 5 \equiv 1\times 2\bmod 5 \equiv 2\bmod 5$.
So, $\dfrac{3}{4}\bmod 5$ is equivalent to $\dfrac{3\times 7}{4\times 7}\bmod 5$
On the other hand, one might also expect $\dfrac{3}{4}\bmod 5$ to be equivalent to $\dfrac{3\cdot 5}{4\cdot 5}\bmod 5$. But $\dfrac{15}{20}\bmod 5$ is meaningless because 20 has no inverse in mod 5.

REFERENCES

Niven, I., & Zuckerman, H. S. (1966). *An introduction to the theory of numbers*. New York, NY: Wiley.

Strayer, J. K. (1994). *Elementary number theory*. Long Grove, IL: Waveland Press.

Weisstein, E. W. (n.d.). *Congruence*. Retrieved from http://mathworld.wolfram.com/Congruence.html

CHAPTER 12

ABSOLUTE VALUE EQUATIONS AND INEQUALITIES

Situation 6 From the MACMTL–CPTM Situations Project

Shari Reed, AnnaMarie Conner, Sarah Donaldson,
Kanita DuCloux, Kelly Edenfield, and Erik Jacobson

PROMPT

A student teacher began a tenth-grade geometry lesson on solving absolute value equations by reviewing the meaning of absolute value with the class. They discussed that the absolute value represents a distance from 0 on the number line and that the distance cannot be negative. The teacher then asked the class what the absolute value symbol indicates about the equation $|x| = 2$. A student responded, "anything coming out of it must be 2." The student teacher stated, "x is the distance of 2 from 0 on the number line." Then on the board, the student teacher wrote

$$|x + 3| = 5$$

Mathematical Understanding for Secondary Teaching: A Framework and Classroom-Based Situations, pages 147–153.

$$x + 3 = 5 \text{ and } x + 3 = -5$$

$$x = 2 \qquad x = -8$$

and graphed the solution on a number line. A puzzled student asked, "Why is it 5 and –5? How can you have –8? You said that you couldn't have a negative distance."

COMMENTARY

The primary issue is to understand the nature of absolute value in the real numbers. A discussion of absolute value in the context of complex numbers appears in Situation 7 (see chapter 13). In working with absolute value, the goal is not simply to know the steps and methods for solving absolute value equations and inequalities but to have a deeper knowledge about the reasons why certain solutions are valid and others are not. The absolute value of a number can be defined as the number's distance from 0. There are several interpretations of absolute value, some of which will be addressed in the Foci that follow. Each view adds a new way to think about absolute value and to solve absolute value equations or inequalities. Focus 1 and Focus 2 consider the absolute value as distance from 0 on a number line and discuss how this characterization allows equations and inequalities to be solved graphically and symbolically. Focus 3 and Focus 4 are based on a two-dimensional view of absolute value as a piecewise-defined function, and Focus 5 addresses how describing the absolute value of x as the square root of x^2 can be employed to solve absolute value equations and inequalities.

MATHEMATICAL FOCI

Mathematical Focus 1

Describing absolute value as distance from 0 on a number line provides a clear representation of the meaning of $|x + 3| = 5$ and related inequalities.

The absolute value of a real number is often defined as the number's distance from 0 on the number line. For example, $|2| = 2$ because 2 is 2 units from 0, and $|-2| = 2$ because –2 is also 2 units from 0. This can be seen on the number line in Figure 12.1.

Absolute value equations can also be solved using number lines. These graphs may be expressed in terms of the value of the variable (e.g., x) that produces the

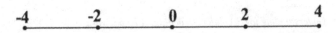

FIGURE 12.1. Locations of 2 and –2 relative to 0 on the number line.

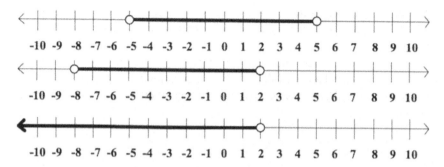

FIGURE 12.2. Graphs of solution sets of three inequalities ($|x| < 5$, $|x + 3| < 5$, and $x + 3 < 5$, respectively) related to $|x + 3| < 5$.

solution, or in terms of the values of a variable expression (e.g., $x + a$) that convey the solution. In these cases, graphs for $x + a$ are translations of the graphs for x.

In $|x + 3| = 5$, x represents a number such that, when 3 is added to it, the result will be 5 units away from 0 on a number line. Consider $|x| = 5$. The solutions to $|x| = 5$ would be 5 and −5. However, translating the solutions three units to the left to reflect the problem in the prompt yields the solutions 2 and −8. Therefore −8 and 2 are solutions to $|x + 3| = 5$.

Likewise, real-number solutions to inequalities can be represented using number lines. Consider the related inequality, $|x + 3| < 5$. One could interpret this as asking for all numbers (written as $x + 3$) that are less than five units from 0. Graphing the set of numbers, x, that are less than five units from 0 yields the first graph shown in Figure 12.2. To compensate for the "+3" and thus have a graph of the values of x that satisfy the inequality $|x + 3| < 5$ requires translating the first graph three units to the left, as shown in the second graph. A graph of $x < 2$ (the solution set of $x + 3 < 5$) is shown as the third graph. Comparing the second and third graphs illustrates that some solutions of $x + 3 < 5$ are not solutions of $|x + 3| < 5$.

Mathematical Focus 2

Describing absolute value as distance from 0 on the number line allows one to solve absolute value equations and inequalities symbolically.

When solving for an unknown (such as x) in an equation, one must list all the possible real solutions (values for x) that make the equation true. Some equations yield only one real solution. For example, in $x + 3 = 5$, the only real number that x could be to make the equation true is 2. Other equations yield more than one solution. For example, if $x^2 = 9$, then x could be either 3 or -3 because both $(3)^2$ and $(-3)^2$ equal 9.

Absolute value equations often yield more than one solution. In $|x| = 5$, for example, there are two values for x that make the equation true, 5 and –5, because both $|5|$ and $|-5|$ are 5.

Extending this to other absolute value equations, such as $|x + 3| = 5$, there are two possibilities for the value of $(x + 3)$. The expression $(x + 3)$ could be 5 or –5, because both $|5|$ and $|-5|$ equal 5. Representing each of these possibilities in equations yields

$$x + 3 = 5 \text{ as well as } x + 3 = -5.$$

So

$$x = 2 \text{ or } x = -8.$$

Each solution can be checked in the original equation to see that it is, indeed, in the solution set of the absolute value equation.

When an inequality involves a numerical constraint on the absolute value of an algebraic expression in one variable, distance from zero can be used to write an extended inequality that can then be solved algebraically. One could interpret the question as asking for all numbers (written $x + 3$) that are less than five units from 0, thus generating the inequality $-5 < x + 3 < 5$, which can be solved algebraically to find the solution set $-8 < x < 2$.

Similarly, if the inequality were reversed, $|x + 3| > 5$, a disjoint statement could be written. One could interpret this question as asking for all numbers (written $x + 3$) that are more than five units (in either direction) from 0. This yields the following:

$$x + 3 < -5 \text{ or } x + 3 > 5$$

$$\Rightarrow x < -8 \text{ or } x > 2.$$

This solution is the complement of what is displayed in the second number line in Figure 12.2.

Mathematical Focus 3

Absolute value equations can be expressed using functions and can be represented graphically. This provides visual representations of domain and range and allows for graphical solutions to absolute value equations and inequalities.

Expressing $y = |x|$ as a function yields $f(x) = |x|$. The domain of the function f with $f(x) = |x|$ typically is taken to be $\{x : x \in \mathbb{R}\}$, and the range is $\{y : y \geq 0\}$. The y-values in the range are never negative (the graph of the function does not appear below the x-axis, where $y < 0$), but the x-values in the domain can be any real numbers. Using absolute value notation, $y = |x|$, what is inside the absolute

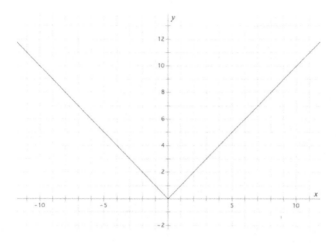

FIGURE 12.3. Graph of $f(x) = |x|$.

value symbols, x, can be negative whereas the absolute value itself, y, cannot be negative (see Figure 12.3).

The solution of an equation can be found by graphing the function related to the left member of the equation and the function related to the right member of the equation and finding the x-value of intersection point(s) of the graphs. Similarly, graphs of related functions can be used to determine the solution of an inequality. To solve the inequality $f(x) < g(x)$, find the values of x for which the graph of f indicates smaller output values than those indicated by the graph of g (one can think of this as the portion of the graph of f that is "below" the graph of g).

Consider the functions $f(x) = |x + 3|$ and $g(x) = 5$. The solution to $|x + 3| = 5$ will be all values in the domain for which $f(x)$ is equal to $g(x)$—the x-values of the intersection points of the graphs of f and g. Similarly, the solution to $|x + 3| < 5$ will be all values in the domain for which $f(x)$ is less than $g(x)$. The graphs of the two functions appear in Figure 12.4. The solution to $|x + 3| < 5$ can be seen by determining the x-values for which the graph of $f(x) = |x + 3|$ is below the graph of $g(x) = 5$. As shown in Figure 12.4, $f(x)$ is less than $g(x)$ exactly when $-8 < x < 2$.

Mathematical Focus 4

Absolute value equations can be expressed as piecewise-defined functions. This representation allows absolute value equations and inequalities to be solved symbolically.

A representation of absolute value as a piecewise-defined function, f, is

$$f(x) = |x| = \begin{cases} x, \text{ if } x \geq 0 \\ -x, \text{ if } x < 0 \end{cases}.$$

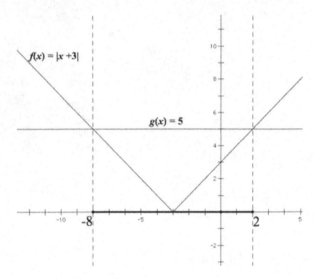

FIGURE 12.4. Graphs of $f(x) = |x + 3|$ and $g(x) = 5$.

Considering the problem in this prompt, this means that

$$g(x) = |x + 3| = \begin{cases} x + 3, \text{ if } x + 3 \geq 0 \\ -x - 3, \text{ if } x + 3 < 0 \end{cases}.$$

To solve the equation $|x + 3| = 5$, one can set each element of the piecewise-defined function g equal to 5 and solve the resulting equations ($x + 3 = 5$ and $-x - 3 = 5$) for x.

$$x + 3 = 5 \Rightarrow x = 2$$

$$-x - 3 = 5 \Rightarrow -x = 8 \Rightarrow x = \text{-}8$$

So, the solutions to the absolute value equation are 2 and -8.

Similarly, the solution set of the inequality $|x + 3| < 5$ is the union of the solution sets of the inequalities generated by these two pieces. Solving the system of inequalities suggested by the first piece, $x + 3 < 5$ and $x + 3 \geq 0$, yields $-3 \leq x < 2$. Solving the system of inequalities suggested by the second piece, $-x - 3 < 5$ and $x + 3 < 0$, yields -8 < x < -3. The union of these two solution sets yields -8 < x < 2.

Mathematical Focus 5

Absolute value of a real number viewed as the positive square root of the square of the number provides another way of understanding absolute value and another tool for solving absolute value equations and inequalities.

Symbolically, absolute value can also be described as the positive square root of the square of that number. That is, $|x| = \sqrt{x^2}$. Use of this characterization to solve $|x + 3| = 5$ is shown here:

$$|x + 3| = 5$$

$$\sqrt{(x+3)^2} = 5$$

$$(x + 3)^2 = 25$$

$$x^2 + 6x + 9 = 25$$

$$x^2 + 6x - 16 = 0$$

$$(x + 8)(x - 2) = 0$$

$$(x + 8) = 0 \text{ or } (x - 2) = 0$$

$$x = \text{-}8 \text{ or } x = 2.$$

Using similar reasoning, $|x + 3| < 5$ implies that $(x + 8)(x - 2) < 0$. Applying to $(x + 8)(x - 2) < 0$ the fact that if the product of two factors is less than 0, one and only one factor is less than 0, either $(x + 8) < 0$ or $(x - 2) < 0$.

If $(x + 8) < 0$, then $x < \text{-}8$. If $x < \text{-}8$, then $(x - 2) < 0$, making both factors less than 0. This is a contradiction, so $(x + 8)$ is not less than 0; it must be greater than 0. Therefore, $(x - 2) < 0$. So $x > \text{-}8$ and $x < 2$. Then, as before, $\text{-}8 < x < 2$.

CHAPTER 13

ABSOLUTE VALUE IN COMPLEX PLANE

Situation 7 From the MACMTL–CPTM Situations Project

Heather Johnson, Shiv Karunakaran, Evan McClintock,
Pawel Nazarewicz, Erik Jacobson, and Kelly Edenfield

PROMPT

A student was asked to produce a function that had certain given characteristics. One of those characteristics was that the function should be undefined for values less than 5. Another characteristic was that the range of the function should contain only nonnegative values. In the process, the student defined $f(x) = \left| \sqrt{x-5} \right|$ and then evaluated $f(-10)$ using his Computer Algebra System (CAS) calculator. The calculator displayed a result of 3.872983346. He looked at the calculator screen and whispered, "How can that be?"

Mathematical Understanding for Secondary Teaching: A Framework and Classroom-Based Situations, pages 155–161.

COMMENTARY

Two major mathematical ideas arise from this Prompt. One is the geometric representation of the complex numbers on a plane. The second is related: the meaning of absolute value in the complex plane. The most commonly used terminology for the absolute value of the complex number, $x + yi$, is *norm, modulus,* or *length of the vector* from 0 to (x, y). Focus 1 and Focus 2 address these ideas. Focus 3 discusses linear absolute value equations in the complex plane. Finally, drawing together the major ideas of the Situation, Focus 4 verifies that $f(-10) \approx 3.873$ by considering the student-generated function $f(x) = \left| \sqrt{x-5} \right|$ as a composition of three functions. The Postcommentary discusses the ambiguous nature of the task in light of the use of a calculator, including a CAS that can function in complex mode.

MATHEMATICAL FOCI

Mathematical Focus 1

Complex numbers can be represented as points on the complex plane.

Representing real numbers requires only a one-dimensional system because all real numbers can be represented on a single line. In this way, a real number, x, can be represented by a unique point on the real number line. However, representing complex numbers requires a plane, in which the real number component is represented on one line and the imaginary number component is represented on another line perpendicular to the real number line. In this way, a complex number $z = x + yi$ can be represented uniquely by a point having coordinates (x, y) on the complex plane. This type of representation is called an *Argand diagram*. For

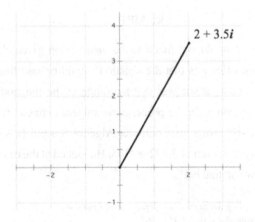

FIGURE 13.1. Representation of $2 + 3.5i$ on the complex plane.

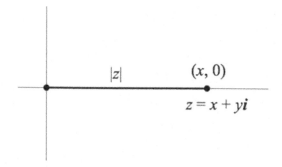

FIGURE 13.2. Graphical representation of $|z|$, where $z = x + yi$ and $y = 0$.

example, the complex number $2 + 3.5i$ is represented by the point $(2, 3.5)$ on the complex plane (see Figure 13.1).

By definition, complex numbers are of the form $z = x + yi$, where x and y are real numbers. Therefore, z is a real number if and only if $y = 0$. In this way, the real numbers are a proper subset of the complex numbers.

Mathematical Focus 2

The absolute value of a complex number, $z = x + yi$, is the number's distance from the origin. This distance is called the modulus or norm and is computed by $|z| = \sqrt{x^2 + y^2}$.

As previously observed, a complex number $z = x + yi$, where x and y are real numbers, can be represented by the point (x, y) on the complex plane. The absolute value of a number, including the "absolute value" of a complex number, is defined as the distance that the number is from zero. In the complex plane, "zero" is the origin, $(0, 0)$. Therefore, the geometric interpretation of $|z|$ is the distance between the point (x, y) and the origin.

Consider a real number expressed in complex form. This means that $y = 0$ and the complex number can be expressed as $z = x + 0i = x$. Plotting this complex number in the complex plane as $(x, 0)$, one can see that the absolute value of z, the distance that z is from the origin, is $|x|$ (see Figure 13.2).

However, consider the general complex number, $z = x + yi$. The distance of z from the origin can be determined by constructing a perpendicular segment from the point (x, y) in the complex plane to the real axis. The resulting diagram suggests a representation of a right triangle with side lengths x and y and hypotenuse of length $|z|$. The Pythagorean Theorem can be used to determine the length of z, the distance the complex number is from the origin, $|z| = \sqrt{x^2 + y^2}$ (see Figure 13.3).

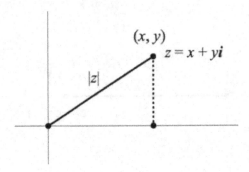

FIGURE 13.3. Graphical representation of $|z|$, where $z = x + yi$ and $y \neq 0$.

Mathematical Focus 3

In the complex plane, there are infinitely many solutions to any linear absolute value equation other than $|z| = 0$, and the graph of these solutions forms a circle.

To address the question in the Prompt, one must understand that many numbers can have the same absolute value in the complex plane. Consider $|x| = 3$. This means that x is 3 units away from 0. If x is a real number, there are two solutions to the equation $|x| = 3$, namely $x = 3$ and $x = -3$. In the complex plane, there are infinitely many solutions for x in $|x| = 3$. Each point on the circle of radius 3 centered at the origin is, by definition, 3 units away from point (0, 0), that is, 3 units away from $0 + 0i$, which is a symbolic representation of the complex number 0.

One can also solve $|x| = 3$ in the following way:

$$|x| = 3$$

$$\Rightarrow |a + bi| = 3$$

$$\Rightarrow \sqrt{a^2 + b^2} = 3$$

$$\Rightarrow a^2 + b^2 = 9$$

Again, consider a circle of radius 3 centered at the origin (see Figure 13.4). Notice that the real number axis (the horizontal axis) intersects the solution circle twice, once at (-3, 0) and again at (3, 0). These points correspond to the real numbers -3 and 3.

This idea can be extended to more interesting absolute value functions such as the one in the equation $|3x + 1| = 5$. There are two real number solutions to the

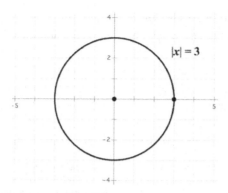

$|x| = 3$

FIGURE 13.4.   Graph of the complex numbers, x, that are solutions to $|x| = 3$.

equation $|3x + 1| = 5$: $x = \dfrac{4}{3}$ and $x = -2$. However, there is an infinite number of complex number solutions. This can be seen in the following calculations.

If $x = a + bi$, then $3x + 1 = 3(a + bi) + 1 = 3a + 3bi + 1$. Grouping, so that $3a + 1$ is the real number component and $3bi$ is the imaginary number component, the following shows the evaluation of $|3x + 1|$.

$$|3x+1| = |3a+1+3bi| = \sqrt{(3a+1)^2 + (3b^2)} = \sqrt{9a^2 + 6a + 1 + 9b^2} \ .$$

Returning to the original equation,

$$|3x+1| = 5$$
$$\Rightarrow \sqrt{9a^2 + 6a + 1 + 9b^2} = 5$$
$$\Rightarrow 9a^2 + 6a + 1 + 9b^2 = 25$$
$$\Rightarrow 9a^2 + 6a + 9b^2 = 24$$
$$\Rightarrow a^2 + \frac{2}{3}a + b^2 = \frac{8}{3}$$
$$\Rightarrow a^2 + \frac{2}{3}a + \frac{1}{9} + b^2 = \frac{8}{3} + \frac{1}{9}$$
$$\Rightarrow \left(a + \frac{1}{3}\right)^2 + b^2 = \left(\frac{5}{3}\right)^2$$

This example suggests that, in the complex plane, solutions to linear absolute value equations are circles. In this case, there is a circle of radius $\dfrac{5}{3}$ that is shifted $\dfrac{1}{3}$ to the left on the real axis. Because there is an infinite number of points on a circle, if solutions to linear absolute value equations are circles then there is an infinite number of solutions to any linear absolute value equation (except $|z| = 0$) in the complex plane.

An alternative, somewhat more general, strategy for solving linear absolute value equations is based in the application of a transformation. First, the solution set of complex numbers, z, to $|z| = r$ is the circle of radius r centered at $(0, 0)$. Translating z to $z - p$ (where p is a real number), yields the complex number solutions to $|z - p| = r$ (for $r > 0$), the circle centered at p with radius r in the complex plane. Applying this idea to $|3x + 1| = 5$, one can write an equivalent equation $\left|x + \frac{1}{3}\right| = \frac{5}{3}$, yielding $\left|x - \left(-\frac{1}{3}\right)\right| = \frac{5}{3}$. So the solution set is the circle centered at complex number $p = -\frac{1}{3}$ (i.e., the point $\left(-\frac{1}{3}, 0\right)$ in the x–y plane) with radius $\frac{5}{3}$.

Mathematical Focus 4

A composite function with the same domain and codomain may be composed of functions with different domains and codomains.

The function represented by $f(x) = \left|\sqrt{x-5}\right|$ is a real function because it has domain and codomain of $\mathbb{R}$, the real numbers. Another way to express $f(x)$ is as the composition of three other functions, r, s, and t, with $r(x) = x - 5$ having domain and codomain $\mathbb{R}$, $s(x) = \sqrt{x}$ having domain $\mathbb{R}$ and codomain $\mathbb{C}$, the complex numbers, and $t(x) = |x|$ having domain $\mathbb{C}$ and codomain $\mathbb{R}$. Note that the usual convention for extending the domain of the square root function from positive to negative reals is that the output is taken to be the product of i and the positive, or principal, square root. This is a choice, not a decision forced by the mathematics itself.

Because f is the composition of the functions r, s, and t, it can be expressed as $f(x) = t \circ s \circ r(x) = t(s(r(x)))$. Thus, the value $f(-10)$ from the Prompt must the same as $t(s(r(-10)))$.

To compute the value of a composite function, evaluate each function in turn, beginning with the innermost, $r(10) = -10 - 5 = -15$. Next, evaluate the function s at $x = -15$: $s(x) = \sqrt{-15} = i\sqrt{15}$. Because the function t has a codomain in the real numbers, the final value of $f(-10)$ will be a real number. The last step is to compute $t(i\sqrt{15}) = \left|i\sqrt{15}\right| = \sqrt{15} \approx 3.873$. Figure 13.5 shows all the steps of the composition. Notice that the computational process begins with the real number -10 that is mapped to the real number -15, but -15 is mapped to $i\sqrt{15}$ in the complex numbers by function s, and then $i\sqrt{15}$ is mapped back to the (positive) real numbers—to $\sqrt{15}$—by the function t.

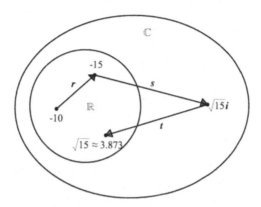

FIGURE 13.5.   Location within the complex numbers and real numbers of the outputs of each of the functions in the composition *t∘s∘r*.

POSTCOMMENTARY

The seventh-grade student whose conundrum is the focus of this Prompt[1] was originally asked to write a function with given characteristics. Given that he had probably not learned about complex numbers at this point, his confusion is justified. In the real number system, the function *f*, with rule $f(x) = \left| \sqrt{x-5} \right|$, is undefined for values of *x* less than 5 and has a nonnegative range. In the complex numbers, however, the function is *not* undefined for values less than 5. Because the student was using a calculator in complex mode, a CAS calculator in this case, he did not get the "undefined" result he expected. Instead, the calculator returned a real answer for an input value for which the student expected the calculator to return an error message.

The issues of domain and codomain raised in Focus 4 become especially salient when a CAS is employed. The technology allows for greater exploration by students, but it also requires teachers and students to be more aware of the issues of domain. Teachers also must be more explicit about the number systems that are intended for use in solving problems of this kind.

NOTE

1.   This incident is described in Heid, Iseri, and Hollebrands (2002).

REFERENCES

Heid, M. K., Iseri, L. W., & Hollebrands, K. F. (2002). Connecting research to teaching: Reasoning and justification, with examples from technological environments. *Mathematics Teacher, 95*, 210–216.

CHAPTER 14

PROPERTIES OF *i* AND OTHER COMPLEX NUMBERS

Situation 8 From the MACMTL–CPTM Situations Project

Erik Tillema, Evan McClintock, M. Kathleen Heid, and Heather Johnson

PROMPT

A teacher of an Algebra 3 course noticed that her students often interpreted *i* in expressions such as $3i$, i^2, and $4i + 2$ as though it were an unknown or variable rather than a number.

COMMENTARY

This Situation provides different interpretations of the imaginary unit and utilizes the multiplication operation involving this unit. Symbolic, graphical, and geometric representations are used in the set of Foci that follow. Focus 1 treats the imaginary unit as a number that is a solution to a quadratic equation with no real solutions. Focus 2 treats multiplication by the imaginary unit as an operation

Mathematical Understanding for Secondary Teaching: A Framework and Classroom-Based Situations, pages 163–170.

163

that rotates points about the origin in the complex plane. Connections are made to vector operations in Focus 3 and Focus 4, and Focus 5 incorporates linear algebra.

MATHEMATICAL FOCI

Mathematical Focus 1

*The imaginary number **i** is a solution to* $x^2 + 1 = 0$ *and is a special case of a complex number.*

Historically, the invention of complex numbers[1] emerged within the context of solving quadratic equations. The quadratic equation $x^2 + 1 = 0$ has no real solutions. Solving for x gives $x = \pm\sqrt{-1}$. By defining a number i, such that $i = \sqrt{-1}$, the solutions to the quadratic equation $x^2 + 1 = 0$ can be expressed as $x = \pm i$. Substituting $x = i$ gives $i^2 = -1$.

To distinguish the number i from a variable x, one can consider i as a special case of a complex number. Because complex numbers are of the form $z = a + bi$, where a and b are real numbers, when $a = 0$ and $b = 1$, $z = i$.

Mathematical Focus 2

*The multiplication operation involving -1 and **i** can be represented as rotations on the real number line and the complex plane.*

Cases involving -1:

Each real number can be represented by a unique point on a line. Multiplying a real number by -1 can be represented as the rotation of a point on the real line 180° counterclockwise about the origin to another point on the real line equidistant from the origin (i.e., a rotation without dilation; see Figure 14.1).

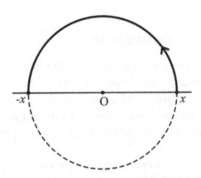

FIGURE 14.1. Illustration of multiplication as rotation about the origin.

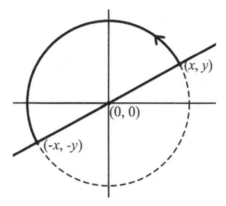

FIGURE 14.2. Illustration of multiplication of a complex number of the form *a +
bi*, (for *b* ≠ 0) by -1 as rotation about the origin.

In contrast, representing complex numbers that have a nonzero imaginary
component requires a coordinate plane, on which one axis is a real-number axis
and an axis perpendicular to the first is an imaginary-number axis. In this way, a
complex number, $z = x + yi$, can be represented uniquely by a point with coordinates (x, y) on the complex plane.

Multiplying a nonzero[2] complex number by -1 can be represented as the rotation of a point on the complex plane 180° counterclockwise about the origin to
another point on the complex plane equidistant from the origin (i.e., a rotation
without dilation; see Figure 14.2).

Cases involving i:

Multiplying by *i* · *i* is equivalent to multiplying by -1. It is consistent to think
about multiplying by *i* · *i* as equivalent to multiplying by *i*, and then multiplying
the resulting product by *i*. So, multiplying a real number by *i* can be thought of as
the rotation of a point on the real-number axis 90° counterclockwise about the origin to a point equidistant from the origin on the imaginary axis (see Figure 14.3).

Applying distributive and associative properties to complex numbers, multiplying a complex number *a* + *bi* by *i* gives:

$$(a + bi)i = ai + bi^2 = ai + b(-1) = -b + ai.$$

It is consistent with this calculation to regard multiplying a nonzero complex
number by *i* as the rotation of a point on the complex plane 90° counterclockwise
about the origin to another point on the complex plane equidistant from the origin
(see Figure 14.4).

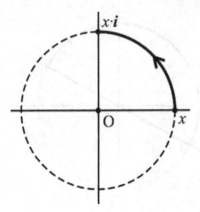

FIGURE 14.3. Illustration of multiplication of a complex number of the form $a + bi$, (for $b \neq 0$) by i as rotation about the origin.

Mathematical Focus 3

> *The multiplication operation involving* i *can be represented as rotations of unit vectors.*

Consider multiplying a positively oriented unit vector on the *x*-axis by i. In this way, the vector 1 is rotated 90 degrees counterclockwise about the origin to the vector i on the *y*-axis (see Figure 14.5). Now consider multiplying the vector i by i. In this way, the vector i is rotated 90 degrees counterclockwise about the origin to the vector $i^2 = -1$ on the *x*-axis (Figure 14.6).

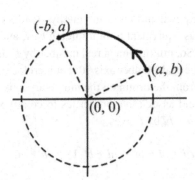

FIGURE 14.4. Multiplying $a + bi$ (for $b \neq 0$) by i is consistent with rotating the point (a, b) in the complex plane to $(-b, a)$ with a 90-degree counterclockwise rotation about the origin.

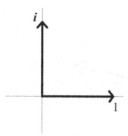

FIGURE 14.5. When multiplied by *i*, the vector **1** is rotated 90 degrees counterclockwise about the origin to the vector *i* on the *y*-axis.

Mathematical Focus 4

The multiplication operation involving real and complex numbers can be represented as rotations, dilations, and linear combinations of vectors.

Consider the multiplication of a complex number by another complex number, for example, $(2i + 3)(4 + i)$. The distributive property of multiplication over addition can be used to determine $(2i + 3)(4 + i)$ because the set of complex numbers satisfies the field properties for addition and multiplication, which include distributivity of multiplication over addition.

The vector $(4 + i)$ is depicted in Figure 14.7.

Determining partial products:

Multiplying the vector $(4 + i)$ by $2i$ rotates it 90 degrees counterclockwise about the origin and dilates it by a factor of 2, as shown by the dashed vector in Figure 14.8. Multiplying the vector $(4 + i)$ by 3 dilates it by a factor of 3, as shown by the solid black vector in Figure 14.8.

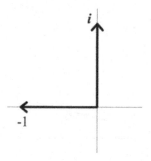

FIGURE 14.6. Multiplying the vector *i* by *i* is equivalent to rotating the vector *i* 90 degrees counterclockwise about the origin to the vector $i^2 = -1$ on the *x*-axis.

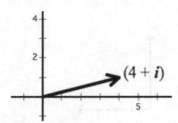

FIGURE 14.7. The vector 4 + *i*.

Summing the partial products:
The sum of the vectors 2*i*(4 + *i*) and 3(4 + *i*) is shown in Figure 14.9. Hence (2*i* + 3)(4 + *i*) = 10 + 11*i*.[3]

Mathematical Focus 5

Complex numbers and their dilations and rotations can be represented with matrices.

A complex number $a + bi$ can be represented as a matrix in the form $\begin{bmatrix} a & -b \\ b & a \end{bmatrix}$ and a rotation matrix is defined as $\begin{bmatrix} \cos\theta & -\sin\theta \\ \sin\theta & \cos\theta \end{bmatrix}$ where θ is the angle of rotation in the counterclockwise direction. The imaginary number bi can be represented by a matrix of the form $\begin{bmatrix} 0 & -b \\ b & 0 \end{bmatrix} = b\begin{bmatrix} 0 & -1 \\ 1 & 0 \end{bmatrix}$. This can be interpreted in terms of the rotation matrix as $\cos\theta = 0$ and $\sin\theta = 1$, giving $\theta = 90°$ followed by a dilation with scale factor b. In this way, multiplying by the imaginary number bi produces

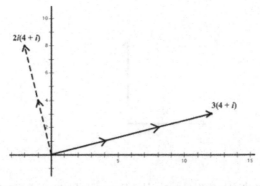

FIGURE 14.8. Illustration of the results of multiplying the vector 4 + *i* by 2*i* and by 3.

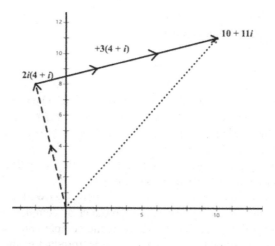

FIGURE 14.9.   The sum of the vectors 2i(4 + i) and 3(4 + i).

a rotation of 90° about the origin in the counterclockwise direction and a dilation with scale factor b. Next, the real number a can be represented by a matrix of the form $\begin{bmatrix} a & 0 \\ 0 & a \end{bmatrix} = a \begin{bmatrix} 1 & 0 \\ 0 & 1 \end{bmatrix}$. This can be interpreted in terms of the rotation matrix as $\cos\theta = 1$ and $\sin\theta = 0$, giving $\theta = 0°$ followed by a dilation with scale factor a.

In this way, multiplying by the real number a produces a rotation of 0° about the origin and a dilation with scale factor a.

The action of left matrix multiplication on 2×1 column vectors representing complex numbers yields the same result as multiplying the complex numbers as binomials. For a, b, x, y real numbers, $(a + bi)•(x + yi) = (ax - by) + (ay + bx)•i$. When complex numbers are represented as 2×1 vectors with the real part in row 1 column 1, and the imaginary part in row 2 column 1, this complex number multiplication can be represented by the matrix multiplication

$$\begin{pmatrix} a & -b \\ b & a \end{pmatrix} • \begin{pmatrix} x \\ y \end{pmatrix} = \begin{pmatrix} ax - by \\ ay + bx \end{pmatrix}.$$

This matrix equation avoids the use of the imaginary number i, but requires two (real) equations. Both ways of representing complex number multiplication are valid, but one must keep in mind that *row 2 column 1* of the 2×1 vector stands for the imaginary part of the complex number.

Writing or thinking of a complex number as an expression of the form $a + bi$, or as a 2-tuple $\begin{pmatrix} a \\ b \end{pmatrix}$, where a and b are real, are both valid. The one logical advantage of the 2-tuple is that one need not worry about what "multiplication" of a real number by the nonreal symbol i means logically.

NOTES

1. The set of complex numbers, of the form $a + bi$ where a and b are real numbers, with addition defined as $(a + bi)+(c + di)=(a + c) + (b + d)i$ and multiplication defined as $(a + bi)\cdot(c + di) = ac +(ad + bc)i + bdi^2$, is a field (see Childs, 2009).

2. Multiplying the zero complex number, $0 + 0i$, by -1 does not lead to "another point"; it leads to the same point.

3. In many introductory complex analysis books, as well as in some secondary mathematics textbooks, the product of complex numbers (say z and w) is described geometrically as the complex number $z \cdot w$ with modulus, norm, or length equal to $|z| \cdot |w|$ (product of the lengths of z and w) and polar angle given by the sum of the polar angles of z and w.

REFERENCE

Childs, L. N. (2009). *A concrete introduction to higher algebra* (3rd ed.). New York, NY: Springer.

CHAPTER 15

SQUARE ROOT OF *i*

Situation 9 From the MACMTL-CPTM Situations Project

Heather Johnson, Shiv Karunakaran, Ryan Fox, and Evan McClintock

PROMPT

Knowing that a Computer Algebra System (CAS) had commands such as **cfactor** and **csolve** to factor complex number expressions and solve complex number equations, a teacher was curious about what would happen if she entered $\sqrt{i}$. The result was $\frac{\sqrt{2}}{2} + \frac{\sqrt{2}}{2}i$. She wondered why a CAS would give a result such as that.

COMMENTARY

When using a CAS, students and teachers can encounter situations that cause them to question why the CAS may give a particular result. Symbolic verification and manipulation can be used to confirm results given by a CAS. Focus 1 accounts for the reasoning behind the symbolic work by confirming that the result makes sense. To address the underlying mathematical logic relating to why $\sqrt{i} = \frac{\sqrt{2}}{2} + \frac{\sqrt{2}}{2}i$, Focus 2, Focus 3, and Focus 4 utilize representations of complex

Mathematical Understanding for Secondary Teaching: A Framework and Classroom-Based Situations, pages 171–177.

numbers on the complex plane. Focus 2 connects powers of i to points of the unit circle on the complex plane and their images under rotations, and Focus 3 uses Euler's formula to represent complex numbers in exponential and trigonometric form. Focus 4 considers the powers of i as elements of cyclic groups.

MATHEMATICAL FOCI

Mathematical Focus 1

Solving the equation $x^2 = i$, *where* $x = a + bi$, *and verifying the solution to the equation provides a representation of the square root of the imaginary number,* i.

Using the fact that any complex number is of the form $a + bi$, where a and b are real numbers, the square roots of i can be determined by solving the equation $(a + bi)^2 = i$ for a and b. Expanding $(a + bi)^2$, the left member of the equation, results in the equivalent equation, $a^2 + 2abi - b^2 = i$. Equating the real components and equating the imaginary components of the equation yield $a^2 - b^2 = 0$ and $2ab = 1$, respectively. Therefore, $a = \pm b$ and either $2b^2 = 1$ or $-2b^2 = 1$. However, because both a and b are real and $-2b^2 = 1$ has no real solutions, then $2b^2 = 1$ must be true. However, if $a = -b$, then $2 \cdot a \cdot b = 2 \cdot (-b) \cdot b = -2b^2 = 1$, which is not possible, meaning that $a = -b$ is not possible, leaving $a = b$ as the only possibility. Solving $2b^2 = 1$ for b gives $b = \dfrac{\sqrt{2}}{2}$ and $b = -\dfrac{\sqrt{2}}{2}$. Therefore the equation $(a + bi)^2 = i$ has two sets of solutions, namely $a = \dfrac{\sqrt{2}}{2}$, $b = \dfrac{\sqrt{2}}{2}$ and $a = -\dfrac{\sqrt{2}}{2}$, $b = -\dfrac{\sqrt{2}}{2}$. In this way, $\dfrac{\sqrt{2}}{2} + \dfrac{\sqrt{2}}{2}i$ and $-\dfrac{\sqrt{2}}{2} - \dfrac{\sqrt{2}}{2}i$ are both square roots of i (see Spencer, 1999).

One way to verify that a complex number is a square root of another number is to square that complex number and verify that the square and the other number are equivalent. By squaring the expression $\dfrac{\sqrt{2}}{2} + \dfrac{\sqrt{2}}{2}i$, it can be verified that $\dfrac{\sqrt{2}}{2} + \dfrac{\sqrt{2}}{2}i$ is a square root of $x^2 = i$. A similar argument verifies that $-\dfrac{\sqrt{2}}{2} - \dfrac{\sqrt{2}}{2}i$ is also a square root of i. It is useful to note that the symbolic manipulation needed to expand the expression $\left(\dfrac{\sqrt{2}}{2} + \dfrac{\sqrt{2}}{2}i \right)^2$ treats it as though it were an algebraic expression of the form $(a + b)^2$ from the real domain.

Expanding $\left(\dfrac{\sqrt{2}}{2} + \dfrac{\sqrt{2}}{2}i \right)^2$ gives

$$\left(\frac{\sqrt{2}}{2} + \frac{\sqrt{2}}{2}i \right)^2 = \left(\frac{\sqrt{2}}{2} \right)^2 + 2\left(\frac{1}{2} \right)i + \left(\frac{\sqrt{2}}{2}i \right)^2 = \frac{1}{2} + i - \frac{1}{2} = i.$$

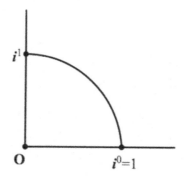

FIGURE 15.1.   First quadrant of the unit circle on the complex plane.

So, $\left(\dfrac{\sqrt{2}}{2}+\dfrac{\sqrt{2}}{2}i\right)^{2}=i$. Because $\left(\dfrac{\sqrt{2}}{2}+\dfrac{\sqrt{2}}{2}i\right)^{2}=i$, $\dfrac{\sqrt{2}}{2}+\dfrac{\sqrt{2}}{2}i$ is a square root of i.

Mathematical Focus 2

Powers of i *can be related to rotations involving the unit circle on the complex plane.*

Consider the unit circle on the complex plane, and, on this circle, consider the point representations of i^{0} and i (see Figure 15.1).[1] Note that the point representing i is the image of the point representing i^{0} under $P_{(0,90°)}$, a counterclockwise rotation of 90° about the origin (O). Thus, if the point for i^{0} could be represented as (1, 0), and if the point for i is represented as $_{(0,90°)}(1, 0) = (0, 1)$ then the point for $\sqrt{i}$ can be thought of as the image of the point for i^{0} under $P_{(0,45°)}$, a rotation of 45° about the origin (see Figure 15.2).[2] Moreover, the point for i can also be thought about as the image of the point for $\sqrt{i}$ under $P_{(0,45°)}$. So, $P_{(0,45°)}$ composed with itself is the same as $P_{(0,90°)}$, that is, $P^{2}_{(0,45°)} = P_{(0,90°)}$.

This analysis is consistent with the definition of product of complex numbers which implies that the product of a complex number, z, with itself, $z \times z$, is a complex number with double the polar angle of z and with modulus or norm equal to the square of the modulus or norm of z. So, the solution of $z^{2} = i$ requires a z with the property that its modulus squared is 1 and its polar angle, when doubled, yields polar angle $\dfrac{\pi}{2}+2\pi n$ (for any integer n). There are two nontrivial solutions given by the points on a unit circle with polar angles $\dfrac{\pi}{4}$ (the one identified here) and $\dfrac{5\pi}{4}$, which doubles to $\dfrac{10\pi}{4}=\dfrac{\pi}{2}+2\pi$. The solution $\dfrac{5\pi}{4}$ gives the square root

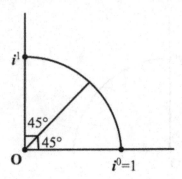

FIGURE 15.2. Images of points representing powers of *i* as rotations.

with negative real part. Additional (trivial) ones can be obtained by adding integer multiples of 2π to the nontrivial ones' polar angles.

Each point on the circle corresponds to the complex number, $\cos(x) + i \sin(x)$. This is shown in Figure 15.3. A point representing $i^{\frac{1}{2}}$ is the image of a point representing i^0 under a rotation of 45° about the origin. Therefore, the coordinates of $i^{\frac{1}{2}}$ have to be $(\cos 45°, \sin 45°) = \left(\dfrac{\sqrt{2}}{2}, \dfrac{\sqrt{2}}{2}\right)$. Thus, $i^{\frac{1}{2}}$.

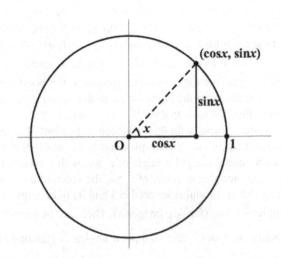

FIGURE 15.3. Coordinates of points on the complex unit circle.

Mathematical Focus 3

*By using Euler's formula, the connection between the trigonometric representation of any complex number and the square root of the imaginary number, **i**, is made more explicit.*

Knowing that every point on the unit circle on the complex plane corresponds to a complex number z, where $z = \cos\theta + i\sin\theta$, Euler's formula, $e^{i\theta} = \cos\theta + i\sin\theta$ can be used to express those complex numbers in the exponential form $z = e^{i\theta}$. For example, letting $\theta = \pi$, $z = e^{i\pi} = \cos\pi + i\sin\pi = -1$, which can be represented by the point (-1, 0) on the unit circle on the complex plane. Similarly, letting $\theta = \dfrac{\pi}{2}$, $z = e^{i\frac{\pi}{2}} = \cos\dfrac{\pi}{2} + i\sin\dfrac{\pi}{2} = i$, which can be represented by the point (0, 1) on the unit circle on the complex plane. Because the task under consideration is to determine $\sqrt{i}$, and because $e^{i\frac{\pi}{2}} = i$, by Euler's formula, it follows that $\sqrt{e^{i\frac{\pi}{2}}} = \sqrt{i}$.

Since $i^{\frac{1}{2}} = \left(e^{i\frac{\pi}{2}}\right)^{\frac{1}{2}}$, using properties of exponents, one can conclude that $i^{\frac{1}{2}} = e^{i\frac{\pi}{4}}$. In this way, letting $\theta = \dfrac{\pi}{4}$, it follows that $z = e^{i\frac{\pi}{4}} = \cos\dfrac{\pi}{4} + i\sin\dfrac{\pi}{4} = \dfrac{\sqrt{2}}{2} + \dfrac{\sqrt{2}}{2}i$, which can be represented by the point $\left(\dfrac{\sqrt{2}}{2}, \dfrac{\sqrt{2}}{2}\right)$ on the unit circle on the complex plane.

Because $e^{i\frac{\pi}{4}} = \dfrac{\sqrt{2}}{2} + \dfrac{\sqrt{2}}{2}i$ and $i^{\frac{1}{2}} = e^{i\frac{\pi}{4}}$, one can conclude that $i^{\frac{1}{2}} = \dfrac{\sqrt{2}}{2} + \dfrac{\sqrt{2}}{2}i$.

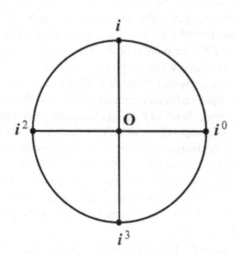

FIGURE 15.4.   Locations in the complex plane of i^0, i^1, i^2, and i^3.

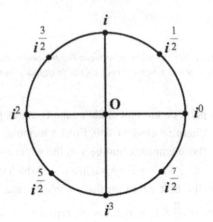

FIGURE 15.5. Location on the complex plane of the elements of the cyclic group (H, ∘).

Mathematical Focus 4

*The value of the square root of the imaginary number, **i**, can be determined by relating this value to cyclic groups.*

This Situation deals with $\sqrt{i}$, which can be written as $i^{\frac{1}{2}}$. So, one strategy is to search for patterns in the powers of i. One can examine the integer powers of i, starting with i^0. The locations of the points in the complex plane representing i^0, i^1, i^2, and i^3 are shown in Figure 15.4. Note that points representing all four powers of i appear on the complex unit circle. Moreover, the four points are positioned at equal increments around the circle (exactly at 90° increments). Furthermore, the fourth power of i can be plotted in the same position as the zero power of i (i.e., $i^4 = i^0 = 1$). Also, every integer power of i greater than 3 is plotted on the four points around the complex unit circle in Figure 15.4. This arrangement of the four powers of i at equal increments around a circle and the cyclic property of the powers as just described, leads to looking at a cyclic group generated by i.

Consider a cyclic group, $(G, \circ)$, of order 4, isomorphic to $(\mathbb{Z}_4, +)$, which can be generated using the imaginary number i as the generator (i.e., $i^{4k} = 1$, where $k \geq 0$ and k is an integer). Note that 1 is the identity element of the group G. Also, one can list all the elements of this group by considering the powers of i (i.e., $G = \langle i \rangle_4 = \{i^0, i^1, i^2, i^3\}$). As previously discussed, the elements of the cyclic group, G, can be illustrated as four symmetric points on the unit circle in the complex plane, as shown in Figure 15.4.

Because the focus of the investigation is $i^{\frac{1}{2}}$, one can extend this discussion of the powers of i by examining $\left\{ i^0, i^{\frac{1}{2}}, i^1, i^{\frac{3}{2}}, i^2, i^{\frac{5}{2}}, i^3, i^{\frac{7}{2}} \right\}$. Thus, the order of the group increases from 4 to 8. That yields the cyclic group, (H, °), isomorphic to $(\mathbb{Z}_8,+)$, given by $H = \langle i \rangle_8 = \left\{ 1, i^{\frac{1}{2}}, i^1, i^{\frac{3}{2}}, i^2, i^{\frac{5}{2}}, i^3, i^{\frac{7}{2}} \right\}$. This group can also be illustrated on the unit circle on the complex plane as shown in Figure 15.5. To obtain the coordinates of these points on the complex unit circle, Focus 2 can be used to obtain $i^{\frac{1}{2}} = \frac{\sqrt{2}}{2} + \frac{\sqrt{2}}{2}i$.

NOTES

1. The convention from the real number system, that r^0 is interpreted as equal to 1 for any nonzero real r, is assumed to extend to complex numbers in exactly the same way. See Situation 12 in chapter 18 for further discussion of the definition of a number raised to the zero power.

2. When used with real numbers, the radical symbol, $\sqrt{\ }$, refers to the *principal* (positive) *square root*. When the radical symbol is used with complex numbers, it also refers to the principal square root, namely, the one with the positive real component. For example, $\sqrt{i} = \frac{\sqrt{2}}{2} + \frac{\sqrt{2}}{2}i$, but $\sqrt{i} \neq -\frac{\sqrt{2}}{2} - \frac{\sqrt{2}}{2}i$, even though $-\frac{\sqrt{2}}{2} - \frac{\sqrt{2}}{2}i$ is one of the square roots of *i*.

REFERENCE

Spencer, P. (1999, April 19). *Question corner—What is the square root of i?* Retrieved from http://www.math.toronto.edu/mathnet/questionCorner/rootofi.html

CHAPTER 16

EXPONENT RULES

Situation 10 From the MACMTL-CPTM Situations Project

**Erik Tillema, Sarah Donaldson, Kelly Edenfield,
James Wilson, Eileen Murray, and Glendon Blume**

PROMPT

In an Algebra 2 class, students had just finished reviewing the rules for exponents. The teacher wrote $x^m \cdot x^n = x^5$ on the board and asked the students to make a list of values for m and n that made the statement true. After a few minutes, one student asked, "Can we write them all down? I keep thinking of more."

COMMENTARY

The relevant mathematics in this Situation reaches beyond the basic rules for exponents into issues of the domains of the variables in those rules. The exponent rule $x^m \cdot x^n = x^{m+n}$ is applicable and is key to deciding how many solutions there will be. However, applying this rule beyond the usual context of positive bases and positive exponents to that of other number systems (such as the set of integers or rational numbers) requires consideration of the domains of the base and the ex-

*Mathematical Understanding for Secondary Teaching: A Framework
and Classroom-Based Situations*, pages 179–183.

ponents. In the following Foci, symbolic, numeric, and graphical representations are used to highlight that there are particular values in the domains of both x and m for which x^m is not a real number. In Focus 2 and Focus 3, the domains of x and m are extended and x^m is examined in these new domains.

MATHEMATICAL FOCI

Mathematical Focus 1

Defining the domains of x *and* m *clarifies the rule* $x^m \cdot x^n = x^{m+n}$ *as applying to expressions with positive bases and real number exponents.*

The exponent rule $x^m \cdot x^n = x^{m+n}$ may be applied to the equation $x^m \cdot x^n = x^5$ if the value of x is restricted to positive numbers. That is, if $x > 0$, and m and n are real numbers, then $x^m \cdot x^n = x^{m+n}$.

If the values of m and n are restricted to natural (counting) numbers, then there is a finite number of solutions for (m, n): (1, 4), (2, 3), (3, 2), and (4, 1). If m and n are not restricted to the natural numbers, then there are infinitely many solutions for (m, n), because there are infinitely many solutions to the equation $m + n = 5$. Thus, m and n could be any two values whose sum is 5.

The restriction $x > 0$ is necessary if the value of each term in the equation is restricted to the set of real numbers. If, for example, $m = 1.5$, then $x^m = x^{1.5} = x^{3/2} = \sqrt{x^3} = |x|\sqrt{x}$. For this result to be a real number, x cannot be negative. Some restriction also may be necessary when x is 0, for example, if $m = -1$ and $x = 0$, then x^m is undefined.

Mathematical Focus 2

When the domain of x *is extended to the negative real numbers or 0, then the domain of* m *is limited to those values for which* x^m *is defined.*

If x^m is to be a real number, then the value of x determines the domain for m. (All statements about x^m also hold for x^n.) Specifically, if x and m are real numbers, then:

a. If $x = 0$, then $m > 0$. Note that 0 raised to a negative power (e.g., 0^{-2}) is undefined. Typically, 0^0 is considered to be indeterminate. However, there are instances (for example, in the Binomial Theorem) in which 0^0 is taken to be 1 by convention.[1]

b. If $x > 0$, then m could be any real number. This is what is assumed in the rule $x^m \cdot x^n = x^{m+n}$ as noted in Focus 1.

c. If $x < 0$, then x^m is defined only when m is a rational number (in lowest terms) whose denominator is odd.

Some examples may clarify Item c. First, consider a subset of the rational numbers with odd denominators: the set of integers.

TABLE 16.1   Values of $(-2)^m$ for Various Integer Values of m

m	$(-2)^m$
-3	-0.125
-2	0.25
-1	-0.5
0	1
1	-2
2	4
3	-8

Let $x = -2$. With this (or any negative real number) as a base, the exponent, m, may be any integer and the result will be a real number (see Table 16.1). The exponent, m, may also be any rational number (in lowest terms) as long as its denominator is odd.

Consider the case when m is a rational number with an even denominator, such as ½; recall that the exponent now indicates an even root. An even root of a negative number yields an imaginary result. If the domains of x and m are restricted to the real numbers, then an even root of a negative number lies outside these domains. If x^m were not limited to the domain of real numbers, then m could be any number as long as $x \neq 0$ (as explained previously in Item a.).

Examining a list of values of $(-2)^m$ and the resulting graph also provides insight into these restrictions. The graph in Figure 16.1 is the result of plotting the points in Table 16.2. These points were chosen because they show that $(-2)^m$ is a real number for certain values of m and not for others. The values for m that result in $(-2)^m$ not

TABLE 16.2   Values of $(-2)^m$ for Various Rational Values of m

m	$(-2)^m$	m	$(-2)^m$	m	$(-2)^m$	m	$(-2)^m$	m	$(-2)^m$
-3	-0.125	-2	0.25	-1	-0.5	0	1	1	-2
-2.9	Ø	-1.9	Ø	-0.9	Ø	0.1	Ø	1.1	Ø
-2.8	0.143...	-1.8	-0.287...	-0.8	0.574...	0.2	-1.148...	1.2	2.297...
-2.7	Ø	-1.7	Ø	-0.7	Ø	0.3	Ø	1.3	Ø
-2.6	-0.164...	-1.6	0.329...	-0.6	-0.659...	0.4	1.319...	1.4	-2.639...
-2.5	Ø	-1.5	Ø	-0.5	Ø	0.5	Ø	1.5	Ø
-2.4	0.189...	-1.4	-0.378...	-0.4	0.757...	0.6	-1.515...	1.6	3.031...
-2.3	Ø	-1.3	Ø	-0.3	Ø	0.7	Ø	1.7	Ø
-2.2	-0.217...	-1.2	0.435...	-0.2	-0.870...	0.8	1.741	1.8	-3.482...
-2.1	Ø	-1.1	Ø	-0.1	Ø	0.9	Ø	1.9	Ø

Note. Ø indicates that the value of $(-2)^m$ is not a real number.

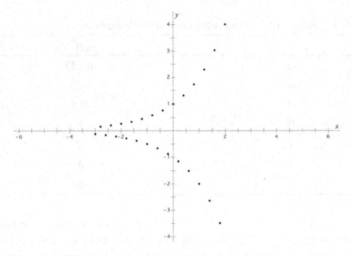

FIGURE 16.1.   Graph of $(-2)^m$ for values of m in Table 16.2.

being a real number (indicated in Table 16.2 by Ø) are those with decimals that, if written as fractions in lowest terms, would have even denominators. Also, the real values of $(-2)^m$ in Table 16.2 alternate between positive and negative. This is because a negative base raised to an even power is positive (such as $(-2)^{-2} = 0.25$), and a negative base raised to an odd power is negative (such as $(-2)^{-3} = -0.125$).

With m on the horizontal axis and $(-2)^m$ on the vertical, the graph resulting from the values in Table 16.2 appears in Figure 16.1.

Mathematical Focus 3

Once the domain of m *is given, the domain of* x *is limited to those values for which* x^m *is defined as a real number.*

Although the value of the base, x, in x^m determines the domain of the exponent, m, it is also true that if x^m is to be defined in the real numbers, then the value of m puts certain limitations on x. (Note, however, that this discussion is still limited to real values, and subsets thereof, of x and m.)

a.   If m is a positive integer, then x may be any real number. For example, x^6 is always defined on the real numbers.

b.   If m is a nonzero rational number with an odd denominator (when written in lowest terms), then x may be any real number. For example, $x^{\frac{3}{5}} = \sqrt[5]{x^3}$ is always defined on the real numbers.

c.   If m is an integer and $m \leq 0$, then x is a real number and $x \neq 0$. This is a modification of the contrapositive of Statement a in Focus 2.

d.   If *m* is an irrational number or a nonzero rational number (in lowest terms) with an even denominator, then $x \geq 0$ (see previous discussion of even denominators). This is the contrapositive of c. in Focus 2.

NOTES

1.   In the binomial theorem, the coefficients for $(a + b)^n$ are given for all whole numbers *n*, and restrictions on *a* and *b* are often ignored. Because Pascal's triangle always starts with 1 at the top, it "encodes" the idea that $(a + b)^0$ is taken to be 1 with virtually no further discussion. If $a + b = 0$ then $(a + b)^0$ becomes 0^0, which is not 1 (see Situation 12, Zero Exponents in chapter 18). If given only the expression 0^0, without knowing the context in which it arose, one does not know a priori whether it might

have arisen from $\lim\limits_{x \to 0^+} 0^x$, which is 0, or a different limit that often is 1. So, it is a matter of convenience that 1 appears in the first row of Pascal's Triangle, and the case when $(a + b) = 0$ often is not addressed.

CHAPTER 17

POWERS

Situation 11 From the MACMTL–CPTM Situations Project

Tracy Boone, Jana Lunt, Christa Fratto, James Banyas,
Eileen Murray, Bob Allen, Sarah Donaldson,
M. Kathleen Heid, Shiv Karunakaran, and Brian Gleason

PROMPT

During an Algebra 1 lesson on exponents, the teacher asked the students to calculate positive integer powers of 2. A student asked the teacher, "We've found 2^2 and 2^3. What about $2^{2.5}$?"

COMMENTARY

The Prompt centers on the extension of the domain of the exponent to selected supersets of the integers. The Foci explore the nature of exponents numerically, graphically, and analytically. The table with integral values in Focus 1 suggests a pattern for a curve and an extended domain that is illustrated in the graphical representation in Focus 2. Although Focus 1 and Focus 2 yield an estimate, the analytical treatment in Focus 3 generates an exact value. These Foci illustrate

*Mathematical Understanding for Secondary Teaching: A Framework
and Classroom-Based Situations*, pages 185–189.
Copyright © 2015 by Information Age Publishing

expansion of the concept of exponentiation beyond repeated multiplication to accommodate the use of some noninteger exponents.

MATHEMATICAL FOCI

Mathematical Focus 1

One method for estimating the value of 2^x, where $x \in \mathbb{R}$, uses linear interpolation and an extension of the properties of the function f with rule f(n) = 2^n, where $n \in \mathbb{Z}$.

If one assumes that the properties of the function f with rule $f(n) = 2^n$, where $n \in \mathbb{Z}$, extend to a function with the same rule but a domain expanded to $\mathbb{R}$, one way to calculate an estimate of the value $2^{2.5}$ is to use linear interpolation between the two known values, 2^2 and 2^3. To use linear interpolation, one can treat the function as if it were linear between the known function values. Using linear interpolation, $2^{2.5}$ is approximately 6 (because 2.5 is halfway from 2 to 3, $2^{2.5}$ is estimated to be halfway from 2^2 to 2^3). This is only an estimate, however, and it is important to understand that the value for $2^{2.5}$ will not be exactly halfway between 2^2 and 2^3 because the function f is not a linear function. Whether the value of $2^{2.5}$ is greater than 6 or less than 6 is determined by the pattern of growth of the function. A linear function has a constant rate of growth, but it can be shown that 2^x does not.

Table 17.1 shows a pattern of increasing growth between successive values of 2^x and illustrates that exponential growth is different from constant growth. In particular, nonlinearity implies that even though 2.5 is the arithmetic mean of 2 and 3, $2^{2.5}$ will not be the arithmetic mean of 4 and 8. Because the differences between the successive values of 2^x are increasing, one can argue that linear approximations will overestimate the value of $2^{2.5}$ (see Figure 17.1). For example, 2^2 is 4, and 2^4 is 16, but 2^3 is less than 10, the arithmetic mean of 4 and 16, which is the value suggested by linear interpolation. Therefore, $2^{2.5}$ should be less than 6.

Thinking graphically can suggest a way to connect the points that will preserve the rate of growth in the table of values. This pattern of points suggests a graph that is increasing and concave up. A graph that is increasing and concave up increases at an increasing rate, whereas the graph of a line increases or decreases at a constant rate. This suggests that the value of $2^{2.5}$ will be closer to 4 than to 8.

TABLE 17.1   Values of 2^x for $x \in \{0, 1, 2, 3, 4\}$

X	2^x
0	1
1	2
2	4
3	8
4	16

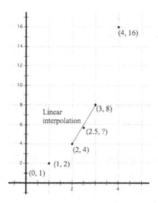

FIGURE 17.1. Linear interpolation yields an overestimate, 6, of $f(2.5) = 2^{2.5}$.

Mathematical Focus 2

Technology-generated graphs of the function f with rule $f(x) = 2^x$, $x \in \mathbb{R}$, can offer a check on one's approximations of 2^x, where $x \notin \mathbb{Z}$.

One can check approximations for $2^{2.5}$ using tool-generated graphics.[1] The graphics themselves arise by numerical approximations that the technology can do far faster than humans. The technology can "check" guesses and refinements of approximations to $2^{2.5}$. For example, $2^{2.5}$ is a number such that its square is 2^5, or 32, and one can guess and check and refine in looking for such a number (but the technology can do it much faster).

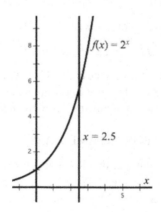

FIGURE 17.2. The value of $f(2.5)$ approximated from the intersection of graphs of $f(x) = 2^x$ and $x = 2.5$.

Moreover, one can estimate the value of $f(2.5)$ from a calculator-generated graph of f using the trace option. Alternatively, one can estimate the point of intersection of the function graph with the vertical line $x = 2.5$ (see Figure 17.2).

Mathematical Focus 3

> By identifying rational numbers as ratios of integers, properties of integral exponents are extended to rational exponents.

For $b \geq 0$, one possible definition for $b^{\frac{m}{n}}$ is that it is the number that when raised to the nth power gives b^m as a result. So, $b^{\frac{m}{n}} = \sqrt[n]{b^m}$ for $m \in \mathbb{Z}$, $n \in \mathbb{N}$, $b \in \mathbb{R}^+$. Using this definition, $\left(b^m\right)^{\frac{1}{n}} = \sqrt[n]{\left(b^m\right)^1} = \sqrt[n]{b^m} = b^{\frac{m}{n}}$. Other properties of rational exponents can be established, such as the following (for m, n, and b as above and $c \in \mathbb{Z}$, $d \in \mathbb{N}$, and $a, b \in \mathbb{R}^+$):

1. $\left(b^{\frac{1}{n}}\right)^m = \left(b^m\right)^{\frac{1}{n}} = b^{\frac{m}{n}}$

2. $\left(b^{\frac{m}{n}}\right)^{\frac{c}{d}} = \left(b^{\frac{c}{d}}\right)^{\frac{m}{n}} = b^{\frac{mc}{nd}}$

3. $b^{\frac{m}{n}} \cdot b^{\frac{c}{d}} = b^{\frac{c}{d}} \cdot b^{\frac{m}{n}} = b^{\frac{m}{n}+\frac{c}{d}}$

4. $(ab)^{\frac{m}{n}} = a^{\frac{m}{n}} \cdot b^{\frac{m}{n}}$

Using the representation $2^{\frac{1}{2}} = \sqrt{2}$, one can analyze $2^{2.5} = 2^{\frac{5}{2}}$ as follows: $2^{\frac{5}{2}} = \left(2^5\right)^{\frac{1}{2}} = \sqrt{2^5} = \sqrt{32} \approx 5.657$ or $2^{\frac{5}{2}} = (2^{\frac{1}{2}})^5 = (\sqrt{2})^5 \approx 1.414^5 \approx 5.657$. In another form: $(2^{\frac{1}{2}})^5 = (\sqrt{2})^5 = \sqrt{2} \cdot \sqrt{2} \cdot \sqrt{2} \cdot \sqrt{2} \cdot \sqrt{2} = 4\sqrt{2} \approx 5.657$.

Thus, 2^x, with $x \in \mathbb{Q}$ and $x \notin \mathbb{Z}$, can be defined in a way that allows the familiar properties of exponents to hold. Once it is known that the properties hold, one can use them to express $2^{2.5}$ in ways that may be easier to understand.

POSTCOMMENTARY

It is also useful to consider the fact that the function f with rule $f(x) = b^x$ would behave differently for different values of b. If $b < 0$, the properties would require further modification. Furthermore, different mathematical discussions would be necessary if $b \notin \mathbb{R}$.

Although it is not strictly necessary, it is common to place the preceding restrictions on the values of m and n as well, depending on the restrictions placed on *b.* For example, $3^{\frac{1}{-4}} = \sqrt[-4]{3}$ is not considered common notation, and an understanding of expressions such as $\sqrt[\pi]{3} = 3^{\frac{1}{\pi}}$ requires more sophisticated techniques. Also, any expression equivalent to 0^0 is not universally defined.

NOTE

1. Use of a technology-generated graph of the function f with rule $f(x) = 2^x$ to approximate the value of $2^{2.5}$ assumes the continuity of f, which can be established by its differentiability.

CHAPTER 18

ZERO EXPONENTS

Situation 12 From the MACMTL-CPTM Situations Project

Tracy Boone, Christa Fratto, Jana Lunt, Heather Johnson,
M. Kathleen Heid, Maureen Grady, and Shiv Karunakaran

PROMPT

In an Algebra 1 class, a student questioned the claim that

$a^0 = 1$ *for all nonzero real number values of* a.

The student asked, "How can that be possible? I know that a^0 is a times itself zero times,[1] so a^0 must be 0."

COMMENTARY

The succinct and mathematically correct answer to the student's question presented in the Prompt is that a^0 is defined to be 1 for specific values of a. The arguments presented in the Foci establish why this definition makes sense mathematically and why defining a^0 in such a way that allows consistency with other

*Mathematical Understanding for Secondary Teaching: A Framework
and Classroom-Based Situations*, pages 191–198.

mathematical facts. The issue that the student raises in the Prompt may be due to viewing a^0 as a numerical value. However, the broader perspective is that what matters is the properties (such as continuity) when one thinks of $y = a^x$ as a function.

MATHEMATICAL FOCI

Mathematical Focus 1

The definition of f(x) = a^x can be extended from having a domain of only nonzero integers to a domain of all real numbers.

The student appears to be drawing on a definition of exponents that is applicable only for exponents that are positive whole numbers. When the values used as exponents are expanded, the following are taken as part of the definition of exponent:

1. $a^0 = 1$, where a is any real number not equal to 0;

2. $a^{-n} = \dfrac{1}{a^n}$, where $n > 0$, and a is any real number not equal to 0; and

3. $a^{m/n} = \sqrt[n]{a^m}$, where m is an integer, n is a positive integer, and a is a positive real number.

The extension of the definition imposes restrictions on the values that may be used for the base, a.

Mathematical Focus 2

The expression 0^0 is defined to be an indeterminate form because the values of $\lim_{x \to 0} x^0$, $\lim_{x \to 0} 0^x$, and $\lim_{x \to 0} x^x$ are not consistent with each other.

In the definition $a^0 = 1$, the restriction that a cannot be equal to 0 can be explained by examining the three functions: $f(x) = x^0$, $f(x) = 0^x$, and $f(x) = x^x$ as the value of x approaches 0. Some evidence that the value of 0^0 should be equal to 1 comes from $\lim_{x \to 0^-}(x^0) = 1$ and $\lim_{x \to 0^+}(x^0) = 1$. However, $\lim_{x \to 0^-}(0^x)$ does not exist because the function does not exist for $x \le 0$, whereas $\lim_{x \to 0^+}(0^x) = 0$. Finally, although $\lim_{x \to 0^+}(x^x) = 1$, the $\lim_{x \to 0^-}(x^x)$ does not exist because the function is not continuous for $x < 0$. Also it is true that $0^n = 0$. If $a = 0$, then $\dfrac{a^n}{a^n} = \dfrac{0}{0}$, is an indeterminate form (see also Situation 1 in chapter 7).

Mathematical Focus 3

For a ≠ 0, a⁰ = 1 can be explained using properties of exponents.

For nonzero a, $a^0 = 1$ can be explained using the multiplication and division properties of exponents. Using the division property of exponents, $\dfrac{a^n}{a^n}$ is equivalent to a^{n-n} or a^0, where n is any nonzero real number. Because of the multiplicative identity field property, $\dfrac{a^n}{a^n} = 1$. Therefore, because of the transitive property, a^0 must equal 1, that is, $1 = \dfrac{a^n}{a^n} = a^{n-n} = a^0$.

A fundamental property of exponentiation is that for $a \neq 0$ and positive integers m and n, $a^{n+m} = a^n \cdot a^m$. Extending this property to m and n being 0 yields $a^n = a^{n+0} = a^n \cdot a^0$, which is consistent with the definition $a^0 = 1$.

Mathematical Focus 4

Defining a⁰ = 1 for a ≠ 0 is consistent with the multiplicative relationship between successive terms in the sequence {aⁿ}, where n is an integer and a is a nonnegative real number.

Examination of a pattern involving the recursive nature of exponential growth can suggest that $a^0 = 1$ is a definition consistent with that pattern. First, consider a specific example using exponents with base 4.

$$4^{-3} = \frac{1}{4^3} = \frac{1}{64}$$

$$4^{-2} = \frac{1}{4^2} = \frac{1}{16}$$

$$4^{-1} = \frac{1}{4^1} = \frac{1}{4}$$

$$4^0 = ?$$

$$4^1 = 4 = 4$$

$$4^2 = 4 \cdot 4 = 16$$

$$4^3 = 4 \cdot 4 \cdot 4 = 64$$

As the exponent increases by 1, each successive term can be obtained by multiplying the preceding term by 4. That is, $4^{n+1} = 4 \cdot 4^n$. For this recursive pattern to hold for all integer values of n, it seems that 4^0 should be equal to 1. It is important to note here that this pattern is developed using only integer values for the value represented by n. The primary reason to set up the pattern using only integer values for the exponents is to examine 4^0 as a part of a sequence of numbers written as 4^n, where n increases by 1. This pattern that each successive term can be obtained by multiplying the preceding term by 4 still holds if the exponents considered are nonintegral rational numbers, as in the following example:

$$4^{-\frac{12}{5}} = \frac{1}{4^{\frac{12}{5}}}$$

$$4^{-\frac{7}{5}} = \frac{1}{4^{\frac{7}{5}}} = \frac{1}{4^{\frac{12}{5}-1}} = \frac{1}{\frac{4^{\frac{12}{5}}}{4}} 4 = 4^{-\frac{12}{5}} \cdot 4$$

$$4^{-\frac{2}{5}} = \frac{1}{4^{\frac{2}{5}}} = \frac{1}{4^{\frac{7}{5}-1}} = \frac{1}{\frac{4^{\frac{7}{5}}}{4}} 4 = 4^{-\frac{7}{5}} \cdot 4$$

$$4^{\frac{3}{5}} = 4^{-\frac{2}{5}+1} = 4^{-\frac{2}{5}} \cdot 4$$

$$4^{\frac{8}{5}} = 4^{\frac{3}{5}+1} = 4^{\frac{3}{5}} \cdot 4$$

$$4^{\frac{13}{5}} = 4^{\frac{8}{5}+1} = 4^{\frac{8}{5}} \cdot 4$$

This pattern can be generalized to all positive values of a. Consider the following list of powers of a.

$$a^{-3} = \frac{1}{a^3}$$

$$a^{-2} = \frac{1}{a^2}$$

$$a^{-1} = \frac{1}{a^1}$$

$$a^0 = 1$$

$$a^1 = a$$

$$a^2 = a \cdot a$$

$$a^3 = a \cdot a \cdot a$$

It can be verified that the pattern holds by looking at a particular definition of a^n, where n is a positive integer. In this case, a^n is defined as a product that consists of a being used as a factor n times and a^{n+1} is a product that consists of a being used as a factor $n + 1$ times, which is the same as a times a product that consists of a being used as a factor n times. So, for all positive values of a and positive integers n, $a^{n+1} = a \cdot a^n$. Extending the domain of n to include 0 suggests that $a^{0+1} = a \cdot a^0$. If $a^{0+1} = a^1 = a$, it follows that a^0 would be equal to 1. However, it is important to realize that the pattern of $a^{n+1} = a \cdot a^n$ will not hold for *all* values of a. For example $a^{\frac{1}{2}}$ is not a real number if a is any negative real number.

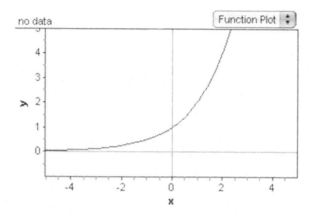

FIGURE 18.1.   The graph of $y = 2^x$.

Mathematical Focus 5

Defining $a^0 = 1$ *for* $a \neq 0$ *is consistent with a pattern established by the graph of* $f(x) = a^x$ *for* $a > 0$ *and* $x \neq 0$.

Another approach to explore the student's statement in the Prompt is through a graphical representation of the function $y = a^x$ for various real values of a. The graph in Figure 18.1 is what is displayed by a graphing utility for the function $y = 2^x$. The value of 2^0 appears to be equal to 1.

The graph in Figure 18.1 shows the behavior of $y = 2^x$. One can examine the behavior of $y = a^x$ at $x = 0$ more generally by graphing $y = a^x$ for several positive values of a. This can be accomplished dynamically by using the slider feature available in dynamical software. The graph in Figure 18.2 represents $y = a^x$ for the value of a indicated by the slider. In Figure 18.2, $a = 3.60$. As the value on the slider is changed, the graph is updated automatically to reflect the change. The value $a^0 = 1$ can be inferred from the graph for any positive value of a. Figure 18.3 shows graphs of $y = a^x$ that are traced for positive values of a. The point $(0, 1)$ appears to be the common point for these graphs.

As mentioned in other Foci, the restrictions placed on the value of a are important. In all graphs in Figure 18.1 to Figure 18.3, it is assumed that a represents a positive real number. These graphs break down if a represents a negative number. For example, the graph of $y = (-2)^x$ is not a continuous and well-defined graph because the function $y = (-2)^x$ is not a real number for any rational number $x = \dfrac{p}{q}$ where $\dfrac{p}{q}$ is in lowest terms and q is even.[2]

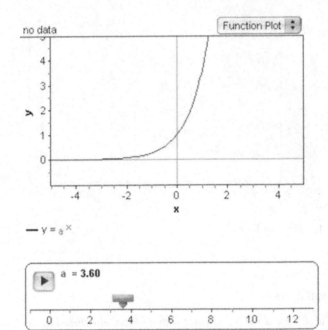

FIGURE 18.2.   Graph of $y = a^x$ for $a = 3.60$.

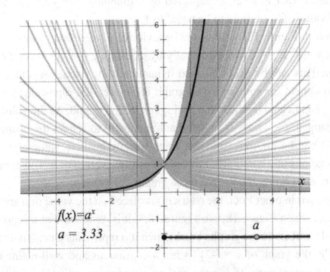

FIGURE 18.3.   Traces of graphs of $y = a^x$ for positive values of a.

TABLE 18.1.   Values of 2^x for Values of x Near 0

x	2^x
-0.0004	0.99972278
-0.0003	0.999792077
-0.0002	0.99986138
-0.0001	0.999930688
0	?
0.0001	1.000069317
0.0002	1.000138639
0.0003	1.000207966
0.0004	1.000277297

Among the assumptions that underlie the preceding graphical argument are that the function $y = a^x$ is well defined and continuous. The inferences drawn from the graphs in Figures 18.1 to 18.3 are based on these assumptions.

Mathematical Focus 6

Defining $a^0 = 1$ *for* $a \neq 0$ *makes the function* f(x) = ax *for* $a > 0$ *continuous everywhere.*

Consider the function $f(x) = 2^x$. For this function to be continuous over all real x, $f(0)$ must be defined. To define $f(0)$, consider $\lim_{x \to 0}(2^x)$. To estimate $\lim_{x \to 0}(2^x)$ numerically, examine values of $f(x) = 2^x$ near $x = 0$ (see Table 18.1). As the values of x approach 0, the values of $f(x) = 2^x$ approach 1; therefore, these computer-generated approximations of 2^x suggest that $\lim_{x \to 0}(2^x) = 1$. This procedure can be expanded to all positive values for a.

Note that if $a = 1$, then $\lim_{x \to 0} a^x = 1$. To prove that $\lim_{x \to 0}(a^x) = 1$ for other positive values of a, show that for each $\varepsilon > 0$ there exists a $\delta > 0$ such that $|a^x - 1| < \varepsilon$ when $0 < |x - 0| < \delta$.

Let $\varepsilon > 0$ such that $|a^x - 1| < \varepsilon$. Now consider the following.

$$|a^x - 1| < \varepsilon$$
$$\Rightarrow -\varepsilon < a^x - 1 < \varepsilon$$
$$\Rightarrow 1 - \varepsilon < a^x < 1 + \varepsilon$$

Case 1: $0 < \varepsilon < 1$

$$1 - \varepsilon < a^x < 1 + \varepsilon$$

$$\Leftrightarrow \ln(1 - \varepsilon) < x \cdot \ln(a) < \ln(1 + \varepsilon)$$

$$\Leftrightarrow \frac{\ln(1 - \varepsilon)}{\ln(a)} < x < \frac{\ln(1 + \varepsilon)}{\ln(a)} \text{ if } a > 1 \ (\therefore \ \ln(a) > 0)$$

and

$$\frac{\ln(1 - \varepsilon)}{\ln(a)} > x > \frac{\ln(1 + \varepsilon)}{\ln(a)} \text{ if } 0 < a < 1 \ (\therefore \ \ln(a) < 0)$$

In this case, choose $\delta = \min\left\{\left|\dfrac{\ln(1 - \varepsilon)}{\ln(a)}\right|, \left|\dfrac{\ln(1 + \varepsilon)}{\ln(a)}\right|\right\}$.

Therefore, if $0 < |x| < \delta$, then $|a^x - 1| < \varepsilon$.

Case 2: $\varepsilon \geq 1$

In this case, choose $\delta = \left|\dfrac{\ln(1 + \varepsilon)}{\ln(a)}\right|$.

Thus, $\lim\limits_{x \to 0} a^x = 1$.

NOTES

1. Lakoff and Núñez (2000, p. 405) address this claim.
2. Another way to express this is that the function $y = (-2)^x$ is well defined in the real number system only over a set of measure 0.

REFERENCE

Lakoff, G., & Núñez, R. E. (2000). *Where mathematics comes from: How the embodied mind brings mathematics into being*. New York, NY: Basic Books.

CHAPTER 19

MULTIPLYING MONOMIALS AND BINOMIALS

Situation 13 From the MACMTL–CPTM Situations Project

Jeanne Shimizu, Tracy Boone, Jana Lunt, Christa Fratto, Erik Tillema,
Jeremy Kilpatrick, Sarah Donaldson, Ryan Fox, Heather Johnson,
Maureen Grady, Svetlana Konnova, and M. Kathleen Heid

PROMPT

The following scenario took place in a high school Algebra 1 class. Most of the students were sophomores or juniors repeating the course. During the spring semester, the teacher asked them to do the following two warm-up items:

1. Are the two expressions, $(x^3y^5)^2$ and x^6y^{10}, equivalent? Why or why not?
2. Are the two expressions, $(a + b)^2$ and $a^2 + b^2$, equivalent? Why or why not?

Roughly one third of the class stated that both pairs of expressions were equivalent because of the distributive property.

Mathematical Understanding for Secondary Teaching: A Framework
and Classroom-Based Situations, pages 199–206.

COMMENTARY

This Situation highlights differences between multiplying monomials and multiplying binomials. The students' incorrect responses to the warm-up problem demonstrate a possible misunderstanding of important differences. The students appear to be misusing the distributive property by applying a procedure, "take the exponent on the outside of the parentheses and multiply it by the exponent of what is inside the parentheses," when that procedure does not apply. Ironically, the students' difficulty with Item 2 may have occurred because they did not use the distributive property. The Foci that follow demonstrate several approaches for exploring the mathematics involved in the two warm-up items, including application of the properties of real numbers, geometric representations, graphical representations, and numerical exploration.

MATHEMATICAL FOCI

Mathematical Focus 1

> *Equivalence of expressions can be explained by application of the properties of real numbers.*

The distributive property of multiplication over addition (hereafter referred to as the *distributive property*) states that $a(b + c) = ab + ac$ and $(b + c)a = ba + bc$ for all real numbers a, b, and c. This applies to multiplication being distributed over addition, and the property does not generalize to all configurations of the form $a*(b @ c)$ or $(b @ c)*a$ where $*$ and $@$ are operations that apply to a, b, and c. The distributive property does not hold for distributing exponentiation over addition.[1] In this particular situation, then, the distributive property cannot be used to claim that $(a + b)^2$ and $a^2 + b^2$ are equivalent.

The distributive property is relevant, however, for Item 2 because $(a + b)^2$ is a product of two binomials. The distributive property can be used as follows to determine the product of binomials:

$$(a + b)(a + b) = (a + b)(a) + (a + b)(b) = a^2 + ba + ab + b^2 = a^2 + 2ab + b^2$$

So, $(a + b)^2 = a^2 + 2ab + b^2$, and unless a or b is equal to 0, $a^2 + 2ab + b^2$—and hence $(a + b)^2$—is not equivalent to $a^2 + b^2$. It is important to note that, as with many pairs of expressions that are not equivalent, the values of the two expressions may be equal for some (but not all) values of the variables in the expressions. In this instance, $(a + b)^2$ and $a^2 + b^2$ have the same value only when $a = 0$, $b = 0$, or both a and b equal 0.

For real numbers, a, b, and c, the commutative property of multiplication states that $ab = ba$ and the associative property of multiplication states that $(ab)c = a(bc)$.

Both of these properties are used in generating expressions that are equivalent to $(x^3y^5)^2$.

$$
\begin{aligned}
(x^3y^5)(x^3y^5) &= x^3(y^5(x^3y^5)) & \text{Associative property of multiplication} \\
&= x^3((y^5x^3)y^5) & \text{Associative property of multiplication} \\
&= x^3((x^3\ y^5)y^5) & \text{Commutative property of multiplication} \\
&= x^3(x^3(y^5y^5)) & \text{Associative property of multiplication} \\
&= (x^3x^3)(y^5y^5) & \text{Associative property of multiplication} \\
&= x^6y^{10} & \text{Property of exponents } (x^a x^b = x^{a+b})
\end{aligned}
$$

Mathematical Focus 2

The distributive property can be illustrated geometrically through the use of an area model and a right triangle inscribed in a circle.

One way to illustrate $(a + b)^2$ is to examine the area of a square with side length $a + b$. As shown in Figure 19.1, the square with side length $a + b$ can be partitioned into four rectangles (two squares having area a^2 and b^2, respectively, and two rectangles, both having area ab). Therefore, by appealing to the area model, $(a + b)^2 = a^2 + 2ab + b^2$, not simply $a^2 + b^2$. It is important to note that, because in this model a and b represent lengths, this area model has limitations in cases in which the variables or variable expressions represent negative values.

It can also be shown that $(a + b)^2 = a^2 + 2ab + b^2$ using a right triangle inscribed in a circle. Let $a > 0$ and $b > 0$, and let $a > b$. Construct the segment AC, so that $AC = a + b$. Locate the point D on the segment AC, so that $AD = a$ and $DC = b$ (see

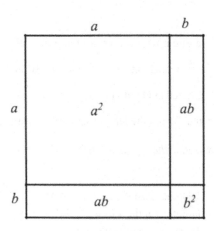

FIGURE 19.1. Area model for $(a + b)^2$ for positive real numbers a and b.

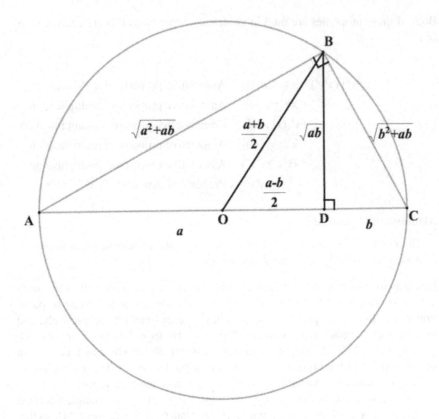

FIGURE 19.2.   Relationships between a, b, and lengths of various segments in the diagram of a right triangle inscribed in a circle.

Figure 19.2). Construct a circle with center at O and diameter $\overline{AC}$. The radius of this circle has length $\frac{a+b}{2}$. Construct the segment DB so that $\overline{DB} \perp \overline{AC}$, and the point B is on the circle centered at O with radius of length $\frac{a+b}{2}$. Figure 19.2 shows the relationships between a, b, and the lengths of various segments in the diagram.

I. First, it can be shown that the length of $\overline{BD}$ is the geometric mean of the lengths of $\overline{AD}$ and $\overline{DC}$.

Consider $\triangle ABC$, $\triangle ABD$, and $\triangle BDC$. Because $\triangle ABC$ is inscribed in the circle O and side $\overline{AC}$ is a diameter of the circle, $\angle ABC$ is a right angle. Because $\overline{DB} \perp \overline{AC}$, $\angle ADB$ and $\angle BDC$ are right angles. Because two pairs of corresponding angles of $\triangle ABC$ and $\triangle ABD$ are congruent, $\triangle ABC$ and $\triangle ABD$ are similar triangles and the lengths of their corresponding sides are proportional. So, $\dfrac{DB}{AD} = \dfrac{BC}{AB}$. Also, because two pairs of corresponding angles of $\triangle ABC$ and $\triangle BDC$ are congruent, $\triangle ABC$ and $\triangle BDC$ are similar triangles and the lengths of their corresponding sides are proportional. So, $\dfrac{BC}{AB} = \dfrac{DC}{DB}$, and, by the transitive property of equality: $\dfrac{DB}{AD} = \dfrac{DC}{DB}$. Because $AD = a$ and $DC = b$, $DB = \sqrt{ab}$.

II. Second, it can be shown that $(a + b)^2 = a^2 + 2ab + b^2$.

Consider $\triangle ADB$. $\angle ADB$ is a right angle, therefore, by the Pythagorean theorem, $AB^2 = AD^2 + DB^2$. Consider $\triangle BDC$. The angle $\angle BDC$ is the right angle, therefore, by the Pythagorean theorem, $BC^2 = DB^2 + DC^2$. Consider $\triangle ABC$. The angle $\angle ABC$ is the right angle, therefore, by the Pythagorean theorem, $AB^2 + BC^2 = AC^2$. Then

$$AC^2 = AB^2 + BC^2 = (AD^2 + DB^2) + (DB^2 + DC^2)$$

$$= a^2 + 2(\sqrt{ab})^2 + b^2$$

$$= a^2 + 2ab + b^2$$

III. The proof developed in Sections I and II can be extended to show that $(a - b)^2 = a^2 - 2ab + b^2$.

Consider $\triangle OBD$. $\angle ODB$ is a right angle, therefore, by the Pythagorean theorem, $OD^2 = OB^2 - BD^2$. Because $OD = \dfrac{a - b}{2}$, $OB = \dfrac{a + b}{2}$, and $BD = \sqrt{ab}$,

$$OD^2 = OB^2 - BD^2$$

$$\Rightarrow \left(\frac{a-b}{2}\right)^2 = \left(\frac{a+b}{2}\right)^2 - ab$$

$$\Rightarrow \frac{(a-b)^2}{4} = \frac{(a+b)^2}{4} - (ab)$$

$$\Rightarrow (a-b)^2 = (a+b)^2 - 4ab$$

$$\Rightarrow (a-b)^2 = a^2 + 2ab + b^2 - 4ab$$

$$\Rightarrow (a-b)^2 = a^2 - 2ab + b^2$$

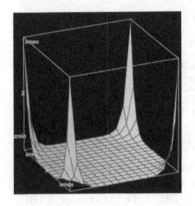

FIGURE 19.3.   Graph of $z = (x^3y^5)^2$.

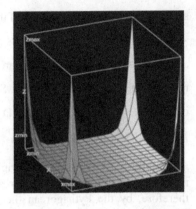

FIGURE 19.4.   Graph of $z = x^6y^{10}$.

Mathematical Focus 3

The equivalence of expressions involving two variables can be explored using three-dimensional graphs.

This Focus uses $(x^3y^5)^2$ and x^6y^{10} to define functions of two variables. That is, let z be a function of two independent variables, x and y. The graph of a function of

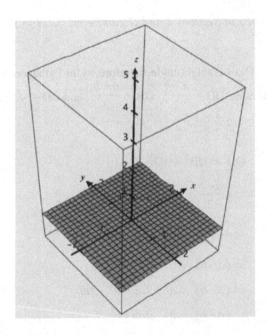

FIGURE 19.5.   Graph of $z = (x^3y^5)^2 - x^6y^{10}$.

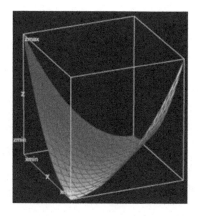

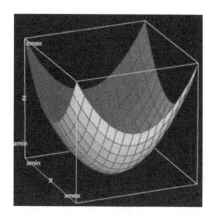

FIGURE 19.6. Graph of $z = (x + y)^2$.   FIGURE 19.7. Graph of $z = x^2 + y^2$.

two variables is a three-dimensional graph. The graphs of $z = (x^3y^5)^2$ and $z = x^6y^{10}$, shown in Figures 19.3 and 19.4, appear to be the same graphs, providing some evidence that $(x^3y^5)^2$ and x^6y^{10} may be equivalent expressions.

When the difference of the two functions is graphed on three-dimensional axes as shown in Figure 19.5, it appears that the result is the function $z = 0$, and hence the two functions appear to be equivalent.

The graphs in Figure 19.6 and Figure 19.7 suggest that the functions $z = (x + y)^2$ and $z = x^2 + y^2$ are not equivalent. In addition, the graph of $z = (x + y)^2 - (x^2 + y^2)$ (see Figure 19.8) supports the conclusion that the functions are not equivalent.

Mathematical Focus 4

A numerical approach can provide some evidence about the equivalence of expressions.

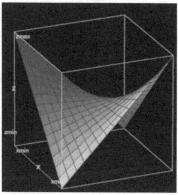

FIGURE 19.8. Graph of $z = (x + y)^2 - (x^2 + y^2)$.

Let $x = 2$ and $y = 3$.

$$(x^3y^5)^2 \qquad x^6y^{10}$$

$$= (2^3 \cdot 3^5)^2 \qquad = 2^6 \cdot 3^{10}$$

$$= (8 \cdot 243)^2 \qquad = 64 \cdot 59{,}049$$

$$= (1944)^2 \qquad = 3{,}779{,}136$$

$$= 3{,}779{,}136$$

FIGURE 19.9. Numerical example for Item 1.

Let $a = 2$ and $b = 3$.

$$(a + b)^2 \qquad a^2 + b^2$$

$$= (2 + 3)^2 \qquad = 2^2 + 3^2$$

$$= (5)^2 \qquad = 4 + 9$$

$$= 25 \qquad = 13$$

FIGURE 19.10. Numerical example for Item 2.

A simple test of a few values for x and y is a way to give evidence that $(x^3y^5)^2$ and x^6y^{10} may be equivalent, whereas testing any nonzero values for a and b will indicate that $(a + b)^2$ and $a^2 + b^2$ are not equivalent. For Item 1, an example is offered in Figure 19.9, using the values $x = 2$ and $y = 3$, for which $(x^3y^5)^2 = x^6y^{10}$. This is merely evidence—not a proof—that $(x^3y^5)^2$ and x^6y^{10} are equivalent, because it is not possible to test all possible values for x and y. For Item 2, however, using the values 2 and 3 for a and b yields a counterexample (see Figure 19.10) that shows that $(a + b)^2$ cannot be equivalent to $a^2 + b^2$.

One example (provided in Item 1) is not sufficient to prove equivalence of the expressions, but one counterexample is sufficient to disprove a statement of equivalence. Because (as provided for Item 2) there exist values of a and b for which the values of $(a + b)^2$ and $a^2 + b^2$ are not equal, it can safely be said that they are not equivalent expressions, because equivalent expressions would have equal values for all real values of a and b.

NOTE

1. For example, $(3 + 5)^2 = 8^2$, or 64, but $3^2 + 5^2 = 34$, so $(3 + 5)^2 \neq 3^2 + 5^2$.

CHAPTER 20

ADDING SQUARE ROOTS

Situation 14 From the MACMTL-CPTM Situations Project

Amy Hackenberg, Eileen Murray, Heather Johnson,
Glendon Blume, and M. Kathleen Heid

PROMPT

Mr. Fernandez was concerned about his ninth-grade algebra students' responses to a recent quiz on radicals, specifically those in response to a question about square roots, in which students added $\sqrt{2}$ and $\sqrt{3}$ and obtained $\sqrt{5}$ as a result.

COMMENTARY

The mathematical basis for determining the appropriateness of the students' work is that the sum of the square roots of two numbers is not, in general, equal to the square root of the sum of the two numbers. Establishing that a statement is not true can be accomplished in different ways, including finding a counterexample and constructing an indirect proof. The students' statement, $\sqrt{2} + \sqrt{3} = \sqrt{5}$, can be disproved using numeric, geometric, and symbolic representations. Connections are also made to linear transformations.

*Mathematical Understanding for Secondary Teaching: A Framework
and Classroom-Based Situations*, pages 207–210.

MATHEMATICAL FOCI

Mathematical Focus 1

Geometric constructions can be used to provide evidence that refutes statements about specific real numbers.

Construct an irrational spiral (see Figure 20.1) by constructing an isosceles right triangle with legs of length 1. This triangle will have a hypotenuse with length $\sqrt{2}$. Then construct another right triangle with legs of length 1 and $\sqrt{2}$. This triangle will have hypotenuse with length $\sqrt{3}$. Figure 20.1 illustrates the continuation of this pattern through the construction of a right triangle with hypotenuse $\sqrt{}$.

If $\sqrt{2}+\sqrt{3}=\sqrt{5}$ is a true statement, then the sum of the measures of the segments with lengths $\sqrt{2}$ and $\sqrt{3}$ must equal the measure of the segment with length $\sqrt{5}$. Figure 20.1 gives evidence to support the conclusion that the sum of the measures of the segments with lengths $\sqrt{2}$ and $\sqrt{3}$ is greater than the measure of the segment with length $\sqrt{5}$, suggesting that $\sqrt{2}+\sqrt{3} \neq \sqrt{5}$.

Mathematical Focus 2

A given statement can be shown to be false by supposing that it is true and showing that the supposition leads to a contradiction.

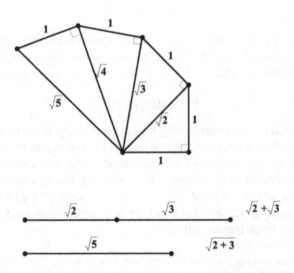

FIGURE 20.1. Irrational spiral suggesting that $\sqrt{2}+\sqrt{3} \neq \sqrt{5}$.

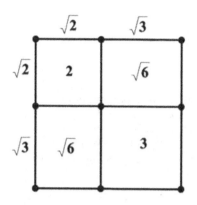 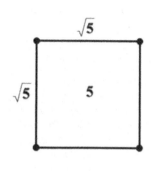

FIGURE 20.2. Squares with sides $\sqrt{2}+\sqrt{3}$ and $\sqrt{5}$.

The following indirect proof uses the property "If $a = b$, then $a^2 = b^2$" and the fact that squares of numbers can be represented geometrically as areas of squares.

If $\sqrt{2}+\sqrt{3} = \sqrt{5}$ is a true statement, then $\sqrt{2}+\sqrt{3} = \sqrt{2+3}$ and $\left(\sqrt{2}+\sqrt{3}\right)^2 = \left(\sqrt{2+3}\right)^2$ would also be true statements. To illustrate, consider a square with side length $\sqrt{2}+\sqrt{3}$. The area of that square would be $\left(\sqrt{2}+\sqrt{3}\right)^2$. The square can be subdivided into four regions, having areas 2, $\sqrt{6}$, $\sqrt{6}$, and 3, as shown in Figure 20.2. Therefore, $\left(\sqrt{2}+\sqrt{3}\right)^2 = 2+\sqrt{6}+\sqrt{6}+3 = 5+2\sqrt{6}$. Now consider a square with side length $\sqrt{2+3}$. The area of that square would be $\left(\sqrt{2+3}\right)^2$, and $\left(\sqrt{2+3}\right)^2 = \left(\sqrt{5}\right)^2 = 5$. Because $5+2\sqrt{6} \neq 5, \left(\sqrt{2}+\sqrt{3}\right)^2 \neq \left(\sqrt{2+3}\right)^2$. Therefore, $\sqrt{2}+\sqrt{3} \neq \sqrt{5}$.

Mathematical Focus 3

A more general question than whether $\sqrt{2}+\sqrt{3} = \sqrt{5}$ is true is whether $\sqrt{a}+\sqrt{b} = \sqrt{a+b}$ is true for all $a, b \in \mathbb{R}^+ \cup \{0\}$. A statement such as $\sqrt{a}+\sqrt{b} = \sqrt{a+b}$ can be disproved by identifying values of a and b that provide a counterexample.

If $\sqrt{a}+\sqrt{b} = \sqrt{a+b}$ were true for all $a, b \in \mathbb{R}^+ \cup \{0\}$, then $\sqrt{9}+\sqrt{16} = \sqrt{25}$ would be true. But $\sqrt{9}+\sqrt{16} = 3+4 = 7$ and $\sqrt{9+16} = \sqrt{25} = 5$, so $\sqrt{9+16} \neq \sqrt{25}$. Therefore it is not true that, for all $a, b \in \mathbb{R}^+ \cup \{0\}, \sqrt{a}+\sqrt{b} = \sqrt{a+b}$.

The counterexample establishes that, as a general rule, $\sqrt{a}+\sqrt{b} = \sqrt{a+b}$ is not true. However, it does not establish that it is never true. To determine values of a and b for which the statement is true, one can solve the equation $\sqrt{a}+\sqrt{b} = \sqrt{a+b}$. Squaring both sides yields $a+2\sqrt{a}\sqrt{b}+b = a+b$, which is true if and only if

$2\sqrt{a}\sqrt{b} = 0$, that is, when $\sqrt{a} = 0$ or $\sqrt{b} = 0$. So, $\sqrt{a} + \sqrt{b} = \sqrt{a+b}$ is true only when $a = 0$ or $b = 0$.

Mathematical Focus 4

An expression such as $\sqrt{2} + \sqrt{3}$ can be thought of as a sum of function values, f(2) + f(3), values of the square root function, $f(x) = \sqrt{x}$, $x \in \mathbb{R}^+ \cup \{0\}$. Some functions satisfy one of the conditions of the linearity property, namely, f(a + b) = f(a) + f(b), and others do not. If a function f does not, in general, satisfy some property, it does not mean that there do not exist some values for which that property might hold.

If $\sqrt{2} + \sqrt{3} = \sqrt{2+3}$, then if, $f(x) = \sqrt{x}$, $x \in \mathbb{R}^+ \cup \{0\}$, f(2 + 3) must be equal to f(2) + f(3). In general, when $f(x) = \sqrt{x}$ and $a, b \in \mathbb{R}^+ \cup \{0\}$, does f(a + b) = f(a) + f(b)?

In general, if a function defined over the nonnegative real numbers has the properties f(a + b) = f(a) + f(b) and f(ca) = c · f(a), then it is said be a linear transformation. Linear functions of the form f(x) = kx that represent a direct variation relationship between variables satisfy the linearity property. This is true because for the linear function f(x) = kx, x ∈ ℝ, for all a, b ∈ ℝ, f(a + b) = k(a + b) = ka + kb = f(a) + f(b). So, f(a + b) = f(a) + f(b) for f(x) = kx, x ∈ ℝ⁺, for all a, b ∈ ℝ.

To determine whether f(a + b) = f(a) + f(b) holds for $f(x) = \sqrt{x}$, x ∈ $\mathbb{R}^+ \cup \{0\}$, one can attempt to find a counterexample. When a = 4 and b = 1, $f(4+1) = \sqrt{4+1} = \sqrt{5}$ and $f(4) + f(1) = \sqrt{4} + \sqrt{1} = 3$. Because $\sqrt{5} \neq 3$, it can be concluded that f(a + b) ≠ f(a) + f(b) for $f(x) = \sqrt{x}$, for all x ∈ $\mathbb{R}^+ \cup \{0\}$. However, this only establishes that f does not (in general) satisfy the linearity property, it does not establish that for every two values of a and b, f(a + b) ≠ f(a) + f(b). For example, $\sqrt{0} + \sqrt{0} = \sqrt{0+0}$. So this linearity property argument does not necessarily establish that $\sqrt{2} + \sqrt{3} \neq \sqrt{2+3}$, it only establishes that the statement in question cannot be deemed true by using an argument that all such statements are true for this particular function. Hence, one can evaluate $\sqrt{2} + \sqrt{3}$ and $\sqrt{2+3}$, as done in Focus 2, to disprove $\sqrt{2} + \sqrt{3} = \sqrt{2+3}$.

CHAPTER 21

SQUARE ROOTS

Situation 15 From the MACMTL–CPTM Situations Project

Tracy Boone, Jana Lunt, Christa Fratto, James Banyas, Sarah Donaldson, James Wilson, Patricia S. Wilson, Heather Johnson, and Brian Gleason

PROMPT

A teacher asked her students to sketch the graph of $f(x) = \sqrt{-x}$. A student responded, "That's impossible! You can't take the square root of a negative number!"

COMMENTARY

This Situation addresses several key concepts that occur frequently in school mathematics: additive inverse, negative numbers, function, domain, and range. Because the symbol "-" has multiple interpretations, it is important to distinguish between a negative number and the additive inverse (i.e., opposite) of a number.[1] Moreover, the domain over which a function is defined determines its range, and a table of values provides an example to illustrate the relationship between domain and range. For a set of points with coordinates $(x, f(x))$ to define the graph of a

---

Mathematical Understanding for Secondary Teaching: A Framework and Classroom-Based Situations, pages 211–215.

function, each first coordinate, x, must correspond to a unique second coordinate, $f(x)$. A graphical representation highlights the univalent relationship between x and $f(x)$.

MATHEMATICAL FOCI

Mathematical Focus 1

The symbols "-" and "−" are used in the representation of several different mathematical entities. Two of these that are commonly confused are negative and additive inverse (also called opposite*).*

A negative real number has a value less than 0, whereas the additive inverse of a real number is the value such that the sum of the number and its additive inverse is the additive identity, 0. For example, *negative 6* indicates a number less than 0, and *the number that is the opposite of positive 6* indicates the additive inverse of +6, which is -6. Every real number has an additive inverse, and only in the case of 0 is a number its own additive inverse.

In the preceding example, -6 and the additive inverse of 6 are the same, but this is not always the case. Take, for example, clock arithmetic. Consider a 24-hour clock in which 24:00 is the same as 0:00. One could ask, "What is the additive inverse of 11:00?" It is not -11:00 because there is no such time. Rather, the additive inverse is 13:00 because 11:00 + 13:00 = 24:00 = 0:00.

Using the variable x to represent a number does not indicate whether the number is positive, negative, or 0. Because $-x$ represents the additive inverse of x, and not necessarily a negative value, it could be positive, negative, or 0, depending upon the value of x.

Mathematical Focus 2

The domain of a function is critical in determining the values in its range.

The implicit assumption that the domain and range of a function are restricted to real numbers could contribute to the statement "You can't take the square root of a negative number." If one assumes that the range of the function f with rule $f(x) = \sqrt{-x}$ is a subset of the real numbers, then the function's domain must be in $\{x \mid x \leq 0\}$. Similarly, if one assumes that the range of the function h with rule $h(x) = \sqrt{x+2}$ is a subset of the real numbers, then the domain must be in $\{x \mid x \geq -2\}$.

Table 21.1 provides an example that illustrates how the domain of the function f with rule $f(x) = \sqrt{-x}$ determines values of the range of f. The values in Table 21.1 are consistent with the ordered pairs for the function f with rule $f(x) = \sqrt{-x}$. If the domain of f is the set of all real numbers, x, such that $x \leq 0$, then the range of f is

TABLE 21.1. Values of $f(x) = \sqrt{-x}$ for Selected Values of x

x	$\sqrt{-x}$
-4	$\sqrt{-(-4)} = 2$
-3	$\sqrt{-(-3)} = \sqrt{3}$
-2	$\sqrt{-(-2)} = \sqrt{2}$
-1	$\sqrt{-(-1)} = 1$
0	$\sqrt{-0} = 0$
1	$\sqrt{-1} = i$
2	$\sqrt{-2} = i\sqrt{2}$
3	$\sqrt{-3} = i\sqrt{3}$
4	$\sqrt{-4} = 2i$

the set of all real numbers, x, such that $x \geq 0$. If the domain of f is the set of all real numbers, then the range of f is a subset of the set of all complex numbers.

Mathematical Focus 3

The interpretation of $-x$ as the additive inverse of x has implications for the graphs of functions. Specifically, the graph of the points $(-x, g(x))$ is a reflection across the vertical axis of the graph of the points $(x, g(x))$.

In the real numbers, the additive inverse of a number is the same as the opposite of the number (see Focus 1). Consider the real number line. The positive and negative numbers are, in a sense, reflections of each other across 0. That is, a number and its additive inverse are on opposite sides of 0 on the number line; they are also the same distance from 0.

In a similar way, for a function f, one can understand values of $f(x)$ in a coordinate plane as being on the opposite side of the vertical axis as the values of $f(-x)$, and the same distance from the vertical axis. The point $(x, g(x))$ is a reflection across the vertical axis of the point $(-x, g(x))$. Applying this to the current Situation, there exists a function defined by the rule $h(x) = \begin{cases} \sqrt{x} \text{ , if } x \geq 0 \\ \sqrt{-x} \text{ , if } x \leq 0 \end{cases}$, for which the graph of the set of points $(x, h(x))$ is a reflection across the vertical axis of

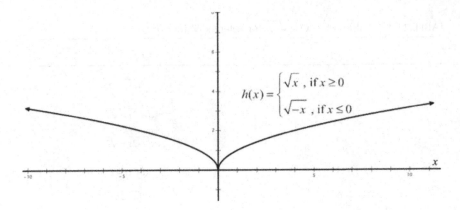

FIGURE 21.1. Graph of $h(x) = \begin{cases} \sqrt{x} \text{ , if } x \geq 0 \\ \sqrt{-x} \text{ , if } x \leq 0 \end{cases}$.

the graph of the set of points $(-x, h(x))$. The graph of the set of points $(x, h(x))$ is identical to the graph of the set of points $(-x, h(x))$ and is shown in Figure 21.1.

POSTCOMMENTARY

The concept of function involves a pairing between two sets. This pairing is often given as a rule (or rule of assignment), such as $f(x) = x + 3$. The terminology involved is far from consistent, so an explanation is offered here.

Typically, the domain of a function refers to the elements that the function "acts on" or "maps from." Choosing a different domain results in a different function. In college mathematics, the domain is often prescribed a priori without any specific rule of assignment. Thus, the notation $h: A \rightarrow B$ could be used to denote the function h with domain A, without explicitly saying how h maps from A to B. In secondary mathematics, however, the rule of assignment is often established first, and the domain is assumed to be the maximal real domain that can be used with the rule of assignment (this assumption is commonly made explicit by the teacher and in the course textbook). For example, a high school algebra class might be given the rule $g(x) = \dfrac{1}{x}$ and then be expected to deduce that the domain is the set of all real numbers except 0. In both cases, however, every element in the domain must be "assignable" to a well-defined element of the range. In the example of the function g, it would be inappropriate in either case to state that g has domain $\mathbb{R}$, because 0 is an element of $\mathbb{R}$ and $\dfrac{1}{0}$ is not well defined.

The terms *range* and *codomain* often need clarification. The range is usually understood to mean the actual set of function values the rule of assignment yields when applied to the domain. Thus, the domain $\mathbb{R} - \{0\}$ and rule $x \mapsto \dfrac{1}{x}$ would result in a range of $\mathbb{R} - \{0\}$. This usage of the word *range* results in all functions being onto (or surjective). The term *codomain* will generally refer to a set that simply *contains* the function values when the rule of assignment is applied to the domain. There are many sets that could serve as the codomain for a function with domain $\mathbb{R} - \{0\}$ and rule $x \mapsto \dfrac{1}{x}$, including $\mathbb{R} - \{0\}$, $\mathbb{R}$, and $\mathbb{C}$. Taken this way, the codomain is generally specified by the one proposing the function, whereas range is used to describe the entire set of function values that result for a given rule of assignment and domain.

Finally, the rule of assignment of a function is constrained. The rule of assignment for the function must be constructed so that each element of the domain maps to precisely one element of the range. If a mapping does not satisfy this constraint it is not termed a *function*; it is referred to as a *relation*. A relation $f: A \to B$ is a function if it satisfies the condition that for all $a \in A$, whenever $f(a) = b_1$ and $f(a) = b_2$, then $b_1 = b_2$. That is, for each "input" there is only one "output." A function need not be determined by a computational formula, for example, a function could map U.S. states to their capital cities (e.g., (Ohio, Columbus), (Maryland, Annapolis), ..., (California, Sacramento)).

NOTES

1. In this and other Situations, a hyphen (-) is used to indicate a negative number (e.g., -17) and an en-dash (–) to indicate either the opposite of a number (e.g., $-b$) or subtraction (e.g., $a - b$).

CHAPTER 22

INVERSE TRIGONOMETRIC FUNCTIONS

Situation 16 From the MACMTL-CPTM Situations Project

**Rose Mary Zbiek, M. Kathleen Heid, Ryan Fox,
Kelly Edenfield, Jeremy Kilpatrick, Evan McClintock,
Heather Johnson, and Brian Gleason**

PROMPT

Three prospective teachers planned a unit of trigonometry as part of their work in a methods course on the teaching and learning of secondary mathematics. They developed a plan in which high school students would first encounter what the prospective teachers called *the three basic trig functions*: sine, cosine, and tangent. The prospective teachers indicated in their plan that students next would work with "the inverse functions," which they identified as secant, cosecant, and cotangent.

COMMENTARY

The Foci draw on the general concept of inverse and its multiple uses in school mathematics. Key ideas related to the inverse are the operation involved, the set

*Mathematical Understanding for Secondary Teaching: A Framework
and Classroom-Based Situations*, pages 217–222.

of elements on which the operation is defined, and the identity element given this operation and set of elements. The crux of the issue raised by the Prompt lies in the use of the term *inverse* with both functions and operations.

MATHEMATICAL FOCI

Mathematical Focus 1

> *An inverse requires three entities: a set, a binary operation on that set, and an identity element given that operation and set of elements.*

Secondary mathematics involves work with many different contexts for inverses. For example, opposites are additive inverses defined for real numbers and with additive identity of 0, and reciprocals are multiplicative inverses defined for nonzero real numbers and with multiplicative identity of 1.

In general, if $*$ is the binary operation, S is the set of elements on which $*$ is defined, $e \in$ S, and e is the identity element, then for c, $d \in$ S, c is an inverse of d in the system if and only if $c*d = d*c = e$.

For functions, including trigonometric functions, the set is often assumed to be $\mathbb{R} \rightarrow \mathbb{R}$ functions (or some subset of $\mathbb{R}$), the operation is composition of functions, and the identity is given by $f(x) = x$. If a function g is the inverse of the sine function, for example, then $g(\sin(x)) = \sin(g(x)) = x$. An observation such as $\csc(\sin(0)) \neq 0$ is sufficient to show that $\csc(x)$ is not an inverse function under composition for $\sin(x)$. In contrast, if the operation is multiplication and the set is real numbers, the multiplicative identity is 1, and for any value of x such that $\sin(x) \neq 0$, the number $\csc(x)$ is the multiplicative inverse of the number $\sin(x)$.

More generally, the functions $f: S \rightarrow T$ and $g: T \rightarrow S$ are inverses if and only if g consists of the ordered pairs $(f(s), s)$ and f consists of the ordered pairs $(g(t), t)$. Support for that conclusion comes from consideration of issues of domain and range. In order for f and g to be inverse functions, $f(g(t)) = t$ and $g(f(s)) = s$. Note that in this case, the identity "entity" actually consists of an identity function on T and an identity function on S. When the domain and range are copies of the same set, the two identities will be equivalent.

For the preceding functions f and g, the range of f must be the domain of g, and the range of g must be the domain of f. For example, the domain of sine is the set of all real numbers, but because it is not one-to-one it does not have an inverse function unless its domain is restricted to a set for which it is one-to-one.

Mathematical Focus 2

> *Although the inverse under multiplication is not the same as the inverse under function composition, the same notation, the superscript −1, is used for both.*

The general function notation, $y = f(x)$, means that y is the image of x under the function f. To indicate the inverse function, the notation $x = f^{-1}(y)$ is used; it means that x is the image of y under the inverse of f. The superscript -1 is used to show that the inverse f^{-1} is a function related to f; the superscript is not to be interpreted as an exponent. In contrast, $z = xy$ means that z is the product of x and y, and to indicate the inverse of the product, the notation $x = zy^{-1}$ is used, where the superscript is interpreted as an exponent. These two usages of the superscript, -1 and -1, are distinct in that one represents a function inverse and the other is usually thought of as an exponent that happens to yield the multiplicative inverse.

The functions secant, cosecant, and cotangent are defined, respectively, as follows: $\sec(x) \equiv \dfrac{1}{\cos(x)}$, $\csc(x) \equiv \dfrac{1}{\sin(x)}$, and $\cot(x) \equiv \dfrac{1}{\tan(x)}$. They are not defined for $\cos(x) = 0$, $\sin(x) = 0$, or $\tan(x) = 0$. They are written using exponents as $\sec(x) \equiv (\cos(x))^{-1}$, $\csc(x) \equiv (\sin(x))^{-1}$, and $\cot(x) \equiv (\tan(x))^{-1}$.

In contrast, the inverse functions of sine, cosine, and tangent are, respectively, $\sin^{-1}(x)$ (sometimes written $\arcsin(x)$), $\cos^{-1}(x)$ (sometimes written $\arccos(x)$), and $\tan^{-1}(x)$ (sometimes written $\arctan(x)$ or $\operatorname{arctg}(x)$).

There are multiple values of x such that $y = \sin(x)$. For example, with $y = 1$, $1 = \sin\left(\dfrac{\pi}{2}\right) = \sin\left(\dfrac{5\pi}{2}\right)$, and so on. So it is not clear what the meaning of $\sin^{-1}(1)$ should be. Should it be $\dfrac{\pi}{2}, \dfrac{5\pi}{2}$, or something else? For $\sin^{-1}$ to be a function, it must map each input value to one and only one corresponding output value. This problem can be overcome by restricting the domain of the sine function so that each output, $\sin(x)$, corresponds to one and only one input, x.

For example, restricting the x-values of $y = \sin(x)$ to $-\dfrac{\pi}{2} \le x \le \dfrac{\pi}{2}$ with $-1 \le y \le 1$ yields an inverse sine function with domain $-1 \le x \le 1$ and range $-\dfrac{\pi}{2} \le y \le \dfrac{\pi}{2}$. Principal values are sometimes denoted with a capital letter; for example, the principal value of the inverse sine may be denoted $\operatorname{Sin}^{-1}(x)$ or $\operatorname{Arcsin}(x)$ (but this capitalization notation is far from universal and may, in fact, be used with the opposite meaning).

So this so-called *principal branch* of the inverse function can be described as the well-defined function with domain [-1, 1] and range $\left[-\dfrac{\pi}{2}, \dfrac{\pi}{2}\right]$. For each input domain value d in [-1, 1], the output $\operatorname{Sin}^{-1}(d) =$ the unique number x in $\left[-\dfrac{\pi}{2}, \dfrac{\pi}{2}\right]$ that satisfies $\sin(x) = d$.

Mathematical Focus 3

When functions are graphed in an xy-coordinate system with y as a function of x, these graphs are reflections of their inverses' graphs (under composition) in the line y = x.

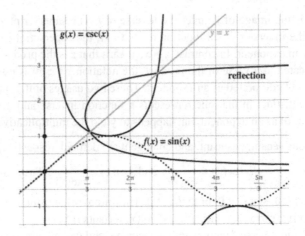

FIGURE 22.1.   Graphs of the sine function, the line given by $y = x$, the cosecant function, and the reflection of the cosecant function in the line $y = x$.

The graph of a function reflected in the line $y = x$ is the graph of its inverse, although without restricting to principal values, the inverse may not be a function. Justifying this claim requires establishing that the reflection of an arbitrary point (a, b) in the line $y = x$ is the point (b, a).

The following argument shows that the reflection of an arbitrary point (a, b) in the line $y = x$ is the point (b, a), using the geometric properties of reflection (the point A' is a reflection of the point A in line l if and only if l is the perpendicular bisector of $\overline{AA'}$). First, it should be noted that the perpendicular bisector of $\overline{AA'}$ consists of precisely those points in the plane that are equidistant from A and A'. So, one can choose an arbitrary point on the line $y = x$, say (c, c). Using the distance formula to compute the distance between (a, b) and (c, c) and between (b, a) and (c, c) yields $\sqrt{(c-a)^2 + (c-b)^2} = \sqrt{(c-b)^2 + (c-a)^2}$, and so the line $y = x$ is the perpendicular bisector of the line segment between (a, b) and (b, a). Therefore, the reflection of the point (a, b) in the line $y = x$ is (b, a).

Suppose that cosecant and sine were inverse functions. Then, a reflection of the graph of $y = \csc(x)$ in the line $y = x$ would be the graph of $y = \sin(x)$. Figure 22.1 shows, on one coordinate system, graphs of the sine function, the line given by $y = x$, the cosecant function, and the reflection of the cosecant function in the line $y = x$. Because the reflection of the graph of the cosecant function in the line $y = x$ does not coincide with the graph of the sine function, sine and cosecant are not inverse functions.

POSTCOMMENTARY

1. When the trigonometric functions are defined in terms of the ratios of the sides of a right triangle, the three sides of the triangle give rise to six ratios. Because calculation with the ratios was so difficult in the days before calculators, names were given to each one and tables were constructed for them. The advent of computers has meant that given one of the functions, the others are easily calculated using trigonometric identities. The secant, cosecant, and cotangent, which were never used much in applications, have consequently diminished somewhat in importance relative to the other three.

    The inverse trigonometric functions—especially the inverse sine, inverse tangent, and inverse secant—turn out to be useful in calculus as antiderivatives for integrals involving quotients and noninteger powers of polynomials. They are often used in precalculus courses, primarily to illustrate the concept of inverse function.

    There are several other methods for proving that the line $y = x$ is the perpendicular bisector of the line segment between (a, b) and (b, a). One alternative is to find the midpoint of the line segment between (a, b) and (b, a), note that it is on the line $y = x$, and then compute the slope of the line segment between (a, b) and (b, a) to confirm that it is the negative (opposite) reciprocal of the slope of the line $y = x$. Another alternative is to use a diagram similar to that in Figure 22.2 to construct an argument involving congruent triangles.

    Some texts use the terms *range* of a function, *image* of a function, and *codomain* of a function. Here, use of *range*, or *image*, indicates the set of values to which the function evaluates and *codomain* to indi-

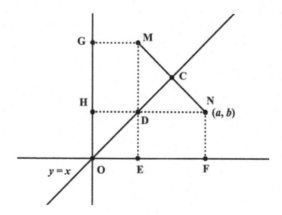

FIGURE 22.2.   Diagram from which a congruent-triangles argument can be developed.

cate a set containing the range or image. Thus, the range, or image, of $f(x) = \sin(x)$ is the interval determined by $-1 \leq f(x) \leq 1$, and a codomain of $f(x) = \sin(x)$ is $\mathbb{R}$.

2. The topics in the Foci represent very important mathematics for secondary and postsecondary mathematics. The idea of inverse functions for real-valued functions extends to higher dimensional analogs in higher mathematics and applications (e.g., solving a differential equation by looking for an "inverting" operator such as integration).

CHAPTER 23

ZERO-PRODUCT PROPERTY

Situation 17 From the MACMTL–CPTM Situations Project

Jeanne Shimizu, Heather Johnson, Ryan Fox, Laura Singletary, and Sarah Donaldson

PROMPT

A student in an Algebra 1 class wrote the following solution to a homework problem:

$$x^2 - 4x - 5 = 7$$

$$(x - 5)(x + 1) = 7$$

$$x - 5 = 7 \qquad x + 1 = 7$$

$$x = 12 \qquad x = 6$$

A different student commented that 6 was a solution to the equation since $6^2 - 4(6) - 5 = 7$, but that 12 was not.

Mathematical Understanding for Secondary Teaching: A Framework and Classroom-Based Situations, pages 223–227.
Copyright © 2015 by Information Age Publishing
All rights of reproduction in any form reserved.

COMMENTARY

In this Prompt, the student uses an overgeneralization of the zero-product property. The zero-product property states that if $ab = 0$, then $a = 0$ or $b = 0$. The only real number, n, for which the property "If $ab = n$, then $a = n$ or $b = n$" holds for all real values a and b, is 0. In this Prompt, the student extends the zero-product property to values of n other than 0. The inaccuracy of this extension is dealt with in Focus 1. Factors are a recurring theme in the study of quadratic polynomials, so integral factors are addressed in Focus 2 and Focus 3. These Foci consider some special cases in which the student's proposed property and solution method do apply. Focus 4 considers this Situation from an abstract-algebra standpoint, using the fact that the set of polynomials forms a mathematical entity known as an *integral domain*.

MATHEMATICAL FOCI

Mathematical Focus 1

A well-chosen counterexample will prove that a "seven-product property" may not be employed in the same manner as the zero-product property.

The zero-product property states that if $ab = 0$, then $a = 0$ or $b = 0$. In the solution given in the Prompt, the student seems to have created a "seven-product property," which would imply that if $ab = 7$, then $a = 7$ or $b = 7$. However, it is not necessarily the case that if the product of two numbers is 7, then one of the numbers must be 7. One counterexample that would disprove this "seven-product property" is $2 \times 3.5 = 7$: The product is 7, but neither of the factors is 7.

The converse of the zero-product property is also true: If $a = 0$ or $b = 0$, then $ab = 0$. So another valuable counterexample related to the "seven-product property" comes from an investigation of its converse: If $a = 7$ or $b = 7$, then $ab = 7$. If one of the factors is 7, and the other factor is not the multiplicative identity (to be discussed in Focus 2), then the product cannot be 7. For example, let $a = 7$ and $b = 10$. The product is 70, not 7, thus disproving the converse of the "seven-product property."

Mathematical Focus 2

A "seven-product property" holds true if and only if one of the factors is the multiplicative identity.

It is important to note the reason that allowed one of the student's answers to be a correct solution even though an incorrect assumption was used. The "seven-product property" works under certain conditions. For example, when multiplying two factors, if one of the factors is 7 and the other is the multiplicative identity, 1, then the product is 7. Consider again the student's solution.

$$x^2 - 4x - 5 = 7$$

$$(x - 5)(x + 1) = 7$$

$$x - 5 = 7 \qquad x + 1 = 7$$

$$x = 12 \qquad x = 6$$

The solution $x = 6$ is correct because $x = 6$ allows one of the factors to equal 7 and the other to equal 1. That is, if $x = 6$, then $(x - 5)(x + 1) = (6 - 5)(6 + 1) = (1)(7) = 7$. The solution $x = 12$ is not correct, however, because although it allows one factor to equal 7, it causes the other factor to equal 13, and the product of 7 and 13 is not 7. It is also important to note that the other correct solution to the quadratic equation is $x = -2$, which allows one factor to be -1 and the other -7. That is, if $x = -2$, then $(x - 5)(x + 1) = (-2 - 5)(-2 + 1) = (-7)(-1) = 7$.

Mathematical Focus 3

Investigating the integral factors of the right member of an equation such as $x^2 - 4x - 5 = 7$ can yield both valid and invalid solutions, highlighting the relevance of and necessity for the zero-product property.

It is possible to determine solutions to the equation $x^2 - 4x - 5 = 7$ in a manner similar to the student's strategy, but in general, the student's method is too complicated to be useful for most polynomials. If used correctly, the method would work like this: Factor the polynomial on the left side of the equation into two binomial factors (in this case, $(x - 5)$ and $(x + 1)$), as the student did. Then factor the integer in the right member of the equation into a pair of its integral factors. In this case, because 7 is prime, the task is not difficult. Now set each factor on the left equal to one of the factors in the right member of the equation and check each possibility to see which one gives correct solutions. The illustration for this particular instance is given in the following:

$$x^2 - 4x - 5 = 7$$

$$(x - 5)(x + 1) = 7$$

(1a) $(x-5)(x+1)=(7)(1)$ (2a) $(x-5)(x+1)=(-7)(-1)$

$x - 5 = 7$ and $x + 1 = 1$ $x - 5 = -7$ and $x + 1 = -1$

$x = 12, 0$ or $x = -2$

or   or

(1b) $x - 5 = 1$ and $x + 1 = 7$ (2b) $x - 5 = -1$ and $x + 1 = -7$

$x = 6$ $x = 4, -8$

In the case of 1a and 2b, x would have to take on two different values at once in order for the equation to hold. Because that would be impossible, those values for x are not solutions. However, the solutions for 1b and 2a are valid. This can be verified by substituting $x = 6$ and $x = -2$ into the original equation:

$$(6)^2 - 4(6) - 5 = 7 \text{ and } (-2)^2 - 4(-2) - 5 = 7$$

This method was relatively manageable in this case because 7 is prime. However, a lengthy process would be required to check all possible solutions if the integer in the right member of the equation had several factors. For example, if the value in the right member of the equation were 16 instead of 7, there would be many more possible solutions to investigate:

$$(x - 5)(x + 1) = 16$$

$x - 5 = 1, x + 1 = 16$	or	$x - 5 = 16, x + 1 = 1$
$x - 5 = -1, x + 1 = -16$	or	$x - 5 = -16, x + 1 = -1$
$x - 5 = 2, x + 1 = 8$	or	$x - 5 = 8, x + 1 = 2$
$x - 5 = -2, x + 1 = -8$	or	$x - 5 = -8, x + 1 = -2$
	$x - 5 = 4, x + 1 = 4$	
	$x - 5 = -4, x + 1 = -4$	

The solutions in this case are found by using two of the factors from above: $x - 5 = 2$ and $x + 1 = -2$. However, it is not obvious that these are the appropriate factors to use to solve the quadratic equation correctly. This highlights the relevance of employing the zero-product property to solve polynomial equations. Because $ab = 0$ implies $a = 0$ or $b = 0$, there are far fewer solutions to check after factoring a particular polynomial. One simply can set each factor equal to 0, resulting in fewer equations to solve. In the case of the equation $x^2 - 4x - 5 = 7$, the solution can be found by first setting the quadratic polynomial equal to zero $(x^2 - 4x - 12 = 0)$ and *then* factoring: $(x - 6)(x + 2) = 0$. Because this implies that $x - 6 = 0$ or $x + 2 = 0$, the solutions are $x = 6$ and $x = -2$.[1]

Mathematical Focus 4

The zero-product property is unique to particular contexts, one of which is the set of polynomials.

The zero-product property may seem obvious, but it is unique to particular sets of numbers. The zero-product property is relevant in this Situation because (in abstract algebra terms) the set of polynomials is an integral domain[2] and therefore contains no zero divisors. Zero divisors are nonzero elements of a set (say, a and b), such that $ab = 0$. For example, sets such as the integers, the real numbers, or (in the case of this Situation) polynomials, have no zero divisors, which is why the zero-product property holds for these sets.

This fact may be appreciated by noting some cases in which the zero-product property does *not* hold. For example, consider the set $\mathbb{Z}_6$, the set of integers modulo 6, whose elements are the integers 0, 1, 2, 3, 4, and 5. According to the zero-product property, $ab = 0$ implies $a = 0$ or $b = 0$. However, in $\mathbb{Z}_6$, it is possible to construct a product of 0 without either of the factors being 0. The product (2)(3) is such an example. In the set of all integers, (2)(3) = 6, but in $\mathbb{Z}_6$, (2)(3) = 0 because (2)(3) = 6, and 6 is congruent to 0 mod 6. That is, 2 and 3 are zero divisors in $\mathbb{Z}_6$. Another nontrivial example of a ring with a zero divisor is, for example, the ring of $n \times n$ matrices. An example of a zero divisor in the ring of 2×2 matrices (over any nonzero ring) is the matrix $\begin{pmatrix} 1 & 1 \\ 2 & 2 \end{pmatrix}$, because, for example,

$$\begin{pmatrix} 1 & 1 \\ 2 & 2 \end{pmatrix}\begin{pmatrix} 1 & 1 \\ -1 & -1 \end{pmatrix} = \begin{pmatrix} -2 & 1 \\ -2 & 1 \end{pmatrix}\begin{pmatrix} 1 & 1 \\ 2 & 2 \end{pmatrix} = \begin{pmatrix} 0 & 0 \\ 0 & 0 \end{pmatrix}.$$

NOTES

1. It is also important to realize that the types of arguments given here will only determine whether there are *integral* solutions. Of course, there are infinitely many polynomial (quadratic) equations over the natural numbers with no integral solutions.

2. An *integral domain* is a nonzero commutative ring in which the product of any two nonzero elements is nonzero.

CHAPTER 24

SIMULTANEOUS EQUATIONS

Situation 18 From the MACMTL-CPTM Situations Project

Dennis Hembree, Erik Tillema, Evan McClintock, Rose Mary Zbiek, Heather Johnson, Patricia S. Wilson, James Wilson, and Ryan Fox

PROMPT

A student teacher in a course entitled *Advanced Algebra/Trigonometry* presented several examples of solving systems of three linear equations in three unknowns algebraically using the method of elimination (linear combinations). She started another example and had written the following

$$3x + 5y = -3$$
$$5x + y - 2z = 5$$

when a student asked, "What if you only have two equations?"

COMMENTARY

Knowing necessary and sufficient conditions for unique solutions to systems of linear equations is important in this Situation. The Foci build from systems of equations in two variables to systems of equations in three variables, and examine why n independent equations are necessary to produce a unique solution to a system of equations in n variables. Systems of linear equations in two or three

Mathematical Understanding for Secondary Teaching: A Framework and Classroom-Based Situations, pages 229–237.

variables may be consistent or inconsistent and dependent or independent. If a set of equations has no solutions in common, the system is referred to as being *inconsistent*. If a set of equations has at least one common solution, the system is referred to as being *consistent*. When a consistent system has exactly one solution, it is referred to as being *independent*; if it has more than one solution, it is referred to as being *dependent*.[1] The Foci use physical models, symbolic representations, graphical representations, and matrix representations to examine systems of linear equations with unique solutions, an infinite number of solutions, and no solutions.

MATHEMATICAL FOCI

Mathematical Focus 1

Solutions to equations are invariant under linear combinations.

If two equations in two variables x and y, $ax + by + c = 0$ and $dx + ey + f = 0$, are in a linear combination, $A(ax + by + c) + B(dx + ey + f) = 0$, then any point (x, y) that satisfies both of the original equations will satisfy the linear combination. Similarly, for a linear combination of two equations in three variables, any solution (x, y, z) to the original system of equations will also satisfy the linear combination (in fact, any solution to a system of two equations in n variables will satisfy a linear combination of those equations).

Thus, if either of the two equations is replaced with a linear combination of the two equations, then the solution set does not change. The set of solutions to the original system of two equations is the same as the set of solutions to the new system in which one of the original equations has been replaced with a linear combination. Of course, one has to be careful about not taking the "trivial" linear combination with A and B both 0. This can be extended to any number of variables (2 equations in n unknowns).

The purpose of the following example is to provide an intuitive sense for why solutions to equations are invariant under linear transformations. The example is similar to those found in the ancient Chinese text *Jiu Zhang Suanshu* [*The Nine Chapters of the Mathematical Arts*] (see Lam, 1994) and many early Babylonian mathematical texts.

Two ropes have different lengths, and the sum of their lengths is 10 meters. To measure the length of a 46-meter bamboo rod requires 3 lengths of the first rope and 7 lengths of the second rope. Determine the length of each rope.

Letting x represent the length of the first rope and y the length of the second rope, the sum of the lengths of the rope can be expressed symbolically as $x + y = 10$. Since many combinations of values for x and y satisfy this equation, this one equation does not supply enough information to find a unique solution.

A second equation is needed to determine a unique set of values for the lengths of the ropes. Because measuring a 46-meter bamboo rod requires 3 lengths of the

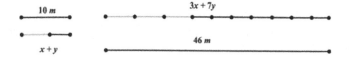

FIGURE 24.1. Length models to represent the equations $x + y = 10$ and $3x + 7y = 46$.

FIGURE 24.2. Length models representing the equations $x + y = 10$ and $3x + 3y = 30$.

first rope and 7 lengths of the second rope, a relationship between the lengths of the two ropes can be expressed symbolically as $3x + 7y = 46$.

Figure 24.1 shows length models that represent the equations $x + y = 10$ and $3x + 7y = 46$.

Because the sum of one length of each rope is 10 m, the sum of three lengths of each rope would be 30 m. This relationship could be expressed symbolically with the equivalent equations $x + y = 10$ and $3x + 3y = 30$. Figure 24.2 shows length models to represent the equations $x + y = 10$ and $3x + 3y = 30$.

Because $3x + 3y = 30$ is equivalent to $x + y = 10$, the lengths of each rope can be expressed with the system of equations $3x + 3y = 30$ and $3x + 7y = 46$. Figure 24.3 shows length models to represent the equations $3x + 3y = 30$ and $3x + 7y = 46$.

By comparing the top rope with the bottom rope in each representation, four lengths of the second rope must equal 16 meters, and therefore one length of the second rope is 4 meters. Expressed symbolically:

$$3x + 7y - 3x - 3y = 46 - 30$$

$$4y = 16$$

$$y = 4$$

Hence, using the information that the sum of the lengths of the two ropes is 10, a unique length for the first rope exists, namely 6 meters.

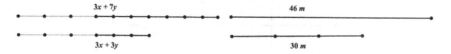

FIGURE 24.3. Length models representing the equations $3x + 3y = 30$ and $3x + 7y = 46$.

Mathematical Focus 2

Two equations with two variables will have a unique solution if the equations are consistent and independent.

A linear equation in two variables is the equation of a line in the plane and therefore has an infinite number of solutions. Figure 24.4 shows an example of the graph of a line.

Two equations have a unique simultaneous solution if the lines they represent intersect and are not coincident. Such equations are consistent and independent. Figure 24.5 shows an example of the graph of a system of two linear equations.

There is no solution if the lines are parallel and not coincident. Such equations are inconsistent; they have no common solution. Figure 24.6 shows a system of inconsistent equations.

There is an infinite number of solutions if the lines are coincident. Such equations are consistent but dependent; they have all solutions in common. Figure 24.7 shows a system of consistent, but dependent, equations.

Mathematical Focus 3

A system of two linear equations in three variables may be consistent or inconsistent. A system of three linear equations in three variables may have no solutions (i.e., be inconsistent), an infinite number of solutions (i.e., be consistent and dependent), or a unique solution (i.e., be consistent and independent).

A graphical representation of the points whose coordinates satisfy a linear equation in three variables is a plane. A graphical representation of the solution of

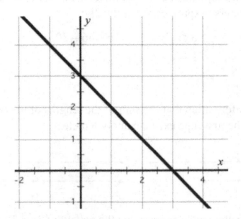

FIGURE 24.4.   Graph of $x + y = 3$.

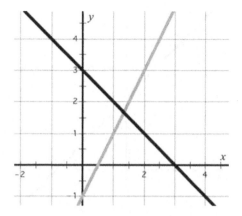

FIGURE 24.5.   Graph of the system of consistent and independent linear equations, $x + y = 3$ and $2x - y = 1$.

a system of two or three linear equations in three variables would be the intersection of the two or three planes representing the solutions of each of the equations.

Two planes will be parallel, intersecting, or coincident. The illustrations for parallel and intersecting planes appear in Figure 24.8 and Figure 24.9, respectively.

When two planes are parallel, as shown in Figure 24.8, the system of two equations is inconsistent and has no solution. When the planes intersect in a line, as shown in Figure 24.9, all the points on the line of intersection of the two planes

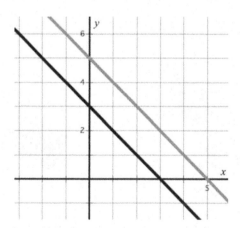

FIGURE 24.6.   Graph of the system of inconsistent linear equations, $x + y = 3$ and $x + y = 5$.

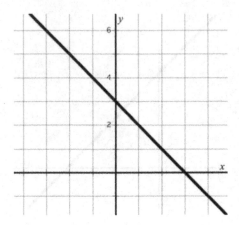

FIGURE 24.7. Graph of the system of consistent and dependent linear equations, $x + y = 3$ and $2x + 2y = 6$.

are solutions. Therefore, the system has infinitely many solutions, and the equations are consistent. Whenever two planes intersect in a line, an equation of one of them cannot be a nonzero multiple of an equation of the other. Therefore, the two equations are independent. Without a third plane to intersect that line of intersection, there is no unique point of intersection. Thus, a system of two equations in three unknowns cannot have a unique solution.

The three planes represented by equations with three variables will either be parallel or they will intersect in several different ways, as illustrated in Figures 24.10 through 24.14. The cases in which two or more of the planes coincide are not shown here.

FIGURE 24.8. Two parallel planes.

FIGURE 24.9. Two intersecting planes.

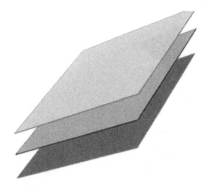

FIGURE 24.10. Three parallel planes.

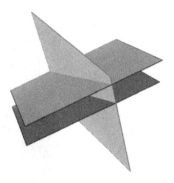

FIGURE 24.11. Three planes, two of which are parallel.

When all three planes coincide, the equations are consistent but not independent. There are an infinite number of solutions. When two of the three planes coincide, the result is depicted in Figure 24.8 (if the third plane is parallel to the two planes that coincide) or in Figure 24.9 (if the third plane intersects the two planes that coincide).

In Figure 24.10, Figure 24.11, and Figure 24.12, the systems of equations have no solution and thus are inconsistent; in Figure 24.13, there is an infinite number of solutions to the system and thus the system is consistent but not independent;

FIGURE 24.12. Three planes intersecting pairwise in three parallel lines.

FIGURE 24.13. Three planes intersecting in a line.

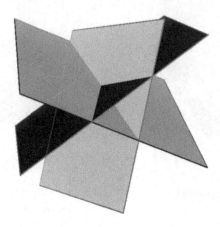

FIGURE 24.14. Three planes intersecting in a point.

and in Figure 24.14, the system of equations has a unique solution, and so it is consistent and independent.

Mathematical Focus 4

Systems of linear equations can be solved by matrix methods. Matrix methods can also be used to determine whether equations are consistent or independent.

One technique using matrix methods to solve systems involves left-multiplying the matrix of constants by the inverse of the coefficient matrix. The determinant of the coefficient matrix must be nonzero for the equations to be consistent and independent and thus have a unique solution.

In the case of a system of two equations with three unknowns, the 2×3 coefficient matrix is not a square matrix. Thus, the coefficient matrix does not have an inverse, and a unique solution does not exist. However, that does not necessarily mean that no solution exists. If the system of equations is dependent, there are infinitely many solutions.

Also, if the system of two equations with three variables is consistent and independent, finding a set of infinite solutions may be accomplished by performing Gaussian elimination on the augmented matrix of coefficients and constants. Consider the general case of a system of two equations in three variables that is independent:

$$a_1 x + b_1 y + c_1 z = k_1$$

$$a_2 x + b_2 y + c_2 z = k_2$$

Performing Gaussian elimination on the augmented matrix of coefficients and constants gives the augmented matrix:

$$\begin{bmatrix} 1 & 0 & r_1 & s_1 \\ 0 & 1 & r_2 & s_2 \end{bmatrix}$$

where r_1, r_2, s_1, and s_2 are constants.

The augmented matrix represents an equivalent system of two equations in three variables:

$$\begin{array}{ccc} x+0y+r_1z=s_1 & & x=s_1-r_1z \\ & \text{or} & \\ 0x+y+r_2z=s_2 & & y=s_2-r_2z \end{array}$$

These equations indicate that, although the values of x and y depend on the value of z, the value of z is arbitrary. Hence, a system of two equations in three variables may have many solutions.

This method, or its matrix equivalent, assumes that $a_1b_2 - a_2b_1$, $b_1c_2 - b_2c_1$, and $a_1c_2 - a_2c_1$ are not all equal to 0 in order for there to be consistent equations.

NOTE

1. In this Situation, systems of equations are called *independent* and *dependent*; linear algebra textbooks typically refer to systems of equations being *linearly independent* or *linearly dependent*.

REFERENCE

Lam, L. Y. (1994). Jiu Zhan Suanshu (Nine chapters on the mathematical art): An overview. *Archive for History of Exact Sciences, 47*, 1–51.

CHAPTER 25

GRAPHING INEQUALITIES CONTAINING ABSOLUTE VALUES

Situation 19 From the MACMTL–CPTM Situations Project

Shari Reed, AnnaMarie Conner, M. Kathleen Heid,
Heather Johnson, Maureen Grady, and Svetlana Konnova

PROMPT

This episode occurred during a course for prospective secondary mathematics teachers. The discussion focused on the graph of $y - 2 \leq |x + 4|$. The instructor demonstrated how to graph this inequality using compositions of transformations, generating the graph in Figure 25.1. Students proposed other methods, which included the two different symbolic formulations and accompanying graphs as seen in Figure 25.2 and Figure 25.3. Students expected their graphs to match the instructor's graph, and they were confused by the differences they noticed.

Mathematical Understanding for Secondary Teaching: A Framework and Classroom-Based Situations, pages 239–248.

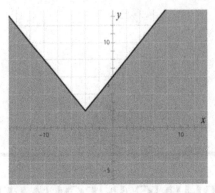

FIGURE 25.1. Instructor's graph of $y - 2 \leq |x + 4|$.

$y - 2 \leq x + 4$ or $-x - 4 \leq y - 2$

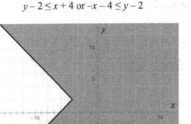

$y - 2 \leq x + 4$ and $-x - 4 \leq y - 2$

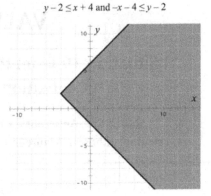

FIGURE 25.2. A student's symbolic formulation of $y - 2 \leq |x + 4|$ as $(y - 2 \leq x + 4)$ or $(-x - 4 \leq y - 2)$ and the student's corresponding graph.

FIGURE 25.3. A student's symbolic formulation of $y - 2 \leq |x + 4|$ as $(y - 2 \leq x + 4)$ and $(-x - 4 \leq y - 2)$ and the student's corresponding graph.

COMMENTARY

This Prompt involves the graph of an absolute value inequality involving two variables. The graph of an absolute value inequality is related to the graph of an absolute value function. Absolute values can be interpreted on the basis of a symbolically stated definition of absolute value as well as based on the conception of absolute value as distance from 0. Composite functions such as absolute value functions can be viewed as transformations of input and/or output values of functions. Working with absolute values entails keeping track of the logic of the conjunctions and disjunctions into which absolute value statements transform, and working with inequalities involves applying an appropriate range of rules governing their transformation. Inequalities in two variables can be interpreted by examining the inequality for each of a sequence of constant values for one of the variables.

MATHEMATICAL FOCI

Mathematical Focus 1

An absolute value inequality that involves a composite function can be graphed by a sequence of graphs of component functions or by a sequence of transformations on the input and/or output of the parent absolute value function.

One way to think about graphing an inequality that contains absolute value expression(s) is to consider the related function as a composition of functions. To graph $y - 2 = |x + 4|$, one can rewrite it as $y = |x + 4| + 2$. This can be viewed as a function rule for a composition of functions, m, for which $m = k \circ h \circ g \circ f$ for functions f, g, h, and k with rules $f(x) = x$, $g(x) = x + 4$, $h(x) = |x|$, and $k(x) = x + 2$. Each successive composition transforms the graph as shown in the sequence of graphs in Figure 25.4.

The portion of the plane to be shaded includes all points (x, y) for which $y - 2$ is less than $|x + 4|$, or, equivalently, all points (x, y) for which y is less than $|x + 4| + 2$,

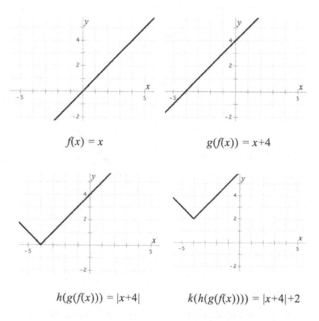

$$f(x) = x \qquad\qquad g(f(x)) = x+4$$

$$h(g(f(x))) = |x+4| \qquad\qquad k(h(g(f(x)))) = |x+4|+2$$

FIGURE 25.4. Successive graphs illustrating the composition, $m = k \circ h \circ g \circ f$.

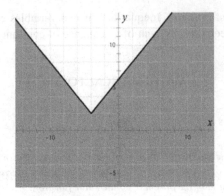

FIGURE 25.5.   Graph of $y - 2 \le |x + 4|$.

so the portion of the plane below the graph of the function m should be shaded. Alternatively, one could test a point on either side of the graphed function to determine which portion of the plane to shade. A rationale for testing a single point is related to what happens on each vertical line of the shaded graph (see Figure

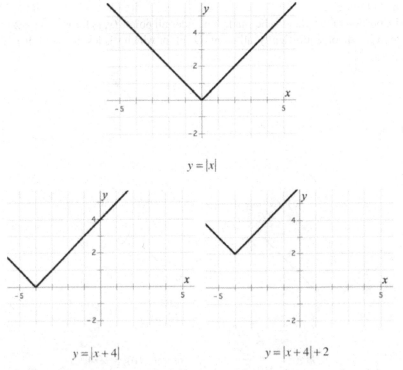

$$y = |x|$$

$$y = |x + 4|$$

$$y = |x + 4| + 2$$

FIGURE 25.6.   A succession of graphs showing the transformation of $y = |x|$ to $y = |x + 4| + 2$.

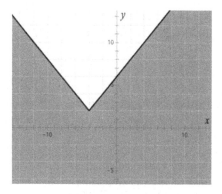

FIGURE 25.7.   Graph of $y - 2 \le |x + 4|$.

25.5 for a graph of $y - 2 \le |x + 4|$). Consider a vertical line $x = k$. For some point (x, y) on that vertical line, $y - 2 = |x + 4|$ or $y = |x + 4| + 2$. The y-values for points on the line $x = k$ above that point will exceed $|k + 4| + 2$, and the y-values below that point will be less than $|k + 4| + 2$.

An alternative way to graph the absolute value inequality $y - 2 \le |x + 4|$ is to begin with the graph of the absolute value function and transform it through a sequence of transformations of either the input or the output. The progression of graphs might look like those shown in Figure 25.6.

In this approach, the parent absolute value function is being transformed, with the input value transformed in the first transformation and the resulting output value being transformed in the second transformation. As previously described, after the related function has been graphed, displaying the graph of the inequality requires shading the appropriate side of the graph of the related equation as shown in Figure 25.7.

Mathematical Focus 2

$a = |b|$ *is not equivalent to* $|a| = b$.

In the Prompt, the students' graphs are both similar to and different from the instructor's graph in two important ways. First, the boundaries of the students' graphs lie on the same lines as the boundaries of the instructor's graph. Second, the shading of the students' graphs is to the left or right of (rather than above or below) these boundaries. This combination of correct boundaries and incorrect shading suggests that the students may have produced the graph that has the absolute value applied to $y - 2$ rather than to $x + 4$. The students' graphs, when analyzed from a transformational perspective, are the graphs of $-|y - 2| \le x + 4$ and $|y - 2| \le x + 4$. Thus, it seems that the students interpreted the original absolute value inequality in a way that "switched" the expression to which the absolute value applied. The student likely applied the absolute value to the expression containing y instead of to the expression containing x.

Mathematical Focus 3

A definition of absolute value can be used to translate an absolute value inequality into a statement composed of conjunctions and disjunctions of statements that involve no absolute values.

One way in which absolute value is defined is:

$$|x| = \begin{cases} x \ if \ x \geq 0 \\ -x \ if \ x < 0 \end{cases}.$$

Using this symbolically stated definition of absolute value, the inequality $y - 2 \leq |x + 4|$ can be interpreted using two cases, one for which the input for the absolute value function is positive or 0 and one for which the input for the absolute value function is negative. This is usually written as

$$y - 2 \leq \begin{cases} x + 4 \ if \ x + 4 \geq 0 \\ -(x + 4) \ if \ x + 4 < 0 \end{cases}.$$

The inequality $y - 2 \leq |x + 4|$ is of the form $a \leq |b|$. If $a \leq |b|$, then $a \leq b$ if $b \geq 0$ and $a \leq -b$ if $b < 0$. Expressing this system entirely in terms of conjunctions and disjunctions, if $a \leq |b|$ then $[(a \leq b \ and \ b \geq 0) \ or \ (a \leq -b \ and \ b < 0)]$. This system of inequalities $y - 2 \leq \begin{cases} x + 4 \ if \ x + 4 \geq 0 \\ -(x + 4) \ if \ x + 4 < 0 \end{cases}$ can be stated using conjunctions and disjunctions as $[y - 2 \leq x + 4 \ and \ x + 4 \geq 0]$ or $[y - 2 \leq -(x + 4) \ and \ x + 4 < 0]$.

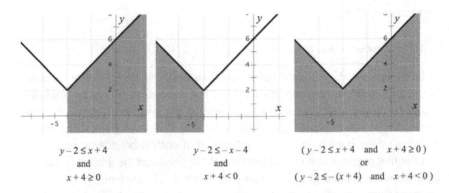

$$y - 2 \leq x + 4$$
and
$$x + 4 \geq 0$$

$$y - 2 \leq -x - 4$$
and
$$x + 4 < 0$$

$$(y - 2 \leq x + 4 \quad and \quad x + 4 \geq 0)$$
or
$$(y - 2 \leq -(x + 4) \quad and \quad x + 4 < 0)$$

FIGURE 25.8.   Sequence of graphs illustrating $(y - 2 \leq x + 4 \ and \ x + 4 \geq 0)$ or $(y - 2 \leq -(x + 4) \ and \ x + 4 < 0)$.

When this system is graphed, it produces the same graph as that produced using the transformational approach to the problem, as shown in the sequence of graphs in Figure 25.8. This system of inequalities is derived directly from the definition of absolute value. Graphing $y - 2 \leq |x + 4|$ by graphing $(y - 2 \leq x + 4$ or $-x - 4 \leq y - 2)$ suggests the misconception that, if $a \leq |b|$ then $(a \leq b$ or $-b \leq a)$ no matter whether b is nonnegative or negative.

Mathematical Focus 4

Inequality is not an equivalence relation.

For $a = |b|$, the symmetric property of equality yields equivalent expressions $|b| = a$ and $a = |b|$. Applying what one might think to be a "symmetric property of inequality" would allow one to write $y - 2 \leq -(x + 4)$ as $-(x + 4) \leq y - 2$. However, the inequality relation "$\leq$" is not an equivalence relation and fails to satisfy the symmetric property of equivalence relations. Applying to $a \leq |b|$ a technique associated with $a = |b|$ without referring to the meaning of the technique is one source of potential error in working with absolute value inequalities.

Mathematical Focus 5

(1) Absolute value can be interpreted as distance from 0.

(2) Inequalities in two variables can be interpreted by examining the inequality for each of a series of constant values for one of the variables.

Alternative mathematical definitions can be equivalent, describing the same mathematical objects. One interpretation of absolute value is based on distance from 0. One can consider the graph of $y - 2 \leq |x + 4|$ to be all points (x, y) such that $x + 4$ is at least $\max\{0, y - 2\}$ units from 0. It is perhaps easier to start by considering the graph of $y \leq |x|$ as all points (x, y) such that, for each nonnegative value of y, x is at least y units from 0. Consider a stacked sequence of rays on aligned and equally spaced parallel number lines (representing the values of x for which $|x| \geq 0, |x| \geq 1, |x| \geq 2, |x| \geq 3, \ldots$). The rays in the diagram in Figure 25.9a represent the pattern that occurs for $|x| \geq y$ for nonnegative integer values of y. For a second number line orthogonal to the first (as in the y-axis), the endpoints of these rays lie on the Cartesian graph of $y = |x|$. One could, in this way, interpret the solution set of ordered pairs, (x, y), for the inequality $y \leq |x|$ (see Figure 25.9b) as the union of the set of all such rays such that $|x| \geq \max\{0, k\}$ for $k \in \mathbb{R}$.[1] This strategy of solving equations or inequalities involving two variables by viewing the system as the union of a set of equations or inequalities involving only one variable is a case of the common and useful strategy of reducing a problem whose solution is unknown to a series of simpler problems whose solution strategy is known.

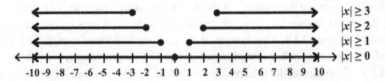

FIGURE 25.9a. A sequence of rays representing the values of x for which $|x| \geq 0$, $|x| \geq 1$, $|x| \geq 2$, and $|x| \geq 3$.

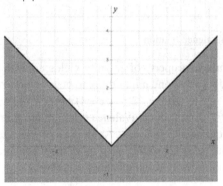

FIGURE 25.9b. Graph of $y \leq |x|$.

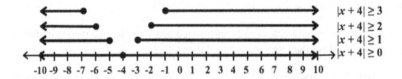

FIGURE 25.10a. A sequence of rays representing the values of x for which $|x + 4| \geq 0$, $|x + 4| \geq 1$, $|x + 4| \geq 2$, and $|x + 4| \geq 3$.

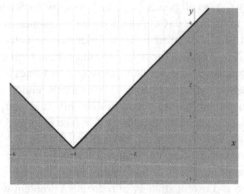

FIGURE 25.10b. Graph of $y \leq |x + 4|$.

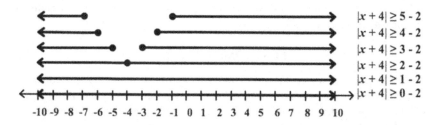

FIGURE 25.11a.   A sequence of rays representing the values of x for which $|x + 4| \geq 0 - 2$, $|x + 4| \geq 1 - 2$, $|x + 4| \geq 2 - 2$, $|x + 4| \geq 3 - 2$, $|x + 4| \geq 4 - 2$, and $|x + 4| \geq 5 - 2$.

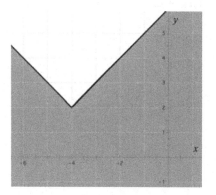

FIGURE 25.11b.   Graph of $y - 2 \leq |x + 4|$.

Similarly, consider the graph of $y \leq |x + 4|$ as all points (x, y) such that $x + 4$ is at least max$\{0, y\}$ units from 0. One can consider a sequence of rays (representing the values of x for which $|x + 4| \geq 0$, $|x + 4| \geq 1$, $|x + 4| \geq 2$, $|x + 4| \geq 3$, ...; see Figure 25.10a) and use the fact that $y = |x + 4|$ is continuous and is piecewise defined with two linear branches to form the Cartesian graph of $y = |x + 4|$. Also $y \leq |x + 4|$ can be interpreted as all points (x, y) such that x is at least max$\{0, y\}$ units from -4 (see Figure 25.10b). The sequence of rays shown in Figure 25.10a and the graph shown in Figure 25.10b result.

Finally, consider the graph of $y - 2 \leq |x + 4|$ as all points (x, y) such that $x + 4$ is at least max$\{0, y - 2\}$ units from 0 (or, alternatively, all points (x, y) such that x is at least max$\{0, y - 2\}$ units from -4). A sequence of rays (representing the values of x for which $|x + 4| \geq 0 - 2$, $|x + 4| \geq 1 - 2$, $|x + 4| \geq 2 - 2$, $|x + 4| \geq 3 - 2$, ...; see Figures 25.11a and 25.11b) can be considered, and the fact can be used that

$y - 2 = |x + 4|$ is continuous and is piecewise defined with two linear functions to form the Cartesian graph of $y - 2 = |x + 4|$.

POSTCOMMENTARY

Different definitions afford the opportunity to approach mathematical concepts in substantially different ways. In these Foci, viewing absolute value through the lens of a symbolic definition led to a significantly different approach from a view of absolute value as distance from 0.

NOTE

1. Allowing k to vary across $\mathbb{R}$ "fills in the spaces" between the rays formed for integral values of k.

CHAPTER 26

SOLVING QUADRATIC EQUATIONS

Situation 20 From the MACMTL–CPTM Situations Project

Jeanne Shimizu, Sarah Donaldson, Kelly Edenfield, and Erik Jacobson

PROMPT

In an Algebra 1 class some students began solving a quadratic equation as follows:
  Solve for x:

$$x^2 = x + 6$$

$$\sqrt{x^2} = \sqrt{x+6}$$

$$x = \sqrt{x+6}$$

They stopped at this point, not knowing what to do next.

Mathematical Understanding for Secondary Teaching: A Framework and Classroom-Based Situations, pages 249–256.
Copyright © 2015 by Information Age Publishing
All rights of reproduction in any form reserved.

COMMENTARY

This Situation highlights some issues concerning solving equations (regarding both quadratic equations and equations in general) that are prevalent in school mathematics. Focus 1 and Focus 2 present two typical methods of solving a quadratic equation: factoring and the quadratic formula. These are included because this Prompt illustrates the importance of having accurate and dependable means by which to attempt to solve quadratic equations. Focus 3 provides a geometric approach for solving $x^2 = x + 6$. Focus 4 provides guidelines for solving any algebraic equation and emphasizes maintaining equivalence. Focus 5 shows the relationship between the solution(s) of an equation and the zero(s) of a function. It contains a graphical approach to solving quadratic equations. The Situation ends with a Postcommentary on the occurrence of extraneous solutions.

MATHEMATICAL FOCI

Mathematical Focus 1

Factoring and using the zero-product property can be used to solve many quadratic equations.

The quadratic equation $x^2 = x + 6$ can be solved by factoring and applying the zero-product property. The zero-product property for real numbers states:

$$\text{If } a \cdot b = 0, \text{ then } a = 0 \text{ or } b = 0 \text{ or } a = b = 0.$$

The zero-product property is a consequence of the axioms of the real number system. The set of real numbers with the operations of addition and multiplication is a *field*[1] and therefore also an *integral domain*[2] because every field is also an integral domain. (Note, however, that the converse is false; the set of integers with the operation of addition and multiplication is an integral domain but not a field because not every integer has a multiplicative inverse.) An important property of integral domains is that there are no *zero-divisors*. Zero-divisors are nonzero elements whose product is 0.

For example, consider the ring $\mathbb{Z}_6$, that is, the set $\{0, 1, 2, 3, 4, 5\}$ with addition and multiplication defined modulo 6. The element 2 and the element 3 are nonzero, but their product is the element 0 (mod 6). Thus, $\mathbb{Z}_6$ has zero-divisors and so is not an integral domain, and a quadratic equation in $\mathbb{Z}_6$ could not be solved using the zero-product property.[3]

Because the set of real numbers is an integral domain, the zero-product property can be used to solve equations involving real numbers. In fact, this property is critical in this Situation because it is the key rule that allows the equation to be solved by factoring. It is worth noting that this property is unique to 0. That is, there is no "two-product property" or "six-product property." A common student error is to generalize the property to something such as:

If $a \cdot b = c$, then $a = c$ or $b = c$.

This error might be used, for example, in the following way:

$$x^2 + 7x + 12 = 2$$

$$(x + 3)(x + 4) = 2$$

$$x + 3 = 2 \quad \text{or} \quad x + 4 = 2$$

$$x = \text{-}1 \quad \text{or} \quad x = \text{-}2$$

Zero is the only number for which the property holds because 0 is the only number that has itself as its only multiple (see Situation 17 in chapter 23 for additional information about the zero-product property). In this example, the property is correctly used by first writing an equation equivalent to $x^2 = x + 6$ that has 0 as a member, factoring, and then setting each factor equal to 0:

$$x^2 = x + 6$$

$$x^2 - x - 6 = 0$$

$$(x - 3)(x + 2) = 0$$

$$x - 3 = 0 \quad \text{or} \quad x + 2 = 0$$

$$x = 3 \quad \text{or} \quad x = \text{-}2$$

Mathematical Focus 2

All quadratic equations can be solved by completing the square or by using the quadratic formula.

Solutions of quadratic equations are not always integers, nor are they necessarily real numbers. For these and other reasons, the quadratic formula, derived by completing the square on $ax^2 + bx + c = 0$, is a useful tool with which to solve a quadratic equation (for a derivation of the quadratic formula using completing the square, see Situation 22 in chapter 28).

$$x^2 = x + 6$$

$$x^2 - x - 6 = 0$$

$$x = \frac{-(-1) \pm \sqrt{(-1)^2 - 4(1)(-6)}}{2}$$

$$x = \frac{1 \pm 5}{2}$$

$$x = 3 \quad \text{or} \quad x = \text{-}2$$

Mathematical Focus 3

A geometric analogy to an area model can be used to represent quadratic equations and their solutions.

It is common to use a geometric analogy to examine products of polynomials, the analogy usually being between an area and a product of binomials. When doing this it is important to consider area as a unitless quantity, because if the variable x represents x centimeters, then in the Prompt, a square with side length x centimeters has an area of x^2 square centimeters, and that area is purported to be equal to $x + 6$ centimeters, an impossibility. On the other hand, one could use units cautiously, thinking of a square with side length x centimeters having an area x^2 square centimeters, and that area being equal to that of a rectangle with side lengths $x + 6$ centimeters and 1 centimeter. A second caution must be offered to indicate that the area model works only for positive quantities.

It is interesting to note that what is termed an "area model" works even when values of the variable expressions are negative. This is an artifact of it being more an organizational tool that accommodates the term-by-term products that result from the multiplication than a model based on area. Further development of the representations of quadratic functions based on area, occur in Situation 22 (see chapter 28).

Mathematical Focus 4

Solving an equation using algebraic manipulation requires that equivalence is maintained between each form of the equation.

In order to solve an algebraic equation, one must determine the value(s) for the unknown(s) that satisfy the equation (i.e., make the equation true). In this Situation, x is an unknown quantity. Simply producing an equation that begins with "$x =$" (such as $x = \sqrt{x+6}$) does not constitute a solution. The x must be expressed in terms of that which does not involve x (that is, x must be "isolated").

A common strategy for solving equations is algebraic manipulation: performing operations to isolate the unknown quantity. Solving equations in this way requires that certain rules be followed. At each step, equivalence must be maintained from one equation to the next. Two guidelines are useful in accomplishing this objective:

1. Keep the equation "balanced" (e.g., by adding 3 to both members of the equation), and
2. Ensure that each equation in the process yields the same solution(s) as the original equation.

For example:

$$2x + 7 = 15$$

$$\underline{-7 \quad -7}$$

$$2x = 8$$

$$\frac{2x}{2} = \frac{8}{2}$$

$$x = 4$$

"Balance" is maintained by applying the same operations (subtracting 7, dividing by 2) to both members of the equation, or, in other words, by operating on the equation using only equivalence-preserving operations (e.g., addition property of equality, multiplication property of equality). Also, each step (in this case, $2x = 8$ and $x = 4$) yields an equation with the same solution set, {4}, as $2x + 7 = 15$.

The quadratic equation example in this Situation is $x^2 = x + 6$. Although the same operation is applied to both sides of the equation, equivalence is not maintained from the first equation to the second equation in the following step:

$$\sqrt{x^2} = \sqrt{x+6}$$

$$x = \sqrt{x+6}$$

because $x = \sqrt{x+6}$ has only one solution, whereas $x^2 = x+6$ has two. Also, it is not true in general that $\sqrt{x^2} = x$. Rather, $\sqrt{x^2} = |x|$. (See the Postcommentary for an explanation of why $x = \sqrt{x+6}$ has only one solution.)

In the following sequence of equations, however, equivalence *is* maintained.

$$x = x +$$

$$\sqrt{x} = \sqrt{x+}$$

$$|x| = \sqrt{x+}$$

$$x = \sqrt{x+6} \text{ or } x = -\sqrt{x+6}$$

In taking these steps, one arrives at an equation involving $\sqrt{x^2}$. It is worth noting here that one of the definitions of absolute value is $|x| = \sqrt{x^2}$. When solving absolute value equations, more than one solution is possible. For example, in the equation $|x| = 3$, the solutions for x are 3 and -3. Similarly, when solving the equation $\sqrt{x^2} = \sqrt{x+6}$, $x = \sqrt{x+6}$ or $x = -\sqrt{x+6}$.

It is not the case, however, that taking the square root of both members of an equation is never a good idea. There are many instances in which taking a root of both members of an equation is a helpful step toward arriving at a solution.

Mathematical Focus 5

Approximate solutions to equations can be found by graphically determining the zeros of the associated function.

The solutions to an equation in which an expression involving x is equal to 0 (such as $x^2 - 4 = 0$) are comparable to the zeros of a function of x (such as $f(x) = x^2 - 4$). This is because a zero, or x-intercept, of a function is an x-value for which the value of $f(x)$ is 0. For example, the solutions to $x^2 - 4 = 0$ are 2 and -2, and the zeros (the x-intercepts) of the graph of f where $f(x) = x^2 - 4$ are 2 and -2. The equation $x^2 - 4 = 0$ and the function f with rule $f(x) = x^2 - 4$ are not the same, as x is an unknown in the equation (representing a specific value), whereas in the function rule, x is a variable (changing quantity). However, the equation and the function in this case are related: The solutions of the equation are the same as the zeros of the function.

Because solutions to equations and zeros of functions are related in this way, graphing a function can be a useful method for solving an equation. However, as noted previously, if the strategy is to find the zeros of the function, the accompanying equation must have one member equal to 0. In this Situation, this will involve performing equivalence-preserving operations on the equation so that one member is 0:

$$x^2 = x + 6$$

$$x^2 - x - 6 = 0$$

The graph of $f(x) = x^2 - x - 6$ (see Figure 26.1) will indicate, by its zeros, approximate solutions of $x^2 - x - 6 = 0$.

A similar method requires graphing the functions f and g with rules $f(x) = x^2$ and $g(x) = x + 6$ (that is, treat each member of the original equation as the rule for a function) and determine the points of intersection of the graphs of the two functions. These are the points at which x^2 and $x + 6$ are equal.

POSTCOMMENTARY

Often, in solving equations, extraneous solutions result. These occur when the original domain is expanded during the course of the solution process.

It was stated previously that $x^2 = x + 6$ and $x = \sqrt{x+6}$ could not be equivalent because they have different solution sets, specifically, the former has two solutions and the latter has only one. To see that $x = \sqrt{x+6}$ has only one solution, consider solving the equation using the principles in Focus 4.

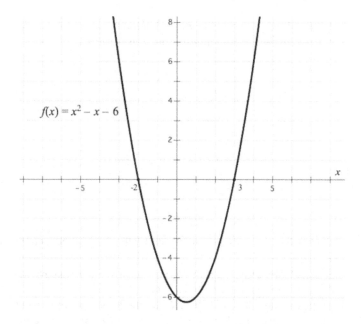

FIGURE 26.1.   A graph of $f(x) = x^2 - x - 6$.

First, however, attention must be given to the domain of x for $x = \sqrt{x+6}$. For $\sqrt{x+6}$ to be defined in the real numbers, $x \geq -6$. But if $x = \sqrt{<\text{some expression}>}$, $x \geq 0$. This means that the domain is the intersection of $x \geq -6$ and $x \geq 0$; thus, the domain is $x \geq 0$.

The inverse operation for taking the square root is squaring. Squaring both members of $x = \sqrt{x+6}$ yields the original equation ($x^2 = x + 6$), and the two solutions are 3 and -2. Note that in this original form of the equation, the domain is the set of real numbers; there are no restrictions on the values of x. However, in $x = \sqrt{x+6}$, $x \geq 0$, so the only solution is $x = 3$. Negative 2 is called an *extraneous solution*. It was introduced by expanding the valid domain from the nonnegative real numbers to all real numbers.

Note that different equations will have extraneous solutions as a result of expanding other domains to the real numbers. Consider $\log_3(5x-12) + \log_3 x = 2$. From the first term, $x > \frac{12}{5}$. From the second, $x > 0$. The intersection of these two restrictions is $x > \frac{12}{5}$. Solving this equation yields two solutions because the problem is converted to solving a quadratic equation whose variable has a domain of all real numbers. This domain expansion introduces a possible extraneous solution.

In fact, the first solution is not in the original domain and is an extraneous solution. The only solution for this logarithmic equation is $x = 3$.

$$\log_3(5x-12)+\log_3 x = 2$$
$$\log_3\big((5x-12)x\big) = 2$$
$$5x^2 - 12x = 9$$
$$5x^2 - 12x - 9 = 0$$
$$(5x+3)(x-3) = 0$$
$$x = -\frac{3}{5} \text{ or } x = 3$$

NOTES

1. A *field* is a commutative ring with identity in which all nonzero elements have inverses.
2. An *integral domain* is a commutative ring in which the cancellation law holds: If $c \neq 0$ and $ca = cb$, then $a = b$.
3. One might wonder how to solve a quadratic in, for example, a commutative ring with identity. For example, $x^2 = (x + n)$ mod 6 where n is one of 0, 1, 2, 3, 4, and 5. In this case, it might be simplest to examine a table of $x^2 = x + n$ for $x = 0, 1, 2, 3, 4,$ and 5. Alternatively, examining zero divisors separately from invertible elements might be a more generalizable approach.

CHAPTER 27

GRAPHING QUADRATIC FUNCTIONS

Situation 21 From the MACMTL–CPTM Situations Project

Ginger Rhodes, Ryan Fox, Shiv Karunakaran, Rose Mary Zbiek, Brian Gleason, and Shawn Broderick

PROMPT

When preparing a lesson on graphing quadratic functions, a student teacher found that the textbook for the class claimed that $x = \dfrac{-b}{2a}$ was the equation for the line of symmetry of a parabola $y = ax^2 + bx + c$. The student teacher wondered how this equation was derived.

COMMENTARY

This Prompt addresses graphing quadratic equations, specifically the derivation of the equation of the line of symmetry of a parabola. The Foci in this Situation deal with the general symbolic representation of a quadratic function, but they differ

Mathematical Understanding for Secondary Teaching: A Framework and Classroom-Based Situations, pages 257–261.

in the approaches used to obtain the equation in question. Focus 1 uses the symmetry of the parabola to find the x-coordinate of the vertex of the parabola. Focus 2 uses the first derivative to find the x-coordinate of the vertex of the parabola. Focus 3 utilizes transformations of the graph of $y = x^2$ to determine the coordinates of the vertex. Focus 4 uses some results about the roots of a polynomial equation, generally known as Viète's formulas, to find the x-coordinate of the vertex of the parabola.

MATHEMATICAL FOCI

Mathematical Focus 1

Knowing the general form of the equation representing the graph of a parabola can enable one to identify the line of symmetry.

A quadratic function can be written in the general form $y = ax^2 + bx + c$, where a, b, and c are real numbers and $a \neq 0$. The graph of this function is a parabola, for which the line of symmetry is the line through the focus that is perpendicular to the directrix. This parabola is symmetric about a line $x = k$, because its directrix is parallel to the x-axis. The symmetry of the graph about the line $x = k$ implies that, for example, the function values $f(k-1)$ and $f(k+1)$ must be equal:

$$a(k-1)^2 + b(k-1) + c = a(k+1)^2 + b(k+1) + c.$$

Expanding the expressions on both sides of the preceding equation yields:

$$ak^2 - 2ak + a + bk - b + c = ak^2 + 2ak + a + bk + b + c.$$

Combining like terms, the equation simplifies to:

$$-2ak - b = 2ak + b.$$

Solving for k reveals that:

$$k = -\frac{b}{2a}$$

Mathematical Focus 2

The first derivative of a polynomial function can be used to obtain the coordinates of the relative extrema of the function. In a parabola, this corresponds to the vertex, the x-coordinate of which gives the x-coordinate of all points on the line of symmetry.

Polynomial functions are differentiable, and one can use the first derivative of the polynomial function to obtain its critical values. These critical values enable one to obtain the coordinates of the vertex of the parabola (the absolute minimum or the absolute maximum).

$$y = ax^2 + bx + c$$

$$y' = 2ax + b$$

Finding the critical values for a quadratic function,[1] and thus the vertex and the equation of the line of symmetry, involves setting the derivative equal to 0 and solving for x:

$$2ax + b = 0$$

$$x = -\frac{b}{2a}.$$

This critical value is the x-coordinate of the vertex. Thus, the equation of the line of symmetry will be of the form $x = -\dfrac{b}{2a}$.

Mathematical Focus 3

Using transformations, the graph of the function $y = x^2$ *can be mapped to the graph of any quadratic function of the form* $y = ax^2 + bx + c$. *The graph of the function given by* $g(x) = v_1 f(h_2(x - h_1)) + v_2$ *is the image of the graph of f under the composition of a horizontal translation, a horizontal stretch or contraction, a vertical stretch or contraction, and a vertical translation related to the values of* $h_1, h_2, v_1,$ *and* v_2, *respectively.*

The point $(0, 0)$ is the vertex of the graph of the function given by $y = x^2$. To find the coordinates of the vertex of the general parabola, the method known as *completing the square* can be applied to the general form of the equation of a parabola, which yields a form given by the transformations noted previously.

$$y = ax^2 + bx + c$$

$$y - c = ax^2 + bx$$

$$\frac{1}{a}(y-c) = x^2 + \frac{b}{a}x$$

$$\frac{1}{a}(y-c) + \left(\frac{b}{2a}\right)^2 = x^2 + \frac{b}{a}x + \left(\frac{b}{2a}\right)^2$$

$$\frac{1}{a}\left(y - c + \frac{b^2}{4a}\right) = \left(x + \frac{b}{2a}\right)^2$$

Solving for y yields a symbolic form of the equation that can be compared to $y = x^2$, the equation of the particular case, to get information about the needed transformation:

$$y = a\left(x + \frac{b}{2a}\right)^2 + \left(c - \frac{b^2}{4a}\right)$$

The presence of $\left(x + \frac{b}{2a}\right)^2$ rather than x^2 implies a horizontal translation through $\frac{b}{2a}$ units in the negative direction (see Situation 25: Translation of Functions, in Chapter 31). This horizontal translation of the graph maps the vertex of the parabola from $(0, 0)$ to $\left(\frac{-b}{2a}, 0\right)$. The rest of the equation suggests a vertical stretch or contraction (by a factor of a) and a vertical translation (through $c - \frac{b^2}{4a}$ units in the positive direction), neither of which affect the axis of symmetry. Thus, the line of symmetry that passes through the vertex of the graph of the general parabola has the equation $x = -\frac{b}{2a}$.

Mathematical Focus 4

General facts about the roots of polynomial equations, known as Viète's formulas, can quickly yield information about the line of symmetry of a parabola.[2]

Given the roots of a quadratic function, r_1 and r_2, a result of François Viète states that $r_1 + r_2 = -\frac{b}{a}$ (and $r_1 r_2 = \frac{c}{a}$).[3] Because the roots of a quadratic are symmetric about the axis of symmetry (which contains the vertex), $\frac{r_1 + r_2}{2} = -\frac{b}{2a}$ is the x-coordinate of the vertex. Therefore, $x = -\dfrac{b}{2a}$ is the equation of the line of symmetry of the parabola.

NOTES

1. For polynomial functions in general, a critical value is tested to see whether a point is an extremum rather than an inflection point. For a parabola, the result is evident (either an absolute maximum or an absolute minimum).

2. Viète's formulas also give a way to show that two possible solutions, r and s, satisfy the equation $ax^2 + bx + c = 0$. One need only show that $r + s = -\dfrac{b}{a}$ and $r \cdot s = \dfrac{c}{a}$.

3. For further discussion of Viète's formula, see http://mathworld.wolfram.com/VietasFormulas.html.

CHAPTER 28

CONNECTING FACTORING WITH THE QUADRATIC FORMULA

Situation 22 From the MACMTL–CPTM Situations Project

Erik Tillema, Heather Johnson, Sharon K. O'Kelley, Erik Jacobson, Glendon Blume, and M. Kathleen Heid

PROMPT

A teacher who had completed a unit on factoring quadratic polynomials began a unit on the quadratic formula. One student asked whether there was a direct connection between factoring quadratic polynomials and the quadratic formula. The teacher wondered about the different ways one might answer the student's question.

COMMENTARY

The topic of solving quadratic equations typically involves three processes: (a) factoring, (b) completing the square, and (c) using the quadratic formula. Although each approach can be used to solve any quadratic equation with real coef-

---

Mathematical Understanding for Secondary Teaching: A Framework and Classroom-Based Situations, pages 263–275.

ficients, sometimes one of the approaches is more efficient than the others. For example, some quadratic equations cannot be solved by factoring expressions over the integers but can be solved by factoring over the real numbers. These equations are more easily solved using the quadratic formula. In addition, not every quadratic can be solved by factoring over the real numbers (e.g., $x^2 + 1 = 0$). In the three Foci, the benefits of each process are presented as well as the connections among them. In addition, a geometric model is offered to support each algebraic explanation.

A quadratic polynomial that can be represented as a product of two binomials over the real numbers can be modeled geometrically using an analogy to the area of a rectangle with side lengths given by the real values of the binomials.[1] Such quadratic polynomials of the form $x^2 + bx + c$ can be represented as a product of the form $(x + m)(x + n)$, where m, n may have integral values or nonintegral real values. Partitioned area models of rectangles emphasize the relationship between the sum $x^2 + bx + c$ and the product $(x + m)(x + n)$.

MATHEMATICAL FOCI

Mathematical Focus 1

If the quadratic expression in a quadratic equation factors over the integers (i.e., $x^2 + bx + c = (x + m)(x + n)$ for some m, n ∈ $\mathbb{Z}$), then the factored expression can be modeled directly with an analogy to a rectangle with sides $(x + m)$ and $(x + n)$.

The most obvious connection between factoring and the quadratic formula is the zero-product property (see Situation 17 in chapter 23). In essence, the solutions generated by the quadratic formula are the roots used in the factors set up using the zero-product property. For example, for the equation $x^2 + 12x + 20 = 0$, the quadratic formula will yield -10 and -2 as the solutions for x. These can be used to produce the equivalent equation $(x + 10)(x + 2) = 0$.

Consider the quadratic polynomial $x^2 + 3x + 2$ and the partitioned area model in Figure 28.1. The process of factoring involves re-expressing a sum as a product. Hence, by composing a rectangle from the partitioned area model, the sum of the areas can be expressed as a product of the lengths of the sides of the rectangle (see

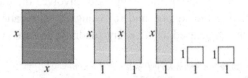

FIGURE 28.1.   Partitioned area model corresponding to $x^2 + 3x + 2$.

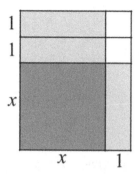

FIGURE 28.2. Rectangle composed from the partitioned area model in Figure 28.1.

Figure 28.2) or as the sum of the partitioned areas, as reflected in the following equation:

$$x^2 + 3x + 2 = (x + 1)(x + 2).$$

Factoring is sometimes more efficient when the roots are integers,[2] but completing the square also works in these cases.

$$x^2 + 3x + 2 = 0$$

Step 1: $x^2 + 3x + \left(\dfrac{3}{2}\right)^2 + [2 - \left(\dfrac{3}{2}\right)^2] = 0$

Step 2: $\left(x + \dfrac{3}{2}\right)^2 - \dfrac{1}{4} = 0$

Step 3: $x = -\dfrac{3}{2} \pm \dfrac{1}{2}$, so $x = -1$ or $x = -2$.

Because -1 and -2 are members of the solution set for x in $x^2 + 3x + 2 = 0$, the root–factor theorem of algebra indicates that $(x + 1)$ and $(x + 2)$ are factors of the polynomial expression $x^2 + 3x + 2$. The result is the same whether one uses factoring or completing the square. Indeed, completing the square relies on the fact that perfect square quadratic trinomials can be factored.[3]

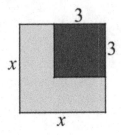

FIGURE 28.3. Partitioned area model for $x^2 - 9$.

FIGURE 28.4. Rectangular region formed from the region in Figure 28.3.

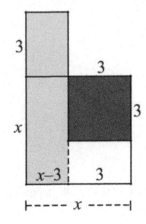

Mathematical Focus 2

If a quadratic expression has one or more real-valued coefficients that are not integers, then the method of completing the square is an efficient method to find factors of the quadratic expression, and partitioned area can geometrically model the expression and its factors.

Completing the square results in a perfect square trinomial less a squared constant that is equivalent to the original expression. The factors of the original expression are the square root of the trinomial plus and minus the constant. Geometrically, these factors correspond to the lengths of the sides of a rectangular region formed by removing from the square region represented by the perfect square trinomial a smaller square region with side lengths equal to the constant. Consider the quadratic polynomial $x^2 - 9$ and the partitioned area model shown in Figure 28.3.

By cutting, rearranging, and pasting back together the region in Figure 28.3, a rectangle of the same area can be made, as shown in Figure 28.4. Because the area of the rectangle can be represented as the product of the lengths of the sides of the rectangle or as the sum of the partitioned areas,

$$x^2 - 9 = (x + 3)(x - 3).$$

This rearrangement of the region in Figure 28.3 into a rectangular region is at the heart of the method of completing the square.

Consider the following quadratic polynomial, $x^2 + 6x + 7$, which cannot be factored over the integers. The roots of the equation $x^2 + 6x + 7 = 0$ can be found by completing the square.

$$x^2 + 6x + 7 = 0$$

Step 1: $x^2 + 6x + \left(\dfrac{6}{2}\right)^2 - \left(\dfrac{6}{2}\right)^2 + 7 = 0$

Step 2: $(x + 3)^2 - 2 = 0$

Step 3: $x = -3 \pm \sqrt{2}$

Using the root–factor theorem once more, leads to the conclusion that

$$x^2 + 6x + 7 = (x + 3 + \sqrt{2})(x + 3 - \sqrt{2}).$$

It is important to note that the process of completing the square is built upon factoring—that the "square" being completed is a perfect square trinomial that is later written in factored form. In the preceding example, the perfect square trinomial is introduced in Step 1 and factored in Step 2.

The method of completing the square can be modeled geometrically. The original quadratic polynomial, $x^2 + 6x + 7$, is represented by the partitioned area model in Figure 28.5. The partitioned area model is re-expressed in Figure 28.6; however, a rectangle has not been composed. Because $x^2 + 6x + 7 = (x^2 + 6x + 9) - 2 = (x + 3)^2 - 2$, the regions in Figure 28.6 can be rearranged into the regions in Figure 28.7. One can think of the area of the region in Figure 28.6 as a rectangular region with area 2 removed from a square region with area $(x + 3)^2$. Now, the area corresponding to the original quadratic expression is, in Figure 28.7, represented by a region that results when a small square region representing two units of area is removed from a square region that models $(x + 3)^2$.

A question arises concerning how one knows that one can dissect a region equivalent to the two upper right-most missing blocks in Figure 28.6 and rear-

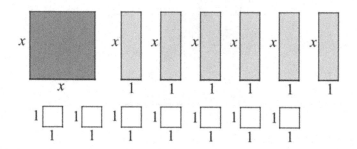

FIGURE 28.5. Partitioned area model for $x^2 + 6x + 7$.

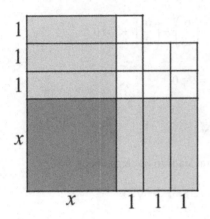

FIGURE 28.6. Nonrectangular rearrangement of the regions in the partitioned area model for $x^2 + 6x + 7$.

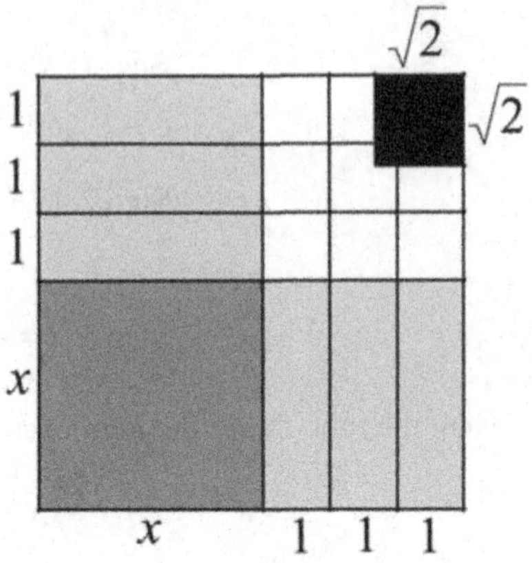

FIGURE 28.7. Rearrangement of the partitioned area model for $x^2 + 6x + 7$ into a region that results when a square region of area 2 square units is removed from a square region with side length $(x + 3)$.

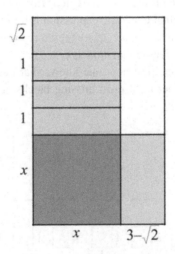

FIGURE 28.8. Rearrangement of the regions in the partitioned area model for $x^2 + 6x + 7$ into a rectangle with dimensions $(x+3+\sqrt{2})\times(x+3-\sqrt{2})$.

range them to get exactly the picture in Figure 28.7, with the black square of side length $\sqrt{2}$.⁴

To factor the quadratic polynomial $x^2 + 6x + 7$, one can compose a rectangle from the partitioned area model (see Figure 28.8). It can be shown that the regions of Figure 28.7 can be decomposed and recomposed to constitute the rectangular region shown in Figure 28.8. Because the area of the rectangle can be represented as the product of the lengths of the sides of the rectangle or as the sum of the partitioned areas,

$$x^2 + 6x + 7 = (x + 3 + \sqrt{2})(x + 3 - \sqrt{2}).$$

This is the same result that was achieved (and by the same process) when completing the square.

If a quadratic equation has complex roots, then although the partitioned area model no longer makes sense, the method of completing the square still gives correct results. Consider the quadratic polynomial $x^2 + 4x + 5$, and the associated quadratic equation $x^2 + 4x + 5 = 0$:

$$x^2 + 4x + 5 = 0$$

$$\text{Step 1: } x^2 + 4x + (2)^2 + [5 - (2)^2] = 0,$$

$$\text{Step 2: } (x + 2)^2 + 1 = 0,$$

$$\text{Step 3: } x = \text{-}2 \pm i.^5$$

Using the quadratic formula would give the same results (see Focus 3).

Mathematical Focus 3

The process of completing the square is used to derive the quadratic formula. In addition, in the case in which the roots are real, the solutions given by the quadratic formula can be represented by a partitioned area model.

Consider a quadratic equation written in general form:

$$ax^2 + bx + c = 0, \text{ with } a \neq 0.$$

The derivation of the quadratic formula from this general equation involves the use of completing the square:

$$1)\ x^2 + \frac{b}{a}x + \frac{c}{a} = 0,$$

$$2)\ x^2 + \frac{b}{a}x + \left(\frac{b}{2a}\right)^2 = -\frac{c}{a} + \left(\frac{b}{2a}\right)^2,$$

$$3)\ \left(x + \frac{b}{2a}\right)^2 = -\frac{c}{a} + \left(\frac{b}{2a}\right)^2,$$

$$4)\ \left(x + \frac{b}{2a}\right)^2 = \frac{b^2 - 4ac}{4a^2},$$

$$5)\ x + \frac{b}{2a} = \pm\sqrt{\frac{b^2 - 4ac}{4a^2}},$$

$$6)\ x = -\frac{b}{2a} \pm \frac{\sqrt{b^2 - 4ac}}{2a},$$

$$7)\ x = \frac{-b \pm \sqrt{b^2 - 4ac}}{2a}.$$

Note that the process of completing the square, as demonstrated in Steps 2 and 3, is built upon the idea of factoring. In turn, it can be argued that the quadratic formula is built upon factoring as well.

The quadratic formula $x = \dfrac{-b \pm \sqrt{b^2 - 4ac}}{2a}$ is used to determine solutions to a

quadratic equation of the form $ax^2 + bx + c = 0$, with $a \neq 0$. When $a = 1$, the quadratic equation becomes $x^2 + bx + c = 0$ and applying the quadratic formula gives the following solutions:

$$x = \frac{-b \pm \sqrt{b^2 - 4c}}{2}$$

$$\Rightarrow x = -\frac{b}{2} \pm \sqrt{\frac{b^2 - 4c}{4}}$$

$$\Rightarrow x = -\frac{b}{2} \pm \sqrt{\frac{b^2}{4} - c}$$

The solutions can be used to write the equation in factored form based upon the zero-product property:

$$x^2 + bx + c = 0$$

$$\left(x + \frac{b}{2} + \sqrt{\frac{b^2}{4} - c}\right)\left(x + \frac{b}{2} - \sqrt{\frac{b^2}{4} - c}\right) = 0.$$

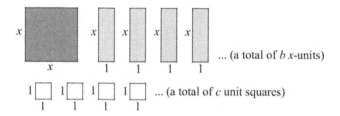

FIGURE 28.9. Partitioned area model for the general, monic, quadratic polynomial $x^2 + bx + c$.

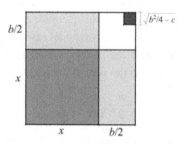

FIGURE 28.10. Rearrangement of the regions in the partitioned area model for $x^2 + bx + c$ into a square with side length $\left(x + \dfrac{b}{2} \right)$ with a square with side length $\sqrt{\dfrac{b^2}{4} - c}$ removed.

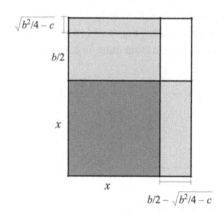

FIGURE 28.11. Rearrangement of the regions in the partitioned area model for $x^2 + bx + c$ into a rectangle with dimensions $\left(x + \dfrac{b}{2} + \sqrt{\dfrac{b^2}{4} - c} \right) \times \left(x + \dfrac{b}{2} - \sqrt{\dfrac{b^2}{4} - c} \right)$.[6]

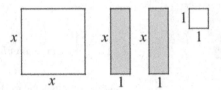

FIGURE 28.12. Partitioned area model for $x^2 - 2x + 1$ with the gray regions signifying regions that must be excised from the white regions.

Consider the quadratic polynomial $x^2 + bx + c$ and a corresponding partitioned area model (see Figure 28.9). To factor the quadratic polynomial $x^2 + bx + c$, compose a rectangle from the partitioned area model (as in Figures 28.10 and 28.11). Because the area of the rectangle can be represented as the product of the lengths of the sides of the rectangle or as the sum of the partitioned areas,

$$x^2 + bx + c = \left(x + \frac{b}{2} + \sqrt{\frac{b^2}{4} - c} \right)\left(x + \frac{b}{2} - \sqrt{\frac{b^2}{4} - c} \right).$$

Note: One can also illustrate this decomposition/recomposition of regions with area $x^2 + bx + c$ with $b < 0$. Although the following is not general, it illustrates, for a specific case how one might handle the necessary decomposition and recomposition. Consider a region with area $x^2 - 2x + 1$. A corresponding partitioned area model is shown in Figure 28.12, with the gray regions signifying regions that must be excised from the white regions. These regions can be rearranged so that the 1-by-1 square replaces a 1-by-1 unit on one of the gray squares. The result is a slightly reconfigured partitioned area model as shown in Figure 28.13, with a region of area x^2 and "negative area" regions of area $1 \cdot x$ and $1 \cdot (x - 1)$. Treating the gray rectangular regions as regions that must be excised from the square region of

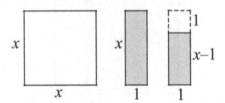

FIGURE 28.13. Slightly reconfigured partitioned area model, with region of area x^2 and "negative area" regions of area $1 - x$ and $1 - (x - 1)$. The combination of a gray region of area 1 and a white region of area 1 "cancel each other out," as shown by the region enclosed by dashed lines. This leaves a gray region of area $x - 1$.

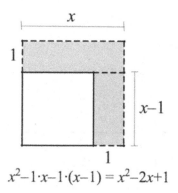

$$x^2 - 1 \cdot x - 1 \cdot (x-1) = x^2 - 2x + 1$$

FIGURE 28.14.   Rearrangement of the partitioned area model for $x^2 - 2x + 1$ into a region that results when rectangles of areas x and $x - 1$ are removed from a square region with side length x.

area x^2, a square of side lengths $x - 1$ can be created with the parts shown in Figure 28.13 (see Figure 28.14).

APPENDIX

This Appendix shows that the areas of the two polygonal regions in Figures 28.6 and 28.7 (identified in Focus 2) are equal. To show that these areas are equal, this demonstration focuses on the smaller region comprised of the 3-by-3 square configuration in the upper right corner of Figure 28.6.

The region ODMN in Figure 28.15, consisting of a configuration of 9 (1-by-1) squares, is intended to combine the features of both the 3-square-by-3-square configuration in the upper right corner of Figure 28.6 and the 3-square-by-3-square configuration in the upper right corner of Figure 28.7. The darkened rectangular region ADHE represents the missing blocks in the upper right corner of Figure 28.6. The square region BDKI represents the darkened $\sqrt{2}$-by-$\sqrt{2}$ square in the upper right corner of Figure 28.7. It is necessary to show that the areas of the two polygons in Figures 28.6 and 28.7 are equal. Because both areas are equal to $x^2 + 6x +$ (the area of the white regions), it is necessary to show only that the areas of the white regions are equal. That is, it is necessary to show only that the area of polygon OAEHMN is the same as the area of polygon OBIKMN.

The area of polygon OAEHMN is 7, because the polygonal region consists of seven unit squares. The area of polygon OBIKMN consists of the area of five unit squares and the areas of parts of several other squares. That is, the area of polygon OBIKMN = 5 + (the area of polygon ABIR) + (the area of polygon KLQI) + (the area of polygon RIQP).

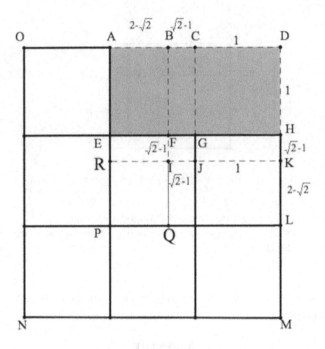

FIGURE 28.15. A configuration of 9 (1-by-1) squares, intended to combine the features of both the 3-square-by-3-square square configuration in the upper right corner of Figure 28.6 and the 3-square-by-3-square configuration in the upper right corner of Figure 28.7.

First, it is helpful to calculate the lengths of various segments in the diagram. Because $AD = 2$ and $BD = \sqrt{2}$, then $AB = AD - BD = 2 - \sqrt{2}$. By an analogous argument, $KL = 2 - \sqrt{2}$. Also, as opposite sides of rectangles with lengths $2 - \sqrt{2}$, $RI = IQ = 2 - \sqrt{2}$.

Using the lengths just computed,

- the area of polygon ABIR $= (AB)(BI) = \left(2 - \sqrt{2}\right)\sqrt{2} = 2\sqrt{2} - 2$,

- the area of polygon KLQI $= (KL)(LQ) = \left(2 - \sqrt{2}\right)\sqrt{2} = 2\sqrt{2} - 2$, and

- the area of polygon RIQP $= (RI)(IQ) = \left(2 - \sqrt{2}\right)\left(2 - \sqrt{2}\right)$
$$= 4 - 4\sqrt{2} + 2 = 6 - 4\sqrt{2}.$$

So, the area of polygon OBIKMN $= 5 +$ (the area of polygon ABIR) $+$ (the area of polygon KLQI) $+$ (the area of polygon RIQP) $= 5 + \left(2\sqrt{2} - 2\right) + \left(2\sqrt{2} - 2\right) + \left(6 - 4\sqrt{2}\right) = 7$.

Thus, the area of polygon OAEHMN is the same as the area of polygon OBIKMN, and so, by the Wallace-Bolyai-Gerwein theorem, one can cut the region enclosed by polygon OAEHMN into finitely many polygonal regions and rearrange the resulting regions to obtain the region enclosed by the polygon OBIKMN.

NOTES

1. Whereas one could argue that binomial factors must represent positive values because of their connection to length, an organizational structure that mimics the area model works for negative-valued factors as well.
2. Note that it is not always true that factoring is more efficient when the roots are integers. For example, it is very easy to find the roots of $x^2 + 232x + 13447 = 0$ using the quadratic formula because one has only to take the square root of 36, but finding the factors of 13447 is not as easy if one factors by hand.
3. In Figure 28.2, the expression $x^2 + 3x + 2$ is said to represent an area. This can be misleading when at least one of the values of x is negative. The strategy of setting $x^2 + bx + c$ equal to 0 is not meant to suggest that the area is 0. It is a strategy for locating zeroes of the function $f(x) = x^2 + bx + c$ because $x^2 + bx + c = (x - r_1) \times (x - r_2)$ for r_1 and r_2 and the resulting expressions for r_1 and r_2 are the usual ones that can be found using the quadratic formula.
4. That this can be done is corroborated by the Wallace-Bolyai-Gerwein theorem, which states that any two simple polygons of equal area are equidecomposable; that is, one can cut the region enclosed by the first into finitely many polygonal regions and rearrange the pieces to obtain the region enclosed by the second polygon. The Appendix shows that the areas of the two polygonal regions in Figures 28.6 and 28.7 are equal.
5. In this case, a geometric interpretation involves the length of complex vectors. If a real, monic quadratic function $x^2 + bx + c$ has a complex root z, then its other root must be z' the complex conjugate of z. Also, because x is real, $x' = x$. So the factorization of $x^2 + bx + c$ can be written $x^2 + bx + c = (x - z)(x - z') = (x - z)(x' - z') = (x - z)(x - z)'$. So $(x - z)(x - z)' = |x - z|^2$, which is the square of the length of the complex vector $x - z$. Equivalently $x^2 + bx + c$ is the area of a square with side length equal to the length of the complex vector $x - z$.
6. One can justify this rearrangement using an argument similar to the ones given with Figures 28.6, 28.7, and 28.8.

CHAPTER 29

PERFECT-SQUARE TRINOMIALS

Situation 23 From the MACMTL–CPTM Situations Project

Bob Allen, Dennis Hembree, Sarah Donaldson, Brian Gleason,
Shawn Broderick, M. Kathleen Heid, and Glendon Blume

PROMPT

While teaching about factoring perfect-square quadratic trinomials, a teacher realized that the students had the impression that a trinomial is a perfect-square quadratic trinomial if and only if the first and last terms of the trinomial are perfect squares. That is, they believed that the "middle term" is irrelevant. The teacher wanted to construct a counterexample that illustrated the importance of the middle term.

COMMENTARY

To generate an appropriate counterexample, the teacher must be able to recognize and factor perfect-square quadratic trinomials, which entails understanding the properties of the components of the trinomial that identify it as a perfect-square

Mathematical Understanding for Secondary Teaching: A Framework and Classroom-Based Situations, pages 277–281.

quadratic trinomial. Then, the teacher must be able to produce a trinomial the components of which lack at least one of those properties.

Using appropriate vocabulary related to perfect-square quadratic trinomials can help one avoid potential ambiguities. A *trinomial* is so named because it is the indicated sum of three (nonzero) monomials,[1] each of which is either a numeral, a variable, or a product of a numeral and one or more variables. A *quadratic trinomial* in one variable is a trinomial in which the greatest of the degrees of its terms is 2. Although $3x^5 + 4x + 7$ is a trinomial, it is not a quadratic trinomial because $3x^5 + 4x + 7$ is a polynomial of degree 5, the greatest of the degrees of its terms. A *perfect-square quadratic trinomial* is the square of a linear binomial, expressed in simplest terms.

It is common to refer to quadratic trinomials as having a *first* term, a *last* term, and a *middle* term. However, it is less ambiguous to refer to terms by their degrees: *quadratic term* (degree 2), *linear term* (degree 1), and *constant term* (degree 0). Using *first, last,* and *middle* is potentially ambiguous, unless the terms of the quadratic trinomial are written in order of descending powers, in which case *middle term* appropriately refers to the linear term.

MATHEMATICAL FOCI

Mathematical Focus 1

> *Factorable quadratic trinomials (factorable over the ring of integers) of the form* $ax^2 + bx + c$ *exhibit a structure in which there exist factors of the product* ac *the sum of which is equal to* b.

For the quadratic trinomial $ax^2 + bx + c$ to be factorable into the product of two linear binomials with integral coefficients, there must exist integer factors of ac that sum to b. For example, the quadratic trinomial $x^2 + 12x + 35$ is factorable because there exist factors of $1 \cdot 35$ that sum to 12, namely, 5 and 7. So $x^2 + 12x + 35 = (x^2 + 5x + 7x + 35)$. Because $(x^2 + 5x + 7x + 35)$ can be rewritten as $x(x + 5) + 7(x + 5)$, $x^2 + 12x + 35 = (x + 7)(x + 5)$. Similarly, the quadratic trinomial $4x^2 - 4x - 3$ is factorable because there exist factors of -12 that sum to -4, namely, 2 and -6. So, $4x^2 - 4x - 3 = (4x^2 - 6x + 2x - 3)$. Because $(4x^2 - 6x + 2x - 3)$ can be rewritten as $2x(2x - 3) + 1(2x - 3)$, $4x^2 - 4x - 3 = (2x + 1)(2x - 3)$.

The following is a proof that if a quadratic trinomial $ax^2 + bx + c$ is factorable into the product of two linear binomials with integral coefficients, then there exist integer factors of ac that sum to b.

Suppose that the quadratic trinomial $ax^2 + bx + c$ is factorable into the product of two linear binomials with integral coefficients, $(a_1x + c_1)(a_2x + c_2)$.[2] Expanding this product results in $(a_1x + c_1)(a_2x + c_2) = a_1a_2x^2 + (c_1a_2 + c_2a_1)x + c_1c_2$. So, $ac = a_1a_2c_1c_2$, and c_1a_2 and c_2a_1 are factors of ac. Thus, there exist integral factors of ac, namely, c_1a_2 and c_2a_1 that sum to $c_1a_2 + c_2a_1$. Because $c_1a_2 + c_2a_1$ and b are both expressions that represent the coefficient of the linear term, they have equal values. Therefore, there exist integral factors of *ac* that sum to *b*.

Mathematical Focus 2

In a quadratic trinomial in x *of the form* ax^2 + bx + c, *a relationship among* a, b, *and* c *can be used to identify whether the trinomial is a perfect square.*

This can be illustrated by forming a perfect-square quadratic trinomial from the square of the binomial $(p + q)$:

$$(p + q)^2 = (p + q)(p + q)$$

$$= p \cdot p + p \cdot q + q \cdot p + q \cdot q$$

$$= p^2 + 2pq + q^2$$

Written in order of descending powers of *p*, the perfect-square quadratic trinomial, $(p + q)^2$ is expressed as $p^2 + 2pq + q^2$, an expression in which the middle term is $2pq$.[3] This indicates that the middle term is twice the product of *p* and *q*. This implies that when a trinomial is written in order of descending powers of a variable, and the absolute value of its middle term is not equal to twice the product of the principal square roots of the first and last terms, it is not a perfect-square trinomial. For example, consider $16x^2 + 24xy + 9y^2$. In this case $\sqrt{16x^2} = |4x|$, $\sqrt{9y^2} = |3y|$, and $|24xy| = 2 \cdot |4x| \cdot |3y|$. Therefore $16x^2 + 24xy + 9y^2$ is possibly a perfect-square trinomial. Note that the trinomial must be expressed as a sum of monomials in order for this test to work. For example, $16x^2 - 24xy + 9y^2 = 16x^2 + (-24xy) + 9y^2$. In this case $\sqrt{16x^2} = |4x|$, $\sqrt{9y^2} = |3y|$, and $|-24xy| = 2 \cdot |4x| \cdot |3y|$. Therefore, $16x^2 - 24xy + 9y^2$ is possibly a perfect-square trinomial. However, complications arise with the trinomial, $16x^2 + 24xy - 9y^2$. In this case $16x^2 + 24xy - 9y^2 = 16x^2 + 24xy + (-9y^2)$, and $\sqrt{16x^2} = |4x|$, $\sqrt{-9y^2} = |y|(3i)$, and $|24xy| \neq 2 \cdot |4x| \cdot |y|(3i)$. Therefore, $16x^2 + 24xy - 9y^2$ is not a perfect-square trinomial.

It is possible to construct a trinomial whose nonmiddle terms are perfect-square monomials, but whose middle term lacks the property presented in Focus 1. A quadratic trinomial in *x* and *y* such as $16x^2 + 23xy + 9y^2$ is not a perfect-square trinomial because, although the first and last terms are perfect squares, the absolute value of the middle term, $23xy$, is not twice the product of the principal square roots of the first and last terms of the trinomial. This reasoning illustrates an important type of mathematical investigation: determining whether a given mathematical object is a mathematical object of a particular type. If one knows that a particular type of mathematical object, O, must possess certain properties, then if one knows that at least one of those conditions is not met by a given mathematical object, that object cannot be an object of type O. The form of reasoning involved is: The truth of the statement *If O, then P* implies that the statement *If not P, then not O* is also true. Thus, if the absolute value of the middle term of a quadratic trinomial written in order of descending powers is not twice the product of principal square roots of the other two terms, then the quadratic trinomial cannot be a perfect-square trinomial.

To produce a trinomial that demonstrates that the middle term (when written in order of descending powers) matters when factoring perfect-square quadratic trinomials, the teacher can simply write a trinomial in the form $m^2 + p + n^2$, where $|p| \neq 2 |m| \cdot |n|$.

POSTCOMMENTARY

There are many subtleties of factoring polynomials that often are ignored, including some in the preceding discussion. For example, factoring occurs over a ring, and this ring will determine which polynomials are factorable and which are irreducible.[4] In secondary curricula, it is conventional to factor polynomials over $\mathbb{Z}$, the set of integers. Additionally, the indeterminates used in polynomials often are not explicitly stated. For example, it is implicitly assumed in the trinomial $ax^2 + bx + c$ that x is the indeterminate in some ring $R[x]$ and that a, b, and c are parameters taken from the ring R. Contrast this with the polynomial discussed in this Situation, $a^2 + 2ab + b^2$, in which a and b are treated more as generalized numbers than parameters. These and other subtleties of factorization and of polynomials add to the complexity of learning to identify particular categories of polynomials and to factor polynomials in that category.

Although it is true that every polynomial with real or complex coefficients does factor over the complex numbers, no proof of that is included in this Situation because existing proofs draw on advanced mathematics. For example, a nonelementary proof exists that is based on estimating the growth of the size of the norm of the polynomial as the variable becomes larger and larger in its distance from 0. Most short proofs of the result that every polynomial with real or complex coefficients factors over the complex numbers use what one might consider to be advanced complex analysis.

NOTES

1. In counting the number of terms in a polynomial, it is common practice to count them only after combining like terms and expressing the polynomial in simplest form.
2. Note that when $a_1 = a_2$ and $c_1 = c_2$, this expression is the square of a binomial.
3. The expression $p^2 + 2pq + q^2$ can be thought of as a quadratic trinomial in p or as a quadratic trinomial in q.
4. Some secondary school mathematics textbooks use the term *prime* for those polynomials that do not factor (over the integers or even the ra-

tionals), however, the term *prime polynomial* seems to occur only in secondary mathematics curricula. College-level mathematics textbooks usually do not refer to prime polynomials; for example, in calculus texts, factoring a quadratic polynomial is always recognized as possible, because complex number roots (which include irrational real roots) are considered part of the possibilities.

CHAPTER 30

TEMPERATURE CONVERSION

Situation 24 From the MACMTL–CPTM Situations Project

Glendon Blume, Heather Johnson, Maureen Grady,
Svetlana Konnova, and M. Kathleen Heid

PROMPT

During a high school Algebra 1 class, students were given the task of producing a formula that would convert Celsius temperatures to Fahrenheit temperatures, given that 0° on the Celsius scale is the temperature at which water freezes and 100° on the Celsius scale is the temperature at which water boils, and given that 32° on the Fahrenheit scale is the temperature at which water freezes and 212° on the Fahrenheit scale is the temperature at which water boils.

The rationale for the task was that if one encounters a relatively unfamiliar Celsius temperature, one could use this formula to convert to an equivalent Fahrenheit temperature that might be more familiar to persons in the United States.

One student developed a formula based on reasoning about the known values from the two temperature scales.

Mathematical Understanding for Secondary Teaching: A Framework and Classroom-Based Situations, pages 283–292.

"Since 0 and 100 are the two values I know on the Celsius scale and 32 and 212 are the ones I know on the Fahrenheit scale, I can plot the points (0, 100) and (32, 212). If I have two points, I can find the equation of the line passing through those two points.

(0, 100) means that the y-intercept is 100. The change in y is (212 − 100) and the change in x is (32 − 0), so the slope is $\frac{112}{32}$. Because $\frac{112}{32} = \frac{7 \cdot 16}{2 \cdot 16}$, if I cancel the 16s, the slope is $\frac{7}{2}$. So the formula is $y = \frac{7}{2}x + 100$."

COMMENTARY

The Situation addresses several key concepts in school mathematics that relate to temperature scales and conversion between scales. The first three Foci highlight the concepts of interval scale, composition of two functions, function, and domain and range of a function. The next three Foci use geometrical and graphical approaches to convert between scales, appealing to the concept of the inverse function and composition of three functions. Additional information about inverse functions appears in Situation 16 (see chapter 22).

MATHEMATICAL FOCI

Mathematical Focus 1

Celsius and Fahrenheit temperature scales are interval scales. Therefore, one can use linear transformations to move between them.

Choosing two reference temperatures on a given temperature scale and partitioning the difference between these two temperatures into intervals of equal length (each of which represents one degree) determine that temperature scale. The temperature scale is an interval scale (Pedhazur & Schmelkin, 1991), because the difference between two scale values can be meaningfully compared with the difference between two other scale values and the differences between numbers reflect the differences among properties in the real world. Because the units on the scale are constant, it is always true that, for example, the difference between 40° and 30° is the same as the difference between 30° and 20°. However, it is not, in general, meaningful to say that 40° is twice as hot as 20°. The reason is that a measure of 0 degrees on the temperature scales does not represent zero amount of heat (except on the Kelvin temperature scale).

The two reference temperatures used for most common scales are the melting point of ice and the boiling point of water. On the Celsius temperature scale, the melting point is taken as 0°C and the boiling point as 100°C, and the difference between them is partitioned into 100 units, each being one degree. On the Fahrenheit temperature scale, the melting point is taken as 32°F and the boiling point as 212°F, and the difference between them is partitioned into 180 units, each being one degree. The zero point on the Celsius or Fahrenheit temperature scales is an agreed-upon convention. For example, the zero point on the Celsius scale has been arbitrarily set at the freezing point of water.

In an interval scale, the relative difference among scale values is unaffected by any linear transformation of the form

$$y = ax + b,$$ (1)

where x is the original scale value, and y is the transformed scale value.

The value of the multiplier constant determines the arbitrary size of the scale unit, and the value of the additive constant b determines the arbitrary location of the zero point on the scale. In other words, the results after transformation retain the same meaning and refer to the same phenomena as the original scale value. For example, a linear transformation of the form (1) maps 0°C to 32°F, and both values refer to the same phenomenon, the freezing point of water.

Mathematical Focus 2

> *To convert from Fahrenheit to Celsius temperature scales, one can apply the concept of composition of functions.*

Consider Figure 30.1, which depicts Celsius, intermediate, and Fahrenheit scales. To make the conversion possible, two features need to match. There needs to be an equal number of degrees between the freezing and boiling points, and the freezing and boiling points have to match. (See CAS-IM Module II (Heid &

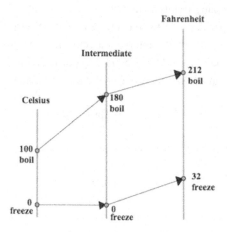

FIGURE 30.1. Celsius, intermediate, and Fahrenheit temperature scales. (Adapted from Heid & Zbiek, 2004.)

Zbiek, 2004) for an exploration of composition of functions via problems involving scale conversions).

The first step is to map values on the Celsius scale to corresponding values on the intermediate scale, in which the difference between freezing point and boiling point is the same as that difference on the Fahrenheit scale. On the Celsius scale the distance between the freezing and boiling points is 100, since $100 - 0 = 100$. On the Fahrenheit scale the distance between the freezing and boiling points is 180, because $212 - 32 = 180$. Therefore, for every 100 degrees on the Celsius scale there are 180 degrees on the intermediate scale. Because $\frac{180}{100} = \frac{9}{5}$, the scale factor should be $\frac{9}{5}$. That means that for every 9 degrees on the intermediate scale there are 5 degrees on the Celsius scale. Consequently, to make the number of degrees between the freezing and boiling points on the intermediate scale equal to the number of degrees between the freezing and boiling points on the Fahrenheit scale, one needs to multiply the Celsius-scale temperatures by $\frac{9}{5}$. This first step corresponds to the function g that takes the temperature in Celsius (w) as the input and multiplies it by the scale factor $\frac{9}{5}$.

$$g(w) = \frac{9}{5}w$$

The second step is to get the freezing points to match. Because the freezing point on the Celsius scale is $0°$ and the freezing point on the Fahrenheit scale is $32°$, to get both freezing points to match, one has to add $32°$ to each of the intermediate-scale temperatures. The freezing temperature moves up from $0°$ on the intermediate scale to $32°$ on the Fahrenheit scale, and the boiling point moves to $212°$ on the Fahrenheit scale. This second step corresponds to the function h, that takes $g(w)$, the output of the function g as the input and adds $32°$. So

$$h(r) = r + 32 \text{ for } r = g(w).$$

The two function rules can be written as follows:

$$g(w) = \frac{9}{5}w \text{ and } h(r) = r + 32.$$

As a result, the function f that converts Celsius temperatures to Fahrenheit temperatures is the composition of the function g followed by the function h:

$$f = h \circ g, \text{ where } g(w) = \frac{9}{5}w \text{ and } h(r) = r + 32.$$

Mathematical Focus 3

The relation that associates the Celsius and Fahrenheit temperature scales is a function.

The student in the Prompt was asked to write the formula that would convert Celsius temperatures to Fahrenheit temperatures. This means that the student had to determine the relation that associates with each element of X (the Celsius temperature value) exactly one element of Y (the Fahrenheit temperature value) and write the corresponding equation. Therefore, by definition, this relation is a function. The set X is called the *domain* of the function. The set Y of all images of the elements of the domain is called the *range* of the function.

According to the Prompt, two scales represent the same phenomena (water freezes and water boils). These facts are useful in determining the relation between the value of the temperature given in Celsius and the corresponding value of the temperature given in Fahrenheit.

As illustrated in Figure 30.2, the domain of the function consists of the Celsius temperature values and the range of the function consists of the Fahrenheit temperature values. Therefore, 0° and 100° belong to the domain of the function and 32° and 212° belong to the range of the function.

What the student has done when using the ordered pairs (0, 100) and (32, 212) is to use the freezing point on a given scale as the input and the boiling point on that scale as the output. For different scales, different function rules would need to be used to calculate the output from a given input, so these two ordered pairs require two different rules rather than a single rule such as $y = \frac{7}{2}x + 100$.

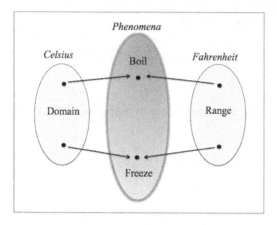

FIGURE 30.2.   For a function that associates the Celsius and Fahrenheit temperature scales, the domain of the function consists of the Celsius temperature values and the range of the function consists of the Fahrenheit temperature values.

TABLE 30.1. Freezing and Boiling Temperatures for the Celsius and Fahrenheit Scales

	Celsius temperature values	Fahrenheit temperature values
	x	y
Freezing point	0°	32°
Boiling point	100°	212°

The equation used for the conversion from the Celsius temperature scale to the Fahrenheit temperature scale is a linear function (see Focus 1). Therefore, two "points" with the coordinates (x_1, y_1) and (x_2, y_2) are needed to find the equation for the line that is the graph of that linear function (see Table 30.1). The first point corresponds to the phenomenon of freezing water. Therefore, the value of the independent variable, x_1, belongs to the Celsius temperature values, so $x_1 = 0°$, and the value of the dependent variable, y_1, belongs to the Fahrenheit temperature values, so $y_1 = 32°$. The second point corresponds to the phenomenon of boiling water. Therefore, the value of the independent variable x_2 is 100°, and the value of the dependent variable, y_2, is 212°.

One can write an equation in slope-intercept form:

$$y - y_1 = m(x - x_1) \tag{2}$$

for the line through the two points (0, 32) and (100, 212).

First, find the slope of the line:

$$m = \frac{y_2 - y_1}{x_2 - x_1} = \frac{212 - 32}{100 - 0} = \frac{180}{100} = \frac{9}{5}.$$

Second, substitute the slope and one of the points (0, 32), rewrite, and simplify the equation (2):

$$y - 32 = \frac{9}{5}(x - 0)$$

$$y = \frac{9}{5}x + 32. \tag{3}$$

Equation (3) is the equation that converts the Celsius temperature values to the Fahrenheit temperature values.

Mathematical Focus 4

A graphical-geometric approach can be used to create the formula to convert Celsius temperature to Fahrenheit temperature.

To find the formula to convert the temperature values in the Celsius temperature scale to temperature values in the Fahrenheit temperature scale, one can use coordinate geometry. In this case the corresponding temperature values in both scales are known:

f and c, the temperature values in the Fahrenheit and Celsius temperature scales, corresponding to one chosen phenomenon (a particular amount of average kinetic energy);

$f_1 = 32°$, the temperature value in the Fahrenheit temperature scale, when water first begins to freeze;

$f_2 = 212°$, the temperature value in the Fahrenheit temperature scale, when water first begins to boil;

$c_1 = 0°$, the temperature value in the Celsius temperature scale, when water first begins to freeze; and

$c_2 = 100°$, the temperature value in the Celsius temperature scale, when water first begins to boil.

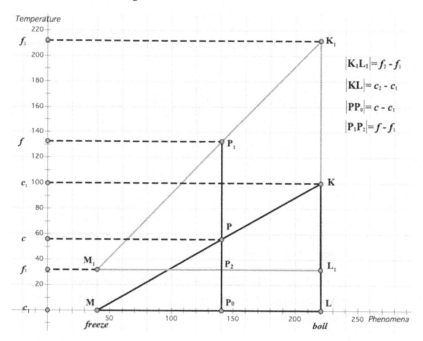

FIGURE 30.3. Corresponding Celsius and Fahrenheit temperatures for various phenomena.

The segment MK represents the Celsius temperature scale, and the segment M_1K_1 represents the Fahrenheit temperature scale (see Figure 30.3).

Choose any point P_0 between points M and L on the Phenomena-axis and draw the line parallel to the Temperature axis. This line intersects the segment MK at point P, and segment M_1K_1 at point P_1. Therefore, the temperature coordinate of the point P represents the Celsius temperature value and the temperature coordinate of the point P_1 represents the corresponding Fahrenheit temperature value.

Consider two pairs of triangles ΔLMK and ΔP_0MP, and $\Delta L_1M_1K_1$ and $\Delta P_2M_1P_1$. Next, apply the parallel side-splitter theorem to prove that

$$\frac{|PP_0|}{|KL|} = \frac{|MP_0|}{|ML|} \text{ and } \frac{|P_1P_2|}{|K_1L_1|} = \frac{|M_1P_2|}{|M_1L_1|} = \frac{|MP_0|}{|ML|}.$$

Therefore,

$$\frac{|P_1P_2|}{|K_1L_1|} = \frac{|PP_0|}{|KL|}$$

$$\frac{f - f_1}{f_2 - f_1} = \frac{c - c_1}{c_2 - c_1}$$

$$f = c\frac{f_2 - f_1}{c_2 - c_1} - c_1\frac{f_2 - f_1}{c_2 - c_1} + f_1 \ .$$

Using substitution of the values c_1, c_2, f_1, f_2, the equation can be rewritten in a simpler form:

$$f = c\frac{212 - 32}{100 - 0} - 0\frac{212 - 32}{100 - 0} + 32$$

$$f = \frac{9}{5}c + 32 \cdot$$

Mathematical Focus 5

By applying the rule for converting from the Celsius scale to the Fahrenheit scale and applying the concept of the inverse function, one can create the rule for converting Fahrenheit temperatures to Celsius temperatures.

Let F be a function[1] whose domain is the set X, and whose range is the set Y. Then, if it exists, the inverse of F is the function F^{-1} with domain Y and range X, defined by the following rule:

If $F(x) = y$, then $F^{-1}(y) = x$.

In the case of functions that map an interval scale to an interval scale, the mapping is a one-to-one correspondence. Therefore, for each nonconstant linear function, one can find the inverse function.

For example, let F be the function that converts a temperature in degrees Celsius (c) to a temperature in degrees Fahrenheit $F(c)$ (see Focus 3):

$$F(c) = \frac{9}{5}c + 32 = f$$

One approach to finding a formula for F^{-1} is to solve the equation $f = F(c)$ for c.

$$c = \frac{5}{9}(f - 32) \qquad (4)$$

Thus, the inverse function converts Fahrenheit degrees to Celsius degrees:

$$F^{-1}(f) = \frac{5}{9}(f - 32)$$

$$C(f) = \frac{5}{9}(f - 32).$$

Mathematical Focus 6

Conversion between scales can be done by applying compositions of functions.

The Kelvin scale is a third standard temperature scale. It is the temperature scale that can be defined theoretically, for which zero degrees (0°) corresponds to zero average kinetic energy. Scientists have determined that the coldest any object of substance can get, in theory, is minus 273° degrees Celsius. The Kelvin temperature scale is an absolute scale having degrees with interval size the same size as those of the Celsius temperature scale.

To write the function rule to convert Celsius temperatures to Kelvin temperatures, the scale factor has to be determined and the freezing points have to match (see Focus 1).

The scale factor of the transformation between the Celsius temperature scale and the Kelvin temperature scale should be equal to 1 (as long as one-degree intervals are the same length on the scales). On the Celsius scale, water freezes at 0°, and on the Kelvin scale, water freezes at 273°. Therefore, the function k takes the temperature in degrees Celsius, c, to a temperature in degrees Kelvin, $k(c)$, with the rule $k(c) = c + 273$.

To find the composition of the functions that will convert from the Fahrenheit scale to the Kelvin scale, one can use the function rule that converts Fahrenheit to Celsius that was established in Mathematical Focus 5

$$C(f) = \frac{5}{9}(f - 32)$$

and compose it with $k(c) = c + 273$. That composition,

$$k(C(f)) = \frac{5}{9}(f - 32) + 273,$$

will convert Fahrenheit to Kelvin.

NOTE

1.   In Focus 5 and Focus 6, F and C are used for the names of the functions, and f and c are used as names of the temperature variables.

REFERENCES

Heid, M. K., & Zbiek, R. M. (2004). *Technology-intensive secondary school mathematics curriculum. Module II: Functions as compositions and inverses.* University Park, PA: The Pennsylvania State University.

Pedhazur, E. J., & Schmelkin, L. P. (1991). *Measurement, design, and analysis: An integrated approach.* Hillsdale, NJ: Lawrence Erlbaum Associates.

CHAPTER 31

TRANSLATION OF FUNCTIONS

Situation 25 From the MACMTL-CPTM Situations Project

Bob Allen, Brian Gleason, and Shawn Broderick

PROMPT

During a unit on functions, the translation of functions was discussed. When the class encountered the function $y = (x - 2)^2 + 3$, one student noted that the vertical translation of 3 upward makes sense when compared to the +3, but the horizontal translation of 2 to the right does not make sense when compared to the -2 in the function rule.

COMMENTARY

Focus 1 and Focus 2 use a quadratic function as the original function for convenience and because it was the function type specified in the Prompt. One may translate any function similarly. Throughout this situation, the term *parent function* is used to refer to functions such as $f(x) = x^2$, $g(x) = x^3$, $h(x) = \cos(x)$, and the term *child function* is used to refer to a transformation, in particular a translation, of a parent

Mathematical Understanding for Secondary Teaching: A Framework
and Classroom-Based Situations, pages 293–297.

function (such as $f_1(x) = x^2 + 3$, $g_1(x) = (x - 1)^3$, and $h_1(x) = \cos\left(x + \dfrac{\pi}{2}\right) - 4$). In general, the Foci address the relationship between $f(x)$ and $f(x - h) + k$.

MATHEMATICAL FOCI

Mathematical Focus 1

Given a function f(x), *graphical representations allow easy comparison of the translations effected by various values of* h *and* k *in the transformed function* f(x − h) + k.

A graphing utility can be used to examine the graphs of parent functions and child functions simultaneously (see Figure 31.1). Simultaneous display of the graph of $f(x - h) + k$ for different values of the parameters h and k illustrates the relationship between the graph of the child function and the graph of the parent function for particular parameter values.

Mathematical Focus 2

A translation to the right of a function f *occurs as a result of the composition* (f ∘ g) *where* g(x) = x − h *for positive* h.

To think about the effect of translating the independent variable, one can ask what makes the translated variable have a value of 0. With no translation of the

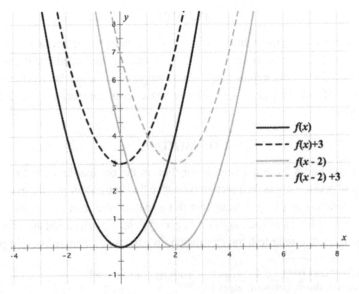

FIGURE 31.1. Graphs of a parent function and related child functions.

TABLE 31.1. Function Values[1] for a Parent Function and a Related Child Function

		Parent function	Child function
x	$x - 2$	$f(x)$	$f(x - 2)$
-5	-7	25	49
-4	-6	16	36
-3	-5	9	25
-2	-4	4	16
-1	-3	1	9
0	-2	0	4
1	-1	1	1
2	0	4	0
3	1	9	1
4	2	16	4
5	3	25	9

"parent" function, $x = 0$ is the y-axis. For the translated variable $(x - h)$, what makes this quantity have value 0 is $x = h$, so that is the "location" of the "y-axis" for the translated function. So, translating the original graph as if the y-axis were now situated at $x = h$ yields the graph of the translated-variable "child" function.

For example, for $f(x) = x^2$, a numerical representation demonstrates that the graph of $(x - 2)^2$ is 2 units to the right of the graph of x^2 as a result of each value of $x - 2$ being 2 less than the corresponding value of x. This difference of 2 means that, in the table, $x - 2$ will produce the same output values as x, just 2 units "earlier." Thus, $f(x - 2)$ will have the same output values as $f(x)$, but 2 units "later" (see Table 31.1). This "2 units later" is along the x-axis and is what yields a horizontal shift to the right by 2 units.

Mathematical Focus 3 [2]

The translation of a function (and its graph) can be thought of as the translation of the underlying coordinate axes.

Let $y = f(x)$ be a function graphed in relation to the axes, X and Y, and let $P = (x_0, y_0)$ be a point on that graph. Suppose there exists another set of axes X' and Y' (parallel to X and Y, respectively) for which the origin has coordinates (h, k) in relation to X and Y. Then the coordinates of P in relation to the axes X' and Y' are $(x_0 - h, y_0 - k)$, as illustrated in Figure 31.2. Thus, the graph of the function f is of the form $y - k = f(x - h)$ in relation to the axes X' and Y', which can be written as $y = f(x - h) + k$.

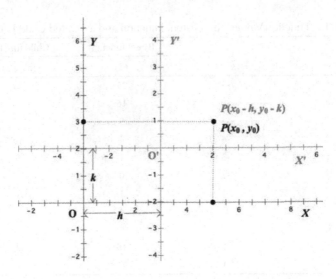

FIGURE 31.2. The coordinates of P in relation to the axes X' and Y'.

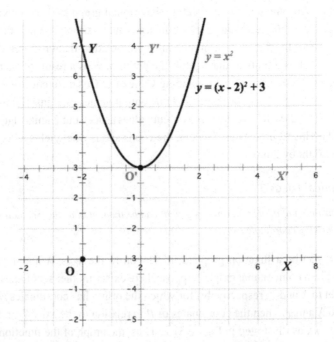

FIGURE 31.3. The graph of $y = (x - 2)^2 + 3$ in relation to a new set of axes with origin at (2, 3).

For the parent function $y = x^2$, the child function $y = (x - 2)^2 + 3$ could be thought of as the formula for $y = x^2$ in relation to a new set of axes (each parallel to its counterpart in the original set) with origin at $(2, 3)$ as illustrated in Figure 31.3.

NOTES

1. Note that Table 31.1 contains the sets of ordered pairs $\{(x, f(x))\}$ and $\{((x - 2), f(x - 2))\}$, whereas the two graphs in Figure 31.1 contain the sets of ordered pairs $\{(x, f(x))\}$ and $\{(x, f(x - 2))\}$.
2. Mathematical Focus 3 draws on Smail (1953).

REFERENCE

Smail, L. L. (1953). *Analytic geometry and calculus*. New York, NY: Appleton-Century-Crofts.

CHAPTER 32

PARAMETRIC DRAWINGS

Situation 26 From the MACMTL–CPTM Situations Project

Rose Mary Zbiek, Eileen Murray, Heather Johnson,
Maureen Grady, Svetlana Konnova, and M. Kathleen Heid

PROMPT[1]

This example, appearing in CAS-Intensive Mathematics (Heid & Zbiek, 2004), was inspired by a student using a dynamic mathematics tool and mistakenly grabbing points representing both parameters (A and B in $f(x) = Ax + B$) and dragging them simultaneously (the difference in value between A and B remains constant). This generated a family of functions that coincided in one point. Interestingly, no matter how far apart A and B were initially, if grabbed and moved together, the graphs of the functions in the family always coincided on the line $x = -1$.

COMMENTARY

In this case, dynamical geometry software was the vehicle that brought mathematical relationships to the fore. When one encounters such a phenomenon, one can enhance the experience by noticing the potential for mathematics in the pat-

Mathematical Understanding for Secondary Teaching: A Framework
and Classroom-Based Situations, pages 299–306.
Copyright © 2015 by Information Age Publishing
All rights of reproduction in any form reserved.

terns that are seen. Focus 1 uses transformations to explain the graphical phenomenon, whereas Focus 2 uses a symbolic proof. Focus 3 extends the phenomenon to quadratic functions (this discussion also appears in CAS-Intensive Mathematics). In addition, Focus 3 considers polynomials of higher degree (which generated another interesting relationship along with its proof). Focus 3 illustrates the decisions needed in designing an extension of a mathematical generalization.

MATHEMATICAL FOCI

Mathematical Focus 1

Graphical phenomena can be explained in terms of transformations.

Let $y = Ax + B$ represent a family of functions for which $A - B = k$, where k is a constant. Then it is also true that $B = A - k$. As the value of k changes, what results is a family of functions each of whose members is generated by a different value of k. Using substitution, $y = Ax + B$ can be rewritten as $y = Ax + (A - k)$, and equivalently, $y = A(x + 1) - k$.

The graphs of the family of lines given by $y = A(x + 1) - k$ result from translating and rotating the graph of $y = x$.

This can be shown by considering the family of lines given by $y = (x + 1) - K$. When the value of K, constant for each family of lines, is changed, it shifts members of the family of lines given by $y = (x + 1) - K$ up or down vertically along the line $x = -1$.

Consider the family of lines given by $y = A(x + 1) - K$, where the value of A varies and K is a constant. Because A represents the slope of any line in the family, when A is changed, it causes the rotation of the graph $y = A(x + 1) - K$ about the point $(-1, -K)$. Thus, the lines of the family $y = A(x + 1) - K$ will intersect in the point $(-1, -K)$.

Figure 32.1 displays a screen capture after the values of A and B have been simultaneously dragged.

Mathematical Focus 2

Graphical phenomena can frequently be explained using symbolic proof.

Figure 32.2 displays a screen capture after A and B have been simultaneously dragged. Notice that the family of lines that appear intersect on the line $x = -1$. To explain why this will be the case for any value of A and B for which their difference remains constant, one can use the following symbolic proof:

Let $y = Ax + B$ and suppose $A - B = k$, a constant. Let $y_1 = A_1x + B_1$ and $y_2 = A_2x + B_2$ be two distinct lines in this family. $A_1 - B_1 = k = A_2 - B_2 \Rightarrow$

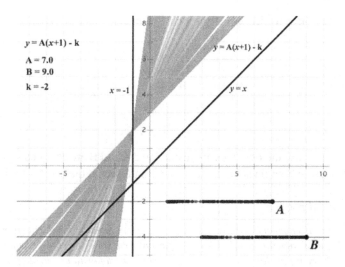

FIGURE 32.1. Screen capture showing trace of $f(x) = Ax + B$ after A and B have been dragged simultaneously.

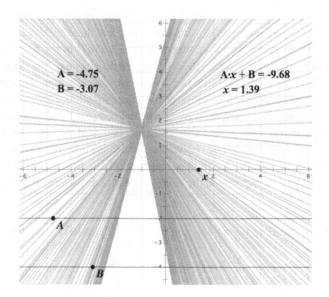

FIGURE 32.2. Screen capture showing trace of $f(x) = Ax + B$ after A and B have been dragged simultaneously.

$A_1 - A_2 = B_1 - B_2$. To determine the point of intersection, y_1 is set equal to y_2. Therefore,

$$A_1 x + B_1 = A_2 x + B_2$$

$$(A_1 - A_2)x = B_2 - B_1$$

$$x = \frac{B_2 - B_1}{A_1 - A_2} = \frac{-(B_1 - B_2)}{A_1 - A_2} = -1 .$$

$A_1 - A_2 \neq 0$ because the two lines are distinct and A and B are changed simultaneously. Hence, any two lines in this family will intersect at the point (-1, –k).

Mathematical Focus 3

Many times phenomena that are observed for certain function families can be extended to other function families with similar properties. For example, phenomena related to linear functions can be extended to polynomial functions with degree > 1.

If one replaces x, the independent variable in the original function defined by $Ax + B$, with any function f, one can consider the family of functions $A \cdot f + B$. If $A - B = k$ is fixed, the corresponding graphs become $y = A \cdot [f(x) + 1] - k$. In this case, every value x_0 that satisfies $f(x_0) = -1$ will yield the same point $(x_0, -k)$ no matter what A and B were (depending only on $A - B = k$). Of course, there can be many such values for x_0.

When considering quadratic functions, additional assumptions must be made to investigate the phenomenon. In a quadratic function, there are three coefficients (A, B, and C in $y = Ax^2 + Bx + C$) rather than two (A and B in $y = Ax + B$). So, for the quadratic function, $y = Ax^2 + Bx + C$, one could consider three possibilities (see Figure 32.3 for graphs of those possibilities) in investigating an extension of the phenomenon observed in the case of the linear function: Case 1: The difference between A and B is constant. C would be held constant with $A - B = k$; Case 2: The difference between A and C is constant. B would be held constant with $A - C = k$; Case 3: The difference between B and C is constant. A would be held constant with $B - C = k$. In each of these cases, symbolic proofs similar to the one in Focus 1 can be developed, and the following conclusions can be drawn.

Case 1:

Consider A_i and B_i as pairs of A- and B-values. If C is held constant,

$$A - B = k \Rightarrow B = A - k$$

and

$$A_1 - B_1 = k = A_2 - B_2 \Rightarrow A_1 - A_2 = B_1 - B_2 .$$

To determine the points of intersection, set $y_1 = A_1 x^2 + B_1 x + C_1$ equal to $y_2 = A_2 x^2 + B_2 x + C_1$.

$$A_1 x^2 + B_1 x + C_1 = A_2 x^2 + B_2 x + C_1$$

$$\Rightarrow (A_1 - A_2) x^2 + (B_1 - B_2) x + C_1 - C_1 = 0$$

$$\Rightarrow (A_1 - A_2) x^2 + (A_1 - A_2) x = 0$$

$$\Rightarrow (A_1 - A_2)(x^2 + x) = 0$$

$$\Rightarrow x^2 + x = 0 \quad \text{because } A_1 - A_2 \neq 0$$

$$\Rightarrow x = 0 \text{ or } x = -1$$

So, if C is held constant and A – B = k then there will be two points of intersection, one at (0, C) and one at (-1, k + C).

Case 2:

Again, consider A_i and B_i as pairs of A- and B-values. If B is held constant,

$$A - C = k \Rightarrow C = A - k$$

and

$$A_1 - C_1 = k = A_2 - C_2 \Rightarrow A_1 - A_2 = C_1 - C_2.$$

To determine the points of intersection, set y_1 equal to y_2.

$$A_1 x^2 + B_1 x + C_1 = A_2 x^2 + B_1 x + C_2$$

$$\Rightarrow (A_1 - A_2) x^2 + (B_1 - B_1) x + C_1 - C_2 = 0$$

$$\Rightarrow (A_1 - A_2) x^2 + (A_1 - A_2) = 0$$

$$\Rightarrow (A_1 - A_2)(x^2 + 1) = 0$$

$$\Rightarrow x^2 + 1 = 0 \quad \text{because } A_1 - A_2 \neq 0$$

$$\Rightarrow x^2 = -1$$

So, if B is held constant and A – C = k then there will be no intersection because the initial equation is equivalent to $x^2 = -1$, for which there is no real number solution.

Case 3:

Again, consider A_i and B_i as pairs of A- and B-values. If A is held constant,

$$B - C = k \Rightarrow C = B - k \text{ and } B_1 - C_1 = k = B_2 - C_2 \Rightarrow B_1 - B_2 = C_1 - C_2.$$

To determine the points of intersection, set y_1 equal to y_2.

$$A_1 x^2 + B_1 x + C_1 = A_1 x^2 + B_2 x + C_2$$

$$\Rightarrow (A_1 - A_1)x^2 + (B_1 - B_2)x + C_1 - C_2 = 0$$

$$\Rightarrow (B_1 - B_2)x + (C_1 - C_2) = 0$$

$$\Rightarrow (B_1 - B_2)x + (B_1 - B_2) = 0$$

$$\Rightarrow (B_1 - B_2)(x + 1) = 0$$

$$\Rightarrow x + 1 = 0 \quad \text{because } B_1 \neq B_2$$

$$\Rightarrow x = -1$$

So, if A is held constant and $A - B = k$ then there will be one intersection at $(-1, A - k)$.

This idea can be extended for n^{th}-degree polynomials. To do this extension, all but two coefficients are held constant and the remaining two have a constant difference. Consider the polynomial $y = A_n x^n + A_{n-1} x^{n-1} + A_{n-2} x^{n-2} + ... + A_2 x^2 + A_1 x + A_0$. Choose two coefficients to vary, but keep their difference constant. All other coefficients will be held constant. Suppose $A_j - A_i = k$, $i < j$, for $i, j \in \{0, \mathbb{N}\}$. For two polyno-

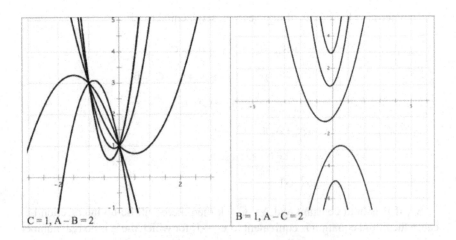

$C = 1, A - B = 2$

$B = 1, A - C = 2$

FIGURE 32.3. Three cases for graphical representations of the family of functions, $Ax^2 + Bx + C$: Case 1 (in which C is held constant), Case 2 (in which B is held constant), and Case 3 (in which A is held constant). (continues)

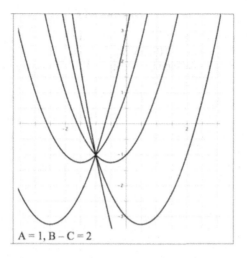

$A = 1, B - C = 2$

FIGURE 32.3.   Continued

mials in this family, one can determine where they intersect by setting them equal to each other. Using a symbolic proof, one can see that the intersections will occur at the following points:

$$x^i = 0, y = A_0$$

$$x^{j-i} = -1, \quad y = A_n(-1)^n + A_{n-1}(-1)^{n-1} + \ldots + A_0$$

However, depending on the values of i and j, these points may or may not be defined. Specifically, because $x^{j-i} = -1$, x^{j-i} for $i, j \in \{0, \mathbb{N}\}$ will be defined as a real number only when $j - i$ is an odd number.

NOTE

1.   This example was developed and used in a research study conducted by Rose Zbiek and colleagues at the University of Iowa during the CAS-Intensive Mathematics project (NSF award number ESI 96-18029). It was further developed as an e-example for the NCTM *Principles and Standards for School Mathematics* and can be accessed through http://www.nctm.org/standards/content.aspx?id=26790.

REFERENCE

Heid, M. K., & Zbiek, R. M. (2004). CAS-Intensive Mathematics Curriculum (Field Test Version, NSF Grant No. TPE 96-18029) [CD]. University Park, PA and Iowa City, IA: The Pennsylvania State University and The University of Iowa.

CHAPTER 33

LOCUS OF A POINT ON A MOVING SEGMENT

Situation 27 From the MACMTL-CPTM Situations Project

Rose Mary Zbiek, James Wilson, Heather Johnson, M. Kathleen Heid, Maureen Grady, and Svetlana Konnova

PROMPT

A high school geometry class was in the middle of a series of lessons on loci. The teacher chose to discuss one of the homework problems from the previous day's assignment.

A student read the problem from the textbook (Brown, Jurgensen, & Jurgensen, 2000, p. 405):

A ladder leans against a house. As A moves up or down on the wall, B moves along the ground. What path is followed by midpoint M? (Hint: Experiment with a meter stick, a wall, and the floor.)

The teacher and two students conducted the experiment in front of the class, starting with a vertical "wall" and a horizontal "floor" and then marking several

Mathematical Understanding for Secondary Teaching: A Framework and Classroom-Based Situations, pages 307–315.
Copyright © 2015 by Information Age Publishing
All rights of reproduction in any form reserved.

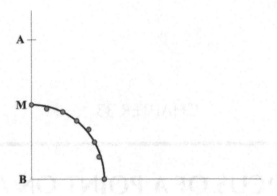

FIGURE 33.1. Data picture produced by teacher and students.

locations of M as the students moved the meter stick. The teacher connected the points. Their work produced the data picture on the board that appears in Figure 33.1.

A student commented, "That's a heck of an arc."

COMMENTARY

The Foci for this Situation utilize a variety of models and representations of the locus. In the class demonstration, the data were generated using a physical model. In two Foci the data have been generated using dynamical geometry software. Other Foci explore the Situation from the standpoints of axiomatic geometry and coordinate geometry.

MATHEMATICAL FOCI

Mathematical Focus 1

Dynamical geometry environments can be used to fit a curve to a set of data points.

Samples of data points can be created using dynamical geometry software. To determine the path traveled by the midpoint of the ladder as it travels from the vertical to the horizontal position, first consider a point above the midpoint of the ladder. Using the marks for a point above the midpoint of the ladder when the ladder is vertical, horizontal and in three other positions, an elliptical arc can be created that appears to pass through all of the other points (see Figure 33.2, in which the midpoint is considered to be the point at which a bucket is hanging from the ladder). This suggests that the path of a point above the midpoint may be an

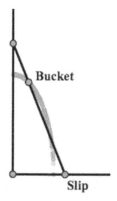

FIGURE 33.2. A trace of points suggesting the path of a bucket hanging at a point above the midpoint of a ladder leaning against a wall as the ladder slips away from the wall along the floor.

elliptical arc. Because an elliptical arc through five or more points is unique, there could be only one such elliptical arc, and it would have been identified.

Now consider a point below the midpoint of the ladder. Using the marks for a point below the midpoint of the ladder when the ladder is vertical, horizontal and three other locations, an elliptical arc can be created that appears to pass through all of the other points (see Figure 33.3). Since an elliptical arc through five or more points is unique, this suggests the path of a point below the midpoint is an elliptical arc.

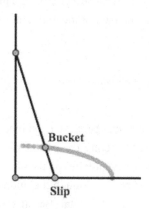

FIGURE 33.3. A trace of points suggesting the path of a bucket hanging at a point below the midpoint of a ladder leaning against a wall as the ladder slips away from the wall along the floor.

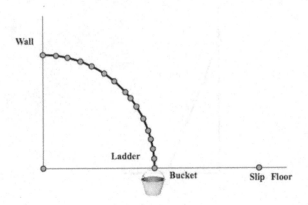

FIGURE 33.4. A trace of points suggesting the path of a bucket hanging at the midpoint of a ladder leaning against a wall as the foot of the ladder slips away from the wall along the floor.

Now consider the paths created by points above and below the midpoint of the ladder as those points move closer to the midpoint of the ladder. The elliptical arcs would converge to a circular arc as the points moved closer to the midpoint of the ladder. This suggests that the path created by the midpoint of the ladder is a circular arc. Moreover, using the marks for the midpoint of the ladder when the ladder is vertical, when the ladder is horizontal, and when the ladder is in one other location, a circular arc can be created that appears to pass through all of the other points (see Figure 33.4). Because a circular arc through three or more points is unique, this suggests that the path of the midpoint is a circular arc.

Mathematical Focus 2

If all the points on a continuous path lie on a circle, then the path created is an arc of the circle.

A dynamical geometry tool can be used to simulate the falling ladder and create a sample of data points depicting the path of the midpoint of the ladder. A circle can be constructed whose center lies on the intersection of the floor and wall (point B in Figure 33.5) and passes through the midpoint of the ladder when the ladder is vertical and when it is horizontal. Because all the data points appear to lie on the circle and every data point from the midpoint of the ladder in its vertical position to the midpoint of the ladder in its horizontal position would form a continuous curve, the path appears to be a circular arc.

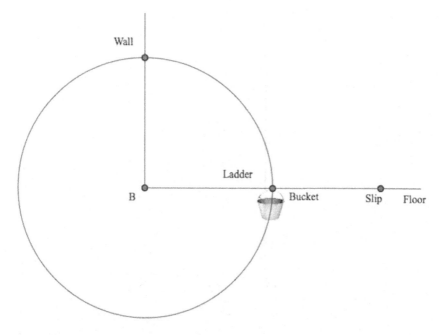

FIGURE 33.5.   Circle with center at the intersection of the floor and wall.

Mathematical Focus 3

When quantities in a situation can be graphed on a coordinate system, coordinate geometry can be used to write an equation that relates the x and y coordinates.

The "Slip" (Point S) is at distance $2a$ from O, the intersection of the floor and the wall. P is at distance $2b$ from O. The length of the ladder is $2c$ (see Figure 33.6). Then the coordinates of S and P are $(2a, 0)$ and $(0, 2b)$, respectively. The coordinates of M are (a, b).

Construct a line perpendicular to $\overline{OS}$ from M. The point where that line intersects $\overline{OS}$ is Q. $\triangle OPS$ is similar to $\triangle QMS$. So, $QM^2 + QS^2 = SM^2$ or $a^2 + b^2 = c^2$. The coordinates of point M satisfy the equation of a circle with radius c and center $(0, 0)$. That is, the coordinates of M are of the form (m_1, m_2) where $m_1^2 + m_2^2 = c^2$. Because these coordinates could be any nonnegative real numbers, the set of all possible points M would lie on an arc of the circle with center $(0, 0)$ and radius c that lies in the first quadrant. In addition, using the diagram in Figure 33.6, it can be proved that any such point M (the midpoint of the hypotenuse of a right triangle) is equidistant from P, S, and O (the vertices of the right triangle). $\triangle QOM$ is congruent to $\triangle QMS$ by SAS ($OQ = QS$; $m\angle MQO = m\angle MQS$; $MQ = MQ$). Because corresponding parts of congruent triangles are congruent, $MO = MS$. Because M is the midpoint of PS, $MP = MS$. So M is equidistant from P, S, and O.

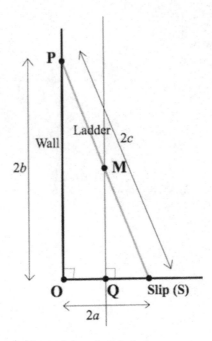

FIGURE 33.6. Falling ladder with lengths labeled.

Mathematical Focus 4

The theorems and definitions of classical geometry can be used to demonstrate that the locus of points for the midpoint in the Prompt is a circle.

The physical situation of a ladder against the wall assumes that the ladder forms the hypotenuse of a right triangle and the legs of the right triangle lie along the wall and the ground at right angles (see Figure 33.7).

The following general result about the midpoint of the hypotenuse of a right triangle that is addressed early in most Geometry courses was proved in Focus 3:

Theorem: The midpoint of the hypotenuse of a right triangle is equidistant from the three vertices of the triangle.

This theorem would lead quickly to the picture in Figure 33.8.

That is, as the ladder moves, the legs of the right triangle change but the hypotenuse (the ladder) remains the same length and the distance of its midpoint from the center of the circle will stay the same. The locus of the midpoint then lies along the arc of a circle with radius determined by the distance of the midpoint of the hypotenuse from the right-angle vertex.

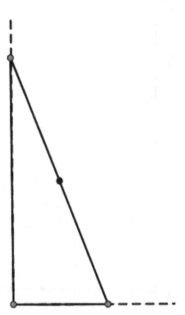

FIGURE 33.7. Ladder as the hypotenuse of a right triangle.

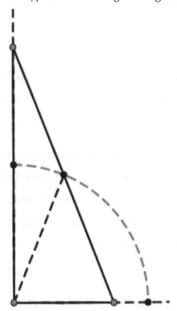

FIGURE 33.8. Diagram illustrating that the midpoint of the hypotenuse of a right triangle is equidistant from the three vertices.

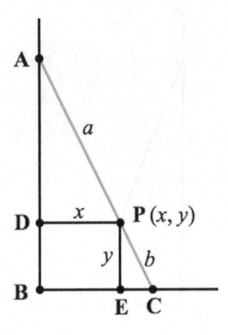

FIGURE 33.9. Coordinatized representation of the falling ladder.

Mathematical Focus 5

Modifying the problem in the Prompt slightly to allow the fixed point on the ladder to be someplace other than the midpoint produces a locus of points that is an elliptical arc.

A rectangular coordinate system can be superimposed on the figure with the origin at the intersection of the wall and the ground. The coordinates of the fixed point can be designated as (x, y), the distance from the top of the ladder to this point can be designated as a, and the distance from the bottom of the ladder to this point can be designated as b (see Figure 33.9).

As illustrated in Figure 33.9, because $\overline{DP}$ is parallel to $\overline{BC}$, and $\overline{PE}$ is parallel to $\overline{AB}$, $\triangle ADP$ is similar to $\triangle PEC$ and $\dfrac{AP}{DP} = \dfrac{PC}{EC}$. By the Pythagorean theorem, $EC = \sqrt{b^2 - y^2}$, so $\dfrac{a}{x} = \dfrac{b}{\sqrt{b^2 - y^2}}$ which yields

$$a\sqrt{b^2 - y^2} = xb$$

$$\Rightarrow a^2(b^2 - y^2) = x^2b^2$$

$$\Rightarrow a^2y^2 + x^2b^2 = a^2b^2$$

$$\Rightarrow \frac{x^2}{a^2} + \frac{y^2}{b^2} = 1.$$

This is the equation of an ellipse with center at the origin and axes of length $2a$ and $2b$. One can then examine the special case of the fixed point as the midpoint of the ladder, which makes $a = b$. By substitution it is evident that the equation is the equation of a circle with center at the origin and radius a.

POSTCOMMENTARY

This problem, the locus of points traced out by a "bucket" on a slipping ladder as the ladder falls, encompasses rich mathematics and is approachable without advanced mathematics when approached with a dynamical geometry tool. Although the results generated by computer software may suggest the correct answer, it is important to note that computer graphics never constitute a mathematical proof of a theorem, even though they may suggest what the correct result is. The mathematics necessary to establish a proof does not rely on computer-generated pictures, which can incorrectly reflect established mathematical results (e.g., on a computer screen one can use elementary operations that appear to trisect any angle one can construct on the computer screen). This issue is addressed in this Situation in that Focus 4 and Focus 3 refer to a plane geometry proof and a coordinate geometry proof.

REFERENCE

Brown, R. G., Jurgensen, J. W., & Jurgensen, R. C. (2000). *Geometry*. Boston, MA: McDougal Littell.

CHAPTER 34

CONSTRUCTING A TANGENT LINE

Situation 28 From the MACMTL–CPTM Situations Project

Pawel Nazarewicz, Sharon K. O'Kelley, Erik Jacobson,
Glendon Blume, and M. Kathleen Heid

PROMPT

A student in a Geometry class was given the following steps describing how to construct a tangent to a circle O from a point A exterior to the circle (see Figure 34.1).

1. First, construct the segment OA and find its midpoint M.
2. Next, using M as the center of a new circle, construct a circle of radius MA.
3. Label one point of intersection of the two circles as B. Construct the line AB that is also a tangent line to circle O.

Mathematical Understanding for Secondary Teaching: A Framework and Classroom-Based Situations, pages 317–322.

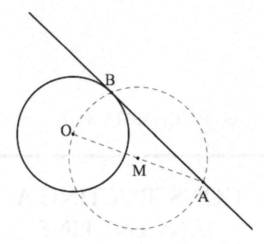

FIGURE 34.1. Construction of a tangent to a circle from a point exterior to the circle.

After seeing this, the student asked how one knows that this is in fact the tangent line, or how one knows that ∠OBA is a right angle.

COMMENTARY

The student's questions are asking why the chosen method of construction is valid. In addressing these questions, it is useful to review the definition of tangent line as well as why a line tangent to a circle is perpendicular to the radius at the point of tangency. The inscribed angle theorem and properties of isosceles triangles can be used to prove that the line of tangency and its corresponding radius form a right angle.

MATHEMATICAL FOCI

Mathematical Focus 1

A line tangent to a circle is perpendicular to the radius at the point of tangency.

This can be proved using an indirect proof supported by the definition of tangent and several previously proven theorems. The definition of *tangent* states that a line is tangent to a circle if it intersects the circle in exactly one point.

Assume that the following two theorems have been proved previously.

1. There exists only one perpendicular to a line from a point not on that line.

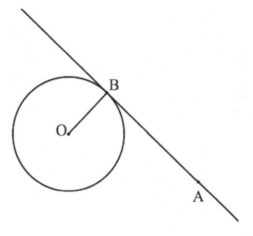

FIGURE 34.2.   Circle O with line AB tangent to O at B.

2.   Given a line and a point not on that line, the perpendicular from the point to the line is the shortest distance from the point to the line.

Let B be the point of intersection of a circle O and a tangent line to circle O, and let point A be another point on the tangent line (see Figure 34.2). For the indirect proof, assume that $\overline{OB}$ is not perpendicular to line AB. By the preceding theorem 1, there is a point X on line AB such that $\overline{OX} \perp \overleftrightarrow{AB}$. By the preceding theorem 2, OX < OB. Because tangent line AB intersects circle O in only one point (the point of tangency), point X is not on circle O. Also, because tangent line AB intersects circle O in only one point, point X is not in the interior of circle O (if it were, line AB would intersect the circle in two points). Therefore, point X is in the exterior of circle O. Because point X is in the exterior of circle O and point B is on circle O, OX > OB, which contradicts the previous conclusion that OX < OB.

Therefore, the assumption that segment OB is not perpendicular to line AB is false. So, $\overline{OB} \perp \overleftrightarrow{AB}$, proving that a tangent line is perpendicular to the radius at its point of tangency.

Mathematical Focus 2

If an inscribed angle in a circle intersects the circle at endpoints of a diameter of that circle, that angle will be a right angle. Conversely, given a right angle, any two points O and A, each on one of the two different rays of the angle, are the endpoints of a diameter of a circle that passes through the vertex of the right angle.

The inscribed angle theorem states that the measure of an angle inscribed in a circle is one half the measure of the arc it intercepts. Consequently, if the intercepted arc is a semicircle, the measure of the inscribed angle is 90 degrees.

Suppose angle ABC is a right angle and segment AC is the diameter of a circle centered at M, the midpoint of segment AC. Drop a perpendicular to segment AC from B. Label the intersection of segment AC and the perpendicular as K (see Figure 34.3). Triangle AKB is similar to triangle BKC. So

$$\frac{AK}{BK} = \frac{BK}{KC}$$

$$\Rightarrow \frac{AM - MK}{BK} = \frac{BK}{CM + MK}$$

$$\Rightarrow \frac{AM - MK}{BK} = \frac{BK}{AM + MK}$$

$$\Rightarrow BK^2 = AM^2 - MK^2$$

$$\Rightarrow BK^2 + MK^2 = AM^2.$$

But $BK^2 + MK^2 = BM^2$ since triangle BKM is a right triangle.

So $AM^2 = BM^2$

$$\Rightarrow AM = BM.$$

So A, B, and C all lie on the circle with center M and diameter AC. Thus, A and C are the endpoints of a diameter of a circle that passes through the vertex of the right angle ABC.

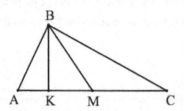

FIGURE 34.3. Right triangle ABC has right angle at B, $\overline{BK} \perp \overline{AC}$, and M is the midpoint of $\overline{AC}$.

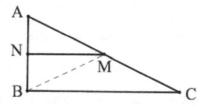

FIGURE 34.4. Right triangle ABC has right angle at B, M is the midpoint of segment AC, and $\overline{MN} \parallel \overline{BC}$.

An alternate proof uses the fact that parallel lines cut transversals proportionally. Let triangle ABC be a right triangle with right angle at B, M is the midpoint of segment AC, and construct segment MN so that $\overline{MN} \parallel \overline{BC}$. Then $\triangle ANM \cong \triangle BNM$ (see Figure 34.4). So $\angle NAM \cong \angle NBM$, and it follows that AM = BM. So A, B, and C all lie on the circle with center M and diameter AC. Thus, A and C are the endpoints of a diameter of a circle that passes through the vertex of the right angle ABC.

Mathematical Focus 3

The radii of the circle constructed following the steps in the Prompt for constructing a tangent to a circle from an exterior point form two isosceles triangles.

This Focus gives another proof that at the point of tangency of the tangent line there is a right angle formed with the radius to that point. Because segments MB,

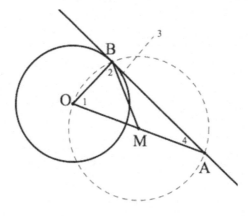

FIGURE 34.5. Triangles formed by radii of circle M in the construction of a tangent to a circle from an exterior point.

MO, and MA are radii of circle M, they are congruent. Thus, triangles OBM and MBA are isosceles triangles. Because these triangles are isosceles, the base angles in each triangle are congruent. Using the notation in Figure 34.5,

$$\angle 1 \cong \angle 2 \text{ and } \angle 3 \cong \angle 4.$$

In addition, triangles OBM and MBA comprise the larger triangle OBA. Because the sum of the interior angles of a triangle is 180 degrees,

$$m\angle 1 + m\angle 2 + m\angle 3 + m\angle 4 = 180.$$

Using substitution,

$$2\,(m\angle 2) + 2\,(m\angle 3) = 180.$$

Therefore,

$$m\angle 2 + m\angle 3 = 90.$$

Using the angle addition postulate, $\angle OBM$ (angle 2) and $\angle MBA$ (angle 3) comprise $\angle OBA$; hence, $\angle OBA$ is a right angle because it measures 90 degrees.

CHAPTER 35

FACES OF A POLYHEDRAL SOLID

Situation 29 From the MACMTL–CPTM Situations Project

Stephen Bismarck, Glendon Blume, Heather Johnson,
Svetlana Konnova, and Jeanne Shimizu

PROMPT

In a 7th-grade mathematics class, in which the lesson was on classification of solids, the teacher held up a rectangular prism and asked the class how many "sides" there were. Two students responded, one with an answer of 12, the other with an answer of 6. The student with the answer of 6 was acknowledged as being correct, and the teacher continued with the lesson.

COMMENTARY

Mathematics dictionary definitions of *solid* and *polyhedron* are not particularly clear. For example, a polyhedron is "a three-dimensional solid which consists of a collection of polygons, usually joined at their edges" (Weisstein, 2003, p. 2304) or "a solid with a surface composed of plane polygonal surfaces (*faces*)" (Nelson, 2003, p. 333). A solid is "a closed, 3-dimensional figure" (Clapham & Nicholson,

Mathematical Understanding for Secondary Teaching: A Framework and Classroom-Based Situations, pages 323–327.

2005, p. 431) or "any limited portion of space bounded by surfaces" (Weisstein, 2003, p. 2751) or "a three-dimensional geometric figure, e.g., a polyhedron or cone" (Nelson, 2003, p. 393). Therefore, the title could use either the word *polyhedron* or *polyhedral solid*; this Situation uses *polyhedral solid*. Also, *polyhedron* refers only to the boundary points, which are the points on the polygonal faces,[1] and *solid* refers to both the boundary points and points interior to the polyhedron. Using the term *polyhedral solid* refers to boundary points and interior points and assumes that the boundary points lie on polygonal faces.

The Prompt raises the issue of how one refers to or names a mathematical entity, in this case, one of the constituting elements that form a prism, or more generally, a polyhedron. Focus 1 addresses the issue of the arbitrariness of conventional mathematical terms, and Focus 2 attends to differences between terms used for two-dimensional figures and those used for three-dimensional figures. Focus 3 uses Euler's formula, which relates the numbers of vertices, edges, and faces of a polyhedron, to determine the number of faces of the rectangular prism. The nature of a polygonal region (one type of bounded, planar region) addressed in Focus 4 offers one mathematical basis for understanding from where the student's answer "12" might have come (other than from counting the 12 edges).

MATHEMATICAL FOCI

Mathematical Focus 1

In mathematics, the words one uses to identify mathematical entities are conventional, but arbitrary. Such conventions are necessary for unambiguous communication.

The term *line segment* refers to the union of two distinct points and the set of collinear points between them. However, one could just as well use the term *line part*, if that were the accepted term for what one typically refers to as a *line segment*. Although, as the teacher did in the Prompt, one could call one of the polygonal regions that constitute a polyhedron (see Figure 35.1) a *side* (or some other term such as *slab* or *dwizzle*), the typical term used for such a polygonal region in the context of a polyhedron is *face*. Using the convention that for a polyhedral solid, *edge* refers to a line segment and *face* refers to a polygonal region, a rectangular prism has 6 faces and 12 edges. The first student may have inferred that the teacher meant *edge* by the term *side*, whereas the second student may have inferred that the teacher meant *face* by the term *side*.

Mathematical Focus 2

Mathematical terms do not necessarily carry the same meaning when moving from one system or dimension to another.[2] When moving between dimensional contexts, a term may take on different meanings that are not communicated by the term alone.

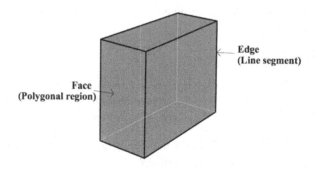

FIGURE 35.1. A face and an edge of a polyhedron.

There are some similarities and some differences between mathematical entities in two dimensions and those in three dimensions. Likewise, there are similarities and differences between the terms used to denote those entities.

Consider the terms used with polygons in two dimensions and the corresponding referents for those terms: *Vertex* refers to a point (a zero-dimensional object) and *side* refers to the infinite set of points on a line segment (a one-dimensional object). The terms used with prisms (or, more generally, polyhedra) in three dimensions and their corresponding referents are *vertex* for a point (a zero-dimensional object), *edge* for the infinite set of points on a line segment (a one-dimensional object), and *face* for the infinite set of points in a bounded, polygonal region (a two-dimensional object).

If the term *side* were used in place of either *edge* or *face*, would it be used to refer to an edge or to a face? If one attempted to map the two-dimensional terms directly onto the three-dimensional objects, one might use *side* in the same way as in two dimensions (i.e., to denote a segment—what one usually calls an *edge*). But, one also could use *side* to denote one of the faces, given that one typically means *face* when referring to things like the north *side* of a building or the left *side* of a box. So, potential confusion could arise if what typically are called *faces* were called *sides*, because one also could use the word *side* to refer to one of the segments forming the polyhedron's edges. In fact, some definitions of *edge* use the word *side* interchangeably with *edge*, for example, "Edge: A LINE SEGMENT where two FACES of a POLYHEDRON meet, also called a SIDE" (Weisstein, 2003, p. 850), and others use *side* for *face*. For example, Clapham and Nicholson (2005) give two definitions for side: "Side: One of the lines joining adjacent vertices in a polygon. One of the faces in a polyhedron" (p. 420). Using *side* to denote a two-dimensional polygonal region (face) as well as a one-dimensional segment (edge) could be confusing. Hence, rather than *side*, the term *edge* typically is used to denote a segment that is the intersection of exactly two faces of a prism, and the term *face* is used to denote a polygonal region bounding a prism (or, more generally, a polyhedron).

Mathematical Focus 3

One can use Euler's formula (the "polyhedron formula")[3] for the number of vertices, faces, and edges of a convex polyhedron to determine the number of faces of a rectangular prism if one knows the number of vertices and edges.

Although one could obtain 6 for the number of faces of a rectangular prism by counting, one also could use Euler's formula,

$$V - E + F = 2 \text{ or } E - V + 2 = F,$$

which relates the numbers of vertices, edges, and faces of a polyhedron.[4] One could subtract the number of vertices (8) from the number of edges (12) and add 2 to get the number of faces (6).

Some mathematics textbooks are not sufficiently careful about conditions under which Euler's formula applies. It is possible to encounter a textbook that indicates that $V - E + F = 2$ is valid for *all* polyhedra. In these cases, the definitions of *polyhedron* often do not include attention to convexity (or to simple connectivity). For example, without a convexity-related assumption, one can have a polyhedron modeled on a torus or on a torus with any number of holes. For those, $V - E + F = 2 - 2 \times$ (number of holes), for which any of the numbers 2, 0, -2, -4, -6, ... can be the result.

Mathematical Focus 4

A polygonal region (a type of bounded, planar region) is a set of points that does not have two distinct "surfaces."

The student who answered "12" may have been thinking that a rectangular prism has 12 faces: 6 faces on its "outside" and 6 faces on its "inside." This is not surprising, given that transparent, plastic models of prisms have "thickness," and one can see one surface on the outside of the model and a different surface on the inside. The fundamental question is, "Does a polygonal region have two 'sides'?"

Although an observer of a polygonal region can be situated in either of two half-spaces formed by the plane in which the polygon lies, the polygonal region consists of only one set of points that is infinitely "thin." There are not two distinct sets of points, one forming a polygonal region in one half-space and another forming a different polygonal region in the other half-space. A polyhedron's face (what the teacher called a *side*) is the same set of points, regardless of the half-space from which it is viewed. Therefore, there are only 6 faces, not 12.

NOTES

1.   A *polygonal region* is the union of the polygon and points interior to the polygon. The term *polygonal face* is used to refer to a polygonal region in the context of a polyhedron.

2.   It is also true that mathematical terms do not necessarily carry the same meaning from one textbook to another.

3.   The polyhedron formula given here is not the only formula referred to as *Euler's formula*. Other formulas attributed to Euler, such as $e^{i\pi} = -1$, may be more commonly thought of as Euler's formula.

4.   See Courant and Robbins (1941, pp. 236–240) for a proof of Euler's polyhedron formula. Also, Eppstein (2005) presents 19 proofs of Euler's polyhedron formula.

REFERENCES

Clapham, C., & Nicholson, J. (2005). *The concise Oxford dictionary of mathematics* (3ʳᵈ ed.). New York, NY: Oxford University Press.

Courant, R., & Robbins, H. (1941). *What is mathematics? An elementary approach to ideas and methods*. London, UK: Oxford University Press.

Eppstein, D. (2005). Geometry junkyard: Nineteen proofs of Euler's formula: $V - E + F = 2$. Retrieved September 28, 2009 from http://www.ics.uci.edu/~eppstein/junkyard/euler/

Nelson, R. D. (2003). *The Penguin dictionary of mathematics* (3rd ed.). London, UK: Penguin.

Weisstein, E. W. (2003). *CRC concise encyclopedia of mathematics* (2nd ed.). Boca Raton, FL: Chapman & Hall/CRC.

CHAPTER 36

AREA OF PLANE FIGURES

Situation 30 From the MACMTL–CPTM Situations Project

Erik Tillema, Tenille Cannon, Kim Johnson, and Rose Mary Zbiek

PROMPT

A teacher in a geometry class introduced formulas for the areas of parallelograms, trapezoids, and rhombi. She removed the formulas from the overhead projector and posed several area problems to her students. One student volunteered the correct answers very quickly. Another student asked, "How did you memorize the formulas so fast?" The first student responded, "I didn't memorize the formulas. I can just see what the area should be."

COMMENTARY

The four Foci for this situation reflect relationships among three classes of figures: parallelograms, trapezoids, and rhombi. In Foci 1 and 2, strategies for a class of figures are applied to one of its subclasses. The two Foci differ in terms of whether the strategy is the application of a known formula or the application of a method that develops the known formulas. Foci 3 and 4 involve decomposition of quadri-

Mathematical Understanding for Secondary Teaching: A Framework and Classroom-Based Situations, pages 329–336.

laterals, with the former emphasizing efficient calculation and the latter targeting the logical development of mathematics from what is known to what is needed.

MATHEMATICAL FOCI

Mathematical Focus 1

Because rhombi, squares, and rectangles are parallelograms, the area for each can be found by A = bh, where b represents the base and h represents the height.

A parallelogram is a *quadrilateral* with two pairs of parallel sides; rhombi, squares and rectangles are special parallelograms. The Venn diagram in Figure 36.1 represents these relationships. One could conclude that the areas of these parallelograms would be found in the same way, the product of the length of a base and the corresponding height.

Mathematical Focus 2

Any parallelogram, trapezoid, or rhombus can be decomposed into two triangles, the sum of whose areas is the area of the original quadrilateral—which suggests a way to generate familiar area formulas.

The method of constructing a diagonal to decompose a parallelogram, rhombus, or trapezoid into two triangles elucidates possible derivations of the area formulas. For example, in Figure 36.2, diagonal $\overline{AC}$ divides parallelogram ABCD into two congruent triangles: $\triangle ABC \cong \triangle CDA$. Using the base and height of the parallelogram, the area of each triangle is $A_{triangle} = \frac{1}{2}bh$. Because the area of parallelogram ABCD is equal to the sum of the areas of the two triangles,

$$A_{parallelogram} = \frac{1}{2}bh + \frac{1}{2}bh = 2\left(\frac{1}{2}bh\right) = bh.$$

Similarly, diagonal $\overline{PS}$ of trapezoid PRST in Figure 36.3 subdivides the trapezoid into two triangles, $\triangle PRS$ and $\triangle PST$. In general, the two triangles are not

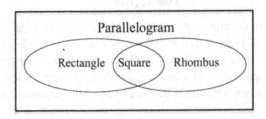

FIGURE 36.1. Venn diagram illustrating relationships among special quadrilaterals.

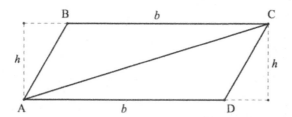

FIGURE 36.2.   Decomposition of parallelogram ABCD into two triangles with side AC.

congruent, but their areas sum to the area of the trapezoid. Using one base and the height of the trapezoid, the area of $\triangle$PRS is $A = \dfrac{1}{2}ah$, and the area of $\triangle$PST is $A = \dfrac{1}{2}bh$. The area of trapezoid PRST is equal to the sum of the areas of $\triangle$PRS and $\triangle$PST: $A = \dfrac{1}{2}ah + \dfrac{1}{2}bh = \dfrac{1}{2}h(a+b)$.

In the case of the rhombus, diagonal $\overline{XZ}$ subdivides rhombus WXYZ into two congruent triangles, $\triangle$WXZ and $\triangle$YZX, as shown in Figure 36.4. Let V be the intersection of $\overline{XZ}$ and $\overline{WY}$, which are diagonals of a rhombus and thus perpendicular bisectors of each other. By defining diagonal $\overline{XZ}$, which has length d_1, to be the base of both $\triangle$WXZ and $\triangle$YZX, segment $\overline{WV}$ with length $\dfrac{1}{2}d_1$ would be the altitude of $\triangle$WXZ. Given $\overline{WY}$, a diagonal of length d_2, segment $\overline{VY}$ with length $\dfrac{1}{2}d_2$ would be the altitude of $\triangle$YZX. The area of each triangle, $\triangle$YZX and $\triangle$WXZ, would be $A_{triangle} = \dfrac{1}{2}d_1\left(\dfrac{1}{2}d_2\right) = \dfrac{1}{4}d_1 d_2$, and thus the area of rhombus WXYZ would be: $A_{rhombus} = \dfrac{1}{4}d_1 d_2 + \dfrac{1}{4}d_1 d_2 = 2\left(\dfrac{1}{4}d_1 d_2\right) = \dfrac{1}{2}d_1 d_2$.

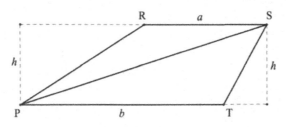

FIGURE 36.3.   Decomposition of trapezoid PRST into two triangles with side PS.

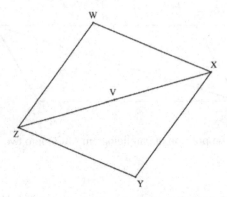

FIGURE 36.4. Decomposition of rhombus into two triangles with side XZ.

Because a rhombus is a parallelogram (see Focus 1), it follows from the preceding discussion of parallelogram ABCD that a rhombus can be decomposed into two triangles the sum of whose areas is equal to the area of the rhombus.

Mathematical Focus 3

A parallelogram, trapezoid, or rhombus can be decomposed into a combination of polygons, the sum of whose areas can be calculated efficiently, and that choice depends both on the original figure and the measures involved.

How one decomposes a quadrilateral to determine its area depends on the specific type of quadrilateral one has. Some decompositions are not possible. For example, one way to decompose a prototypical scalene trapezoid such as that in Figure 36.5 involves two triangles and a rectangle. In contrast, it makes no sense to use two triangles and a rectangle in a decomposition of a right trapezoid such as ABCD in Figure 36.6.

In some cases, the general nature of the decomposition might be the same but the calculation can be done more efficiently based on observations about the numbers

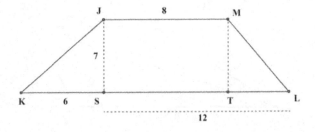

FIGURE 36.5. Decomposition of scalene trapezoid.

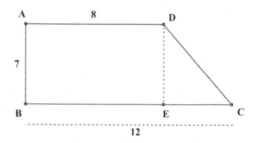

FIGURE 36.6.   Decomposition of right trapezoid.

involved. For example, the decomposition of a prototypical trapezoid into one rectangle and two triangles works with an isosceles trapezoid. To determine the area of the isosceles trapezoid JKLM in Figure 36.7, one can decompose the figure into rectangle JSTM and congruent triangles JSK and MTL. The calculation is slightly easier if one recognizes the equal areas of the two right triangles: The area of trapezoid

$$JKLM \text{ is } 2\left[\left(\frac{1}{2}\right)\left(\left(\frac{1}{2}\right)(12-8)\right)7\right]+(8)(7)=(1)(2)(7)+(8)(7)=(10)(7)=70$$

square units.

For figures of the same type, one might use slightly different computations or decompositions depending on the numbers involved, particularly in the absence of convenient formulas. For example, to determine the areas of the rhombi in Figure 36.8, thinking in terms of two triangles or four triangles yields easy calculations. A slight change in one measure creates the rhombus in Figure 36.9, for which only one of the three options seems to produce a slightly easier mental calculation.

Mathematical Focus 4

Area formulas of rectangles, parallelograms, and triangles and trapezoids can be logically developed in that order.

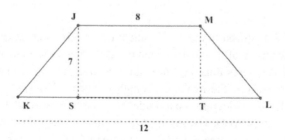

FIGURE 36.7.   Decomposition of isosceles trapezoid.

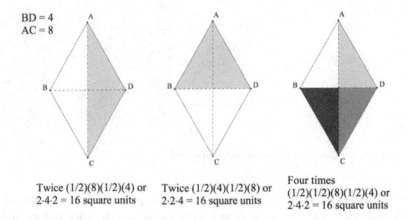

FIGURE 36.8. Rhombus with diagonals of length 4 and 8 decomposed into triangles.

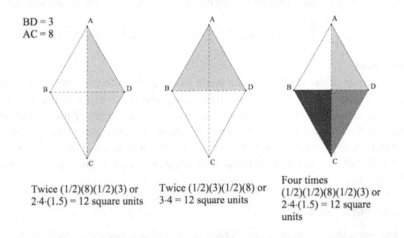

FIGURE 36.9. Rhombus with diagonals of length 3 and 8 decomposed into triangles.

A deductive approach to the development of area formulas is consistent with the systematic nature of Euclidean geometry. Covering a rectangular region with an array of unit squares illustrates the multiplicative nature of the area formula, $A = lw$. The formula for the area of a rectangle can be used to develop the formula for the area of a parallelogram. An altitude from vertex A in parallelogram ABCD creates right triangle AED. Translating $\triangle$AED along a vector from A to B creates rectangle ABFE whose area is equal to the area of parallelogram ABCD, as in Figure 36.10. Thus, $A_{ABCD} = bh$.

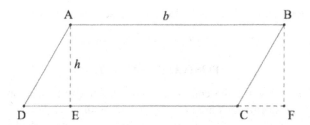

FIGURE 36.10. Parallelogram ABCD and associated rectangle ABFE.

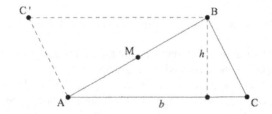

FIGURE 36.11. 180° rotation of a triangle about the midpoint of one of its sides.

The area formula for a parallelogram can be used to find the formulas for the area of a triangle and the area of a trapezoid. Both cases involve rotating a polygon 180° about the midpoint of one side. Figure 36.11 shows the 180° rotation of △ABC about midpoint M of side AB, which creates quadrilateral AC'BC with opposite sides congruent. So, AC'BC is a parallelogram the area of which is twice the area of original triangle; thus $A_{\text{AC'BC}} = bh$ implies $A_{\triangle \text{ABC}} = \frac{1}{2}bh$, where b is the base of △ABC and h is its height.

Similarly, Figure 36.12 illustrates how rotating trapezoid ABCD 180° about midpoint M of a nonparallel side creates quadrilateral C'D'CD with opposite an-

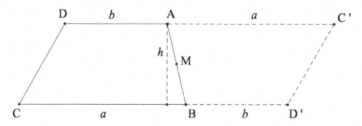

FIGURE 36.12. 180° rotation of trapezoid ABCD about midpoint M of a nonparallel side.

gles congruent. C'D'CD is a parallelogram whose area is twice the area of the original trapezoid; thus $A_{C'D'CD} = (a+b)h$ implies $A_{ABCD} = \frac{1}{2}(a+b)h$.

POSTCOMMENTARY

Foci 1, 2, and 4 involve the conservation of area under translations and rotations. These transformations are not as apparent as those in Focus 3. However, one can interpret the calculation string for the area of the isosceles trapezoid,

$2\left\{\left(\frac{1}{2}\right)\left[\frac{1}{2}(12-8)\right](7)\right\}$, as the area of two congruent triangles and the equivalent

term, $1\left[\frac{1}{2}(12-8)\right](7)$, as the area of the rectangle created by the pairing of one

triangle and the image of the second triangle after a translation and reflection are consecutively applied, as illustrated in Figure 36.13.

All the Foci have implications for interpreting area formulas and related expressions. For example, Focus 1 suggests $A = lw$ as a product of base length l and height w. In Focus 2, the expressions are expanded or simplified versions of the sum of the areas of two triangles. The numerical expressions in Focus 3 are manipulated for ease of calculation. Focus 4 involves seeing formulas as derivations or sources of other area formulas.

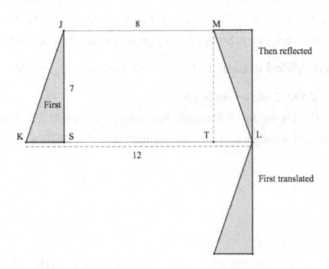

FIGURE 36.13. Transformation of part of trapezoid JMLK to produce a rectangle with area equal to that of the trapezoid.

CHAPTER 37

AREA OF SECTORS OF A CIRCLE

Situation 31 From the MACMTL–CPTM Situations Project

**Dennis Hembree, Sharon K. O'Kelley,
Shawn Broderick, and James Wilson**

PROMPT

An honors geometry class had the following problem for homework.

Complete the following (see Figure 37.1) for a circle with radius 3:

Central angle x	30°	60°	90°	120°	150°
Area of sector					

FIGURE 37.1.   Homework problem posed in an honors geometry class.

*Mathematical Understanding for Secondary Teaching: A Framework
and Classroom-Based Situations*, pages 337–341.

A student asked simply, "How do you do this?"

COMMENTARY

What could this problem afford beyond a series of simple calculations? If the question is passed off as simply a chance to practice computations, then perhaps something is missed. The problem could have been chosen to provide an opportunity to use a spreadsheet, write a program for a handheld calculator, or to look for patterns in the completed table. The completed table could suggest exploring a graph of the five number pairs. This Prompt allows for the introduction of radian measure as a more efficient representation of the sector's area formula. Students can also explore possible values for the area using both the table and programming features of a graphing calculator. In addition, a spreadsheet can be useful in calculating the area. Such uses of technology support students as they attempt to find values within the domain and range of the continuous function represented by the formula. All of these are means to explore the underlying functional relationship of sector areas and their central angles.

MATHEMATICAL FOCI

Mathematical Focus 1

The formula for the area of a sector can use either degree measure or radian measure.

The formula for the area of a sector, $A_x = \dfrac{x}{360}\left(\pi r^2\right)$, is straightforward, although cumbersome. Converting the central degree measure, x, to radians, θ, gives $A_\theta = \dfrac{\theta}{2}r^2$. This formula is less cumbersome and has the same form as the familiar area formula for a circle. The use of radian measure would require the students not only to use a modified version of the original formula but also to convert the central angles noted in the chart to radian measure. Using both degree and radian measure could also confirm that each approach leads to the same set of area measures.

Mathematical Focus 2

Plotting an appropriate set of number pairs (angle measure, sector area) in the Cartesian plane often yields a picture of the trend represented by the data that is not necessarily highlighted by the table of values.

Using either formula, the areas of the five chosen sectors can be approximated readily with a scientific calculator (see Table 37.1).

TABLE 37.1. Area of Sectors for Five Selected Central Angles

Central angle x	30°	60°	90°	120°	150°
Area of sector	2.356 in²	4.712 in²	7.069 in²	9.425 in²	11.781 in²

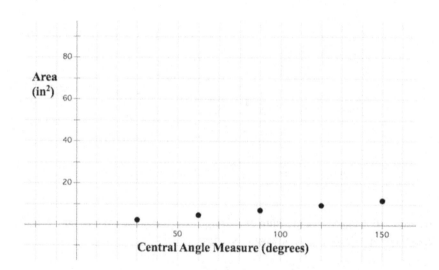

FIGURE 37.2. Graph of pairs of values (central angle in degrees, area of sector) in Table 37.1.

Plotting these points on a graph (see Figure 37.2) can be done either by hand or with technology.

Mathematical Focus 3

The area of a sector of a circle with fixed radius is a continuous linear function of the measure of the sector's central angle.

Superimposing a line on the graph displayed in Focus 2 (see Figure 37.3) may suggest the need to confirm that the relationship is linear.

Considering the slope between pairs of points can accomplish this or it can be done by examining either of the formulas and writing them as either

$$A_x = \frac{x}{360}\left(\pi r^2\right)$$

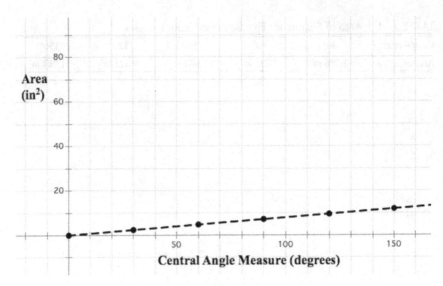

FIGURE 37.3. Line superimposed on the graph in Figure 37.2.

where $x \geq 0$ is the degree measure of the angle, r is the (fixed) radius, and A_x is the area when the central angle measure is x degrees; or

$$A_\theta = \frac{\theta}{2} r^2$$

where $\theta \geq 0$ is the radian measure of the angle, r is the (fixed) radius, and A_θ is the area when the central angle measure is θ radians. Each of these represents a linear function for a fixed r.

Because the familiar formula $A = \pi r^2$, contains a squared variable, students might mistakenly expect a quadratic relation. The relation between the area of a sector and the measure of its central angle is linear when the independent variable is central angle measure (in degrees) and r is fixed. Students may observe that the area of a 120° sector equals 4 times the area of a 30° sector. It should be noted that the fact that this is true depends on the fact that the function has a linearity property (i.e., $f(a) + f(b) = f(a + b)$). Not all functions, and not all linear functions, have this property. For the linear function with rule $f(x) = ax + b$, f has the linearity property only when $b = 0$.

POSTCOMMENTARY

Various technological tools can be useful for completing the table in the Prompt. In addition to completing the table, those tools can help the user to identify mathematical relationships in the data.

One can use the graphing and table features of a graphing tool to find the area of any sector and to plot the table of values (angle measure, sector area). These features of graphing tools are useful when rapid and accurate calculations are needed to explore mathematical relationships. One exploration, for example, might be to examine how the slope of the graph of the linear function changes as the radius of the circle is changed. The slope of the graph of each of the two linear functions depends on whether degree or radian measure was used in the derivation of the function. However, the domains of the two functions are the same, namely, all real numbers. Because the domain is the same, the two functions for a given value of r can be graphed on the same coordinate axes and their slopes compared visually; the units on the coordinate axis will then be coordinate units, not degrees or radians.

This problem also could be an opportunity to write a program for a handheld calculator, to use an existing program, or to generate and explore the data in the table. With a programmable calculator, one can write a simple program to output the area of a sector of a circle with a given central angle and a given radius. The program can be structured to accept either degree measure or radian measure as input for the measure of the central angle.

A spreadsheet also could be used to generate and explore the data in the table. The functionality of a spreadsheet to repeat operations using a Fill command makes the spreadsheet particularly attractive for this problem. Furthermore, once the data in the table are generated, the spreadsheet features for displaying graphs can be used to examine whether a linear relationship is suggested by the graph.

CHAPTER 38

SIMILARITY

Situation 32 From the MACMTL–CPTM Situations Project

Evan McClintock, Susan Peters, Donna Kinol, Shari Reed,
Heather Johnson, Erik Tillema, Rose Mary Zbiek,
M. Kathleen Heid, Sarah Donaldson,
Eileen Murray, and Glendon Blume

PROMPT

In a geometry class, students were given the diagram in Figure 38.1 depicting two acute triangles, $\triangle ABC$ and $\triangle A'B'C'$, and students were told that $\triangle ABC \sim \triangle A'B'C'$

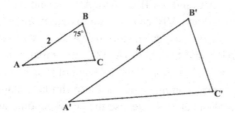

FIGURE 38.1.   Triangles with indicated measures that were given to students.

Mathematical Understanding for Secondary Teaching: A Framework
and Classroom-Based Situations, pages 343–350.
Copyright © 2015 by Information Age Publishing
All rights of reproduction in any form reserved.

with a figure (Figure 38.1) indicating that $A'B' = 2AB$ and $m\angle B = 75$. From this, a student concluded that $m\angle B' = 150$.

COMMENTARY

By definition, two polygons are similar if and only if their corresponding angles are congruent and their corresponding side lengths are proportional. Thus, similar figures may have different sizes, but they have the same shape. The Foci for this Situation incorporate a variety of approaches (geometric, graphical, and symbolic) to shed light on the concept of similarity. The first Focus refutes the claim made in the Prompt by appealing to the definition of similar triangles, and the second Focus refutes the claim using an indirect proof that considers the impact of doubling the measures of each of the angles of the original triangle. In Focus 3 and Focus 4, similarity is examined in terms of transformations in general, and dilations in particular. Under a geometric similarity transformation, angle measure is preserved and the ratio of the measures of corresponding distances is constant. Finally, a geometric construction and proof are included that lend further insight into the definition of similarity. In each Focus, the concept of ratio is emphasized, because common ratio lies at the heart of why size, but not shape, may be different for similar figures.

MATHEMATICAL FOCI

Mathematical Focus 1

Length measures of corresponding parts of similar figures are related proportionally with a constant of proportionality that may be other than 1, but angle measures of corresponding angles of similar figures are equal.

By definition, similar triangles have corresponding angles that are congruent and corresponding sides that are proportional.[1] Figure 38.1 in the Prompt depicts two similar triangles, $\triangle ABC$ and $\triangle A'B'C'$, with $AB = 2$ and $A'B' = 4$. A ratio of the lengths of the sides AB and $A'B'$ can be used to determine the corresponding lengths of the sides of $\triangle A'B'C'$ as scaled sides of $\triangle ABC$ or vice versa. In particular, because $A'B' = 2AB$ and $\triangle ABC$ and $\triangle A'B'C'$ are similar, it must be true that $A'C' = 2AC$, and $B'C' = 2BC$. Although the constant of proportionality, 2, can be used to find lengths of corresponding sides of these similar triangles, it does not apply to the measures of the angles, because the corresponding angles must, by definition, be congruent.

Mathematical Focus 2

The degree measures of a triangle cannot be some nonunit multiple of the corresponding degree measures in a similar triangle.

Suppose that one doubled the degree measures of each of the angles of ΔABC. This would result in the sum of the degree measures of the angles of ΔA′B′C′ being 360. But this is not possible, because the sum of the degree measures of the angles of any triangle is 180. So, the degree measures of the angles A′, B′, and C′ cannot each be double the corresponding degree measure in ΔABC.

More generally, consider ΔABC ~ ΔA′B′C′ with A′B′ = k × AB where $k > 0$ and $k \neq 1$. If one supposed that each degree measure in ΔA′B′C′ were k times the corresponding degree measure in ΔABC, then the sum of the degree measures of the angles of ΔA′B′C′ would be 180k with k > 0 and k ≠ 1. This would contradict the fact that the sum of the degree measures of a triangle must be 180. Therefore, the degree measures of the angles of ΔA′B′C′ cannot be k times the degree measures of the corresponding angles of ΔABC, where $k > 0$ and $k \neq 1$.

Mathematical Focus 3

Similar triangles have corresponding angles that are congruent and have corresponding sides whose lengths are proportional. Given that either one of these properties is true, the other must be true.

Each of the two conditions for triangle similarity—congruence of corresponding angles of two triangles and proportionality among their corresponding sides—can be shown to imply the other. So, if the lengths of corresponding sides of the similar triangles in the Prompt are proportional, then the corresponding angles must be congruent.

First, it can be shown that, given that the three angles of one triangle are congruent to the three corresponding angles of another triangle, the lengths of the corresponding sides of the two triangles are proportional. This is Proposition 4 in Book VI of Euclid's *Elements* (Densmore, 2002).

Consider triangles ΔABC and ΔDEF such that ∠A ≅ ∠D, ∠B ≅ ∠E, and ∠C ≅ ∠F (see Figure 38.2). Construct a circle centered at A with radius $\overline{DE}$. Let G be the point of intersection of the circle and $\overline{AB}$. Construct a line through G parallel to $\overleftrightarrow{BC}$. Let H be the point of intersection of the parallel line and $\overline{AC}$. Since $\overleftrightarrow{GH} \parallel \overleftrightarrow{BC}$, corresponding angles are congruent, so ∠AGH ≅ ∠E. Therefore, ΔAGH ≅ ΔDEF by the ASA theorem of congruence. Because $\overleftrightarrow{GH} \parallel \overleftrightarrow{BC}$, and parallel lines divide transversals proportionally, $\dfrac{GB}{AG} = \dfrac{HC}{AH} \Rightarrow \dfrac{AB}{AG} = \dfrac{AC}{AH} \Rightarrow \dfrac{AB}{AC} = \dfrac{AG}{AH}$, which is equivalent to $\dfrac{AB}{AG} = \dfrac{AC}{AH}$. Because AG = DE and AH = DF, by substitution, $\dfrac{AB}{DE} = \dfrac{AC}{DF}$. Using similar reasoning, it can be shown that $\dfrac{BC}{EF} = \dfrac{AB}{DE}$. Therefore, by the transitive property of equality: $\dfrac{BC}{EF} = \dfrac{AC}{DF}$. Thus, it can be concluded that $\dfrac{AB}{DE} = \dfrac{BC}{EF} = \dfrac{AC}{DF}$.

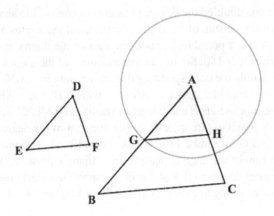

FIGURE 38.2. Triangles ABC and DEF with congruent corresponding angles.

Conversely, given that the lengths of the corresponding sides of two triangles are proportional, it can be shown that the angles of one triangle are congruent to the corresponding angles of the other triangle. This is Proposition 5 in Book VI of Euclid's *Elements* (Densmore, 2002).

Consider triangles $\triangle ABC$ and $\triangle DEF$ such that $\dfrac{AB}{DE} = \dfrac{BC}{EF} = \dfrac{AC}{DF}$, as illustrated in Figure 38.3. At point D on $\overrightarrow{DF}$ copy $\angle CAB$, and at point F on $\overrightarrow{FD}$ copy $\angle ACB$. Let X be the point of intersection of the two nonconcurrent rays of the copied angles. (Note: A more complete proof might also establish that point X exists, namely, that the nonconcurrent rays of the two copied angles must intersect.) Because two angles of $\triangle ABC$ are congruent to two angles of $\triangle DXF$, the remaining angles, $\angle B$ and $\angle X$, are congruent. By the preceding proof (Euclid's Proposition 4), because the angles of $\triangle ABC$ and $\triangle DXF$ are congruent, the lengths of the corresponding sides of $\triangle ABC$ and $\triangle DXF$ are proportional. So, $\dfrac{AB}{AC} = \dfrac{DX}{DF}$ and from the given proportion, $\dfrac{AB}{AC} = \dfrac{DE}{DF}$. So, $\dfrac{DE}{DF} = \dfrac{DX}{DF}$, implying that DE = DX. Similarly, EF = XF, because $\dfrac{BC}{XF} = \dfrac{AC}{DF}$ from the proportional sides of $\triangle ABC$ and $\triangle DXF$, and $\dfrac{BC}{EF} = \dfrac{AC}{DF}$ from the given proportion. Because $\overline{DF}$ is a side of both $\triangle DEF$ and $\triangle DXF$, $\triangle DXF \cong \triangle DEF$ by the SSS triangle congruence postulate. So, $\angle X \cong \angle E$, $\angle EDF \cong \angle XDF$, and $\angle EFD \cong \angle XFD$. Because the angles of $\triangle ABC$ and the corresponding angles of $\triangle DXF$ were already established as being congruent, and because the corresponding angles of $\triangle DXF$ and $\triangle DEF$ are congruent, by transitivity of congruence the corresponding angles of $\triangle ABC$ and $\triangle DEF$ are congruent.

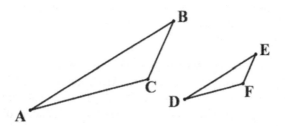

FIGURE 38.3. Triangles ABC and DEF with proportional lengths of corresponding sides.

A more concise proof can be constructed using the law of cosines for the corresponding sides of the two triangles. For example, again considering $\triangle ABC$ and $\triangle DEF$ such that $\dfrac{AB}{DE} = \dfrac{BC}{EF} = \dfrac{AC}{DF}$ as illustrated in Figure 38.3, it follows that $DE = k(AB)$, $EF = k(BC)$, and $DF = k(AC)$. The law of cosines yields

$$AB^2 = BC^2 + AC^2 - 2(BC)(AC)\cos(C)$$

and

$$DE^2 = EF^2 + DF^2 - 2(EF)(DF)\cos(F)$$

from which it follows that

$$(kAB)^2 = (kBC)^2 + (kAC)^2 - 2(kBC)(kAC)\cos(F)$$

$$\Rightarrow (AB)^2 = (BC)^2 + (AC)^2 - 2(BC)(AC)\cos(F).$$

However, $(AB)^2 = (BC)^2 + (AC)^2 - 2(BC)(AC)\cos(C)$.

So, $\cos(C) = \cos(F)$, and since the measures of angles C and F are less than π,

$$m\angle C = m\angle F \text{ and } \angle C \cong \angle F.$$

A similar argument can be used to show that $\angle A \cong \angle D$ and $\angle B \cong \angle E$.

Mathematical Focus 4

When comparing corresponding parts of figures and their images under similarity transformations, angle measure is preserved and the length of an element in the image is the product of the ratio of similitude and the length of the corresponding element in the original figure. Isometries, for which the ratio of similitude is 1, are a subset of similarity transformations.

Transformations in which shape is preserved but size is not necessarily preserved are similarity transformations. Given △ABC, consider △A′B′C′ to be the image of a similarity transformation of △ABC. A geometric similarity transformation is an angle-preserving function such that all corresponding distances are scaled by a constant ratio, $k \neq 0$. For the given similarity transformation in the Prompt, the lengths of sides of △A′B′C′ are double the length of the corresponding sides of △ABC, and, because angles are preserved by the similarity transformation, the angles of △A′B′C′ are congruent to the corresponding angles of △ABC.

Dilations whose center is the origin in a coordinate plane are similarity transformations of the form $F((x, y)) = (kx, ky)$ for some constant ratio $k \neq 0$. In general, if $|k| < 1$, the mapping results in a contraction, for which the resulting image is smaller than its preimage. If $|k| > 1$, the mapping results in an expansion, for which the resulting image is larger than its preimage. If $k = 1$, the mapping is the identity transformation under composition of transformations, for which the resulting image is the same size and shape as its preimage.

Transformations in which shape and size are preserved are known as *isometries*, for which the resulting image is congruent to the preimage. There are five distinct types of isometries: identity, reflection, nonidentity rotation, nonidentity translation,[3] and glide reflection. In the coordinate plane, any figure may be mapped to a similar figure by a composition of dilations and isometries. The constant of proportionality is the product of the ratios of similarity. When one figure is mapped to its image by a composition consisting only of isometries, the product of the ratios is a power of 1 and the figures are congruent.

Mathematical Focus 5

For a triangle inscribed in a circle and its dilation through the center of the circle, the relationship between the inscribed angle and the length of the intercepted arc shows that angle measure is preserved under dilation.

Using dynamical geometry software, a dynamic diagram (see Figure 38.4) can be created to illustrate that shape and angle measure are preserved for similar triangles for which one triangle can be represented as the expansion or contraction of the other triangle (with the center of the circle that circumscribes that triangle as the center of dilation). Consider a triangle inscribed in a circle centered at the origin (in Figure 38.4, △ABC is inscribed in circle c_1 and is mapped to △A′B′C′ under a dilation centered at the origin). Using polar coordinates, the ordered pair of coordinates of any point on the circle is (r, θ), where r is the radius of the circle and θ is the measure of the angle in standard position formed by the x-axis and a ray from the origin to a point on the circle. For any angle in the triangle, the mea-

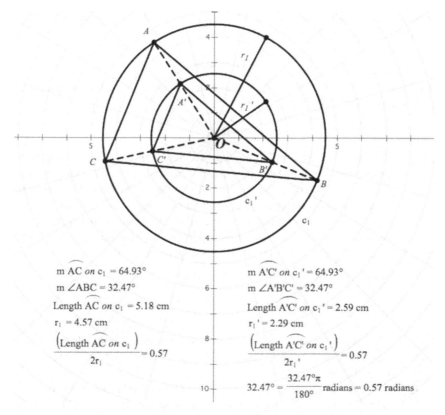

$m \stackrel{\frown}{AC} \text{ on } c_1 = 64.93°$

$m \angle ABC = 32.47°$

Length $\stackrel{\frown}{AC}$ on $c_1 = 5.18$ cm

$r_1 = 4.57$ cm

$\dfrac{\left(\text{Length } \stackrel{\frown}{AC} \text{ on } c_1\right)}{2r_1} = 0.57$

$m \stackrel{\frown}{A'C'} \text{ on } c_1{}' = 64.93°$

$m \angle A'B'C' = 32.47°$

Length $\stackrel{\frown}{A'C'}$ on $c_1{}' = 2.59$ cm

$r_1{}' = 2.29$ cm

$\dfrac{\left(\text{Length } \stackrel{\frown}{A'C'} \text{ on } c_1{}'\right)}{2r_1{}'} = 0.57$

$32.47° = \dfrac{32.47°\pi}{180°} \text{ radians} = 0.57 \text{ radians}$

FIGURE 38.4. Shape and angle measure of triangle ABC are preserved when triangle ABC is mapped to triangle A′B′C′ by a contraction.

sure of the angle is equal to half the length of the intercepted arc divided by the radius (in Figure 38.4, the measure of angle BCA is $\dfrac{\frac{1}{2}m\left(\text{arcAB}\right)}{r_1}$). Thus, angle measure is a function of arc length a and radius r, namely, $m(a,r) = \dfrac{a}{2r}$. Because the length of an arc is equal to the product of the radian measure of the central angle that subtends the arc and the length of radius of the circle (in Figure 38.4, length of arc ABC $= r_1 \cdot m(\angle AOB)$, the ratio of half the length of the arc to the radius of the circle will be equal to half the radian measure of the central angle, which is the radian measure of the inscribed angle intercepting that arc, regardless of the length of the radius.[4] Therefore, as a circle centered at the origin is expanded or contracted, the measure of any angle in the inscribed triangle remains constant. Thus, angle measure is preserved.

REFERENCE

Densmore, D. (Ed.). (2002). *Euclid's elements* (The Thomas L. Heath Translation). Santa Fe, NM: Green Lion Press.

NOTES

1. For triangles (and not for polygons with four or more sides), having congruent angles implies that the triangles are similar (with the corresponding sides in the same proportion). An analogous statement is false for polygons with four or more sides (neither congruent angles nor proportional sides alone implies the other). All rectangles have four right angles but not all are similar; and a rectangle and a nonrectangular parallelogram can have proportional sides without being similar.

2. A rationale for $\dfrac{AB}{AG} = \dfrac{AC}{AH}$ is the following:

$$\frac{AB}{AG} = \frac{GB+AG}{AG} = \frac{GB}{AG}+1 = \frac{HC}{AH}+1 = \frac{HC+AH}{AH} = \frac{AC}{AH}.$$

3. The identity rotation and the identity translation are not included in this list because including them would duplicate the identity mapping already listed.

4. This also can be established using the constant dilation factor to show that, for every circle, the ratio of the arc length to the radius of the circle is constant. Hence, the ratio of half the arc length to the radius of the circle is constant, and regardless of the length of the radius of the circle, the measure of the angle is preserved under dilation.

CHAPTER 39

PYTHAGOREAN THEOREM

Situation 33 From the MACMTL–CPTM Situations Project

Patrick Sullivan, M. Kathleen Heid, Maureen Grady, and Shiv Karunakaran

PROMPT

In both an Algebra 1 course and an Advanced Algebra course, students were given transparency cutouts of graph paper squares with side lengths from 1 unit to 25 units. Students were asked to create triangles whose sides had the side-lengths of three of the squares (see Figure 39.1). Students began to notice the squares that would create right triangles and the relationship involving the areas of those squares. A student asked, "Does this work for every right triangle?"

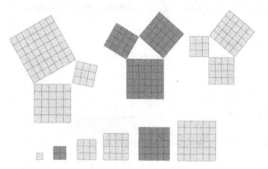

FIGURE 39.1 Triangular configurations of squares created by students.

Mathematical Understanding for Secondary Teaching: A Framework and Classroom-Based Situations, pages 351–363.
Copyright © 2015 by Information Age Publishing

COMMENTARY

The Pythagorean theorem relates the squares of the lengths of the sides of a right triangle. The law of cosines establishes a more general relationship between the lengths of the sides that holds for all triangles. Using algebra and geometry together it is possible to prove both the Pythagorean theorem and the law of cosines.

MATHEMATICAL FOCI

Mathematical Focus 1

Visual inspection alone is insufficient for drawing mathematical conclusions.

The generalization drawn by the student is based on what the student observed using physical models. Although such observations are important for mathematical discovery, they cannot replace mathematical proof. The diagram in Figure 39.2 is an example of a case in which the physical representation is illusory. That diagram makes it appear that two right triangles have the same base and same height but different areas. However, close examination reveals that neither of the figures is actually a right triangle. Calculating the slopes of segments AC, CB, DF, and FE reveals that points A, C, and B are not collinear and that points D, F, and E are not collinear. The figures shown are quadrilaterals: one convex and one concave. The diagram provides a misleading illusion. The existence of many such illusions provides a compelling rationale for the importance of mathematical proof.

Mathematical Focus 2

Algebraic meanings of area formulas for geometric configurations involving right triangles can be used to prove the Pythagorean theorem.

The question posed by the student seems to warrant the need for a justification that encompasses all cases. In other words, how would one prove the Pythagorean

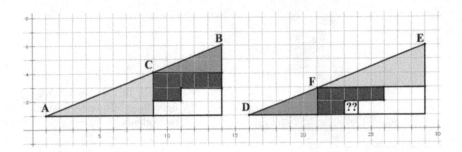

FIGURE 39.2.   What appear to be two right triangles with the same base and same height but with different total areas.

theorem? This Focus presents only two of the many proofs of the Pythagorean theorem. There are various resources available that have compiled many of these proofs. One such resource is *The Pythagorean Proposition* by Elisha Scott Loomis (1968). The following proofs utilize both algebra and geometry.

Proof 1:

Begin with a right triangle that has legs with lengths a and b (with $a > b$) and hypotenuse of length c. Make three copies of the same triangle rotated 90°, 180°, and 270°. Place the four triangles together so that their hypotenuses form a square with side length c as shown in Figure 39.3. [The angles at the vertices of the quadrilateral formed have measure 90 because the acute angles of a right triangle are complementary and each angle of the quadrilateral is formed by noncorresponding acute angles of congruent right triangles.] The area of each of these right triangles is $\dfrac{ab}{2}$. The inner quadrilateral is also a square. The angles of the inner quadrilateral are all right angles because they are supplementary to right angles. The sides of the inner quadrilateral are congruent because they are all equal to $a - b$.

The inner square has side length $(a - b)$, so the area of the inner square is $(a - b)^2$. Also, the sum of the areas of the four triangles is $4 \times \dfrac{ab}{2}$, and the area

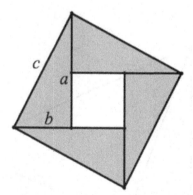

Inner square has side length a - b

FIGURE 39.3. Figure formed by juxtaposing four rotations (0°, 90°, 180°, and 270°) of a right triangle to form a square.

of the outer square is c^2. Knowing that the area of the outer square is equal to the area of the inner square plus the area of the four right triangles, one can conclude:

$$c^2 = (a-b)^2 + 2ab$$

$$\Rightarrow c^2 = a^2 - 2ab + b^2 + 2ab$$

$$\Rightarrow c^2 = a^2 + b^2$$

Therefore, for any right triangle, the square of the length of its hypotenuse is equal to the sum of the squares of the lengths of its legs.

Proof 2:

An alternative proof, attributed to U.S. President James Garfield in 1876, involves both geometry and algebra. Consider the arrangement in Figure 39.4 that includes two copies the same right triangle.

The noncorresponding legs, $\overline{QR}$ and $\overline{RS}$, of the two congruent right triangles are aligned. Note that isosceles triangle PRT formed with legs of length c is a right triangle. This is true because the acute angles of a right triangle are complementary and $\angle PRQ$ and $\angle TRS$ are noncorresponding acute angles of congruent right triangles. Consider the area of trapezoid PQST, formed by the three triangles. This area can be calculated in two different ways: (a) using a formula for the area of a trapezoid, and (b) using the area of the three triangles that constitute the trapezoid.

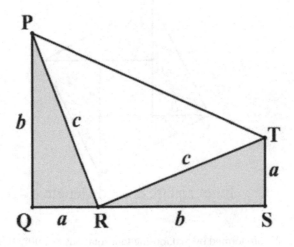

FIGURE 39.4. Diagram for President Garfield's proof of the Pythagorean theorem.

1. Using a formula for the area of a trapezoid,

Area of trapezoid = $\left(\dfrac{1}{2}\right) \cdot$ (height of trapezoid) $\cdot$ (sum of the lengths of the bases)

$$= \left(\dfrac{1}{2}\right) \cdot (a+b) \cdot (a+b)$$

$$= \left(\dfrac{1}{2}\right) \cdot (a^2 + 2ab + b^2).$$

2. Using the area of the three triangles that constitute the trapezoid,

Area of trapezoid = $(2) \cdot$ (area of right triangle with sides a, b, and c)

+(area of isosceles right triangle)

$$= (2)\left(\dfrac{1}{2}\right) \cdot (a) \cdot (b) + \left(\dfrac{1}{2}\right) \cdot (c) \cdot (c)$$

$$= \left(\dfrac{1}{2}\right) \cdot (2ab + c^2).$$

Equating the two expressions for the area of the trapezoid and simplifying the resulting equation results in the familiar equation for the Pythagorean theorem.

$$\left(\dfrac{1}{2}\right) \cdot (a^2 + 2ab + b^2) = \left(\dfrac{1}{2}\right) \cdot (2ab + c^2)$$

$$\Rightarrow (a^2 + 2ab + b^2) = (2ab + c^2)$$

$$\Rightarrow a^2 + b^2 = c^2.$$

Mathematical Focus 3

The law of cosines can be used to prove that if the lengths of the sides of a triangle are a, b, and c and a² + b² = c², then the triangle is a right triangle.[1]

The observation that the sum of the squares of the lengths of the legs of a right triangle is equal to the square of the length of the hypotenuse leads to the question of whether the relationship can apply to other types of triangles. So the question is one of whether every triangle with sides of length a, b, and c (where c is the length of the longest side) for which $a^2 + b^2 = c^2$ is a right triangle.

Using the law of cosines (proven in Focus 4 and Focus 5), for any triangle with side lengths a, b, and c: $c^2 = a^2 + b^2 - 2ab \cos C$. If $a^2 + b^2 = c^2$, then $c^2 = c^2 - 2ab \cos C$. Therefore, $-2ab \cos C = 0$. If $2ab \cos C = 0$, then C is a 90° angle because a and b are nonzero and when cos C = 0 and C is an angle whose degree measure must be less than 180°, the degree measure of angle C is 90°. So triangle ABC is a right triangle.

Mathematical Focus 4

The law of cosines can be used to describe a relationship between a, b, *and* c *for any triangle with sides of length* a, b, *and* c.

If, for triangles with sides of lengths a, b, and c (where c is the length of the longest side), $a^2 + b^2 = c^2$ holds for right triangles and only right triangles, the question arises as to whether a similar relationship holds for other types of triangles. The law of cosines (proven later in this Focus), $c^2 = a^2 + b^2 - 2ab \cos C$, provides such a relationship.

If angle C is obtuse, then cos C will have a negative value and $a^2 + b^2 - 2ab \cos C$ will be greater than $a^2 + b^2$. So, $c^2 > a^2 + b^2$.

If angle C is acute, then cos C will have a positive value and $a^2 + b^2 - 2ab \cos C$ will be less than $a^2 + b^2$. So, $c^2 < a^2 + b^2$.

The law of cosines can be proved using the Pythagorean theorem. The proof involves three cases, based on types of triangles.

Case 1. Triangle ABC is acute.

Side BC is opposite angle A and has length a, side AC is opposite angle B and has length b, and side AB is opposite angle C and has length c. Construct the circles with diameters BC and AC, calling them circles BC and AC, respectively (see

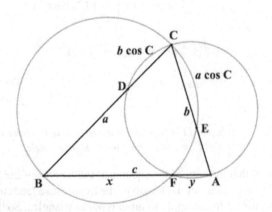

FIGURE 39.5. Diagram for Case 1 of Focus 4. △ABC is acute.

Figure 39.5).[2] One intersection of circle AC and line BC is C; call the other intersection D. One intersection of circle BC and line AC is C; call the other intersection E. One intersection of circle BC and circle AC is C; call the other intersection F. It is clear that F lies on segment AB because $\angle CFB$ and $\angle CFA$ are right angles (they are inscribed in semicircles). Let $x = BF$ and $y = FA$. $\triangle CFA$ and $\triangle CFB$ are right triangles (with right angles at vertex F), so $y^2 + (CF)^2 = b^2$ and $x^2 + (CF)^2 = a^2$. Therefore, $b^2 - y^2 = a^2 - x^2$. Substituting $c - x$ for y and simplifying yields:

$$b^2 - (c - x)^2 = a^2 - x^2$$

$$\Rightarrow b^2 - c^2 + 2cx - x^2 = a^2 - x^2$$

$$\Rightarrow b^2 = a^2 + c^2 - 2cx$$

Substituting $x = a \cos B$ (because $\triangle CFB$ is a right triangle) yields:

$$b^2 = a^2 + c^2 - 2ac \cos B.$$

Case 2. Triangle ABC is obtuse and angle C is obtuse.

Side BC is opposite angle A and has length a, side AC is opposite angle B and has length b, and side AB is opposite angle C and has length c. Again, construct the circles with diameters BC and AC, calling them circles BC and AC, respectively (see Figure 39.6).

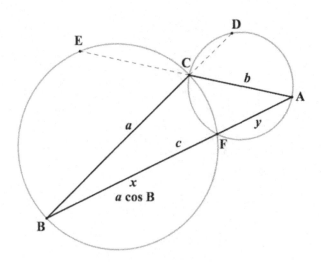

FIGURE 39.6. Diagram for Case 2 of Focus 4. $\triangle ABC$ is obtuse and angle C is obtuse.

One intersection of circle AC and line BC is C; call the other intersection D. One intersection of circle BC and line AC is C; call the other intersection E. One intersection of circle BC and circle AC is C; call the other intersection F. It is clear that F lies on segment AB because $\angle$CFB and $\angle$CFA are right angles (they are inscribed in semicircles). Let $x = $ BF and $y = $ FA. $\triangle$CFA and $\triangle$CFB are right triangles (with right angles at vertex F), so $y^2 + (CF)^2 = b^2$ and $x^2 + (CF)^2 = a^2$. Therefore, one can conclude that $b^2 - y^2 = a^2 - x^2$. Substituting $y = c - x$ and simplifying yields:

$$b^2 - (c - x)^2 = a^2 - x^2$$

$$\Rightarrow b^2 - c^2 + 2cx - x^2 = a^2 - x^2$$

$$\Rightarrow b^2 = a^2 + c^2 - 2cx.$$

Substituting $x = a \cos B$ (because $\triangle$CFB is a right triangle) yields:

$$b^2 = a^2 + c^2 - 2ac \cos B.$$

Case 3. Triangle ABC is obtuse and angle C is acute.

Side BC is opposite angle A and has length a, side AC is opposite angle B and has length b, and side AB is opposite angle C and has length c. Again, construct

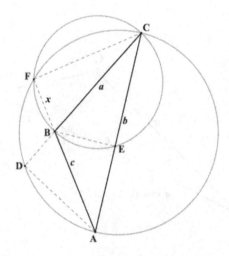

FIGURE 39.7. Diagram for Case 3 of Focus 4. $\triangle$ABC is obtuse and angle C is acute.

the circles with diameters BC and AC, calling them circles BC and AC, respectively (see Figure 39.7).

One intersection of circle AC and line BC is C; call the other intersection D. One intersection of circle BC and line AC is C; call the other intersection E. One intersection of circle BC and circle AC is C; call the other intersection F. It is clear that F lies on line AB because $\angle$CFB and $\angle$CFA are right angles (they are inscribed in semicircles) and because through a point on a line there is exactly one perpendicular. Let BF = x and FA = y.

Assume that B is the obtuse angle. Repeat the preceding constructions and note that F will be in the exterior of triangle ABC (to its left in Figure 39.7). Extend AB to the left and it passes through the point F for the same reasons. Next, construct CF. So

$$(CF)^2 + x^2 = a^2$$

$$\text{and } (CF)^2 + (x + c)^2 = b^2$$

$$\Rightarrow a^2 + c^2 + 2xc = b^2.$$

Also, $\cos(\pi - B) = \dfrac{x}{a}$ and $\cos(\pi - B) = -\cos B$. So, $\dfrac{x}{a} = -\cos B$. Substituting $x = -a \cos B$ (because $\triangle$CFB is a right triangle) yields

$$b^2 = a^2 + c^2 - 2ac \cos B.$$

Mathematical Focus 5

The law of cosines can be proven without reference to the Pythagorean theorem using the Power of the Point theorem.

This proof of the law of cosines involves three cases, based on the shape of different triangles (see Proof of the law of cosines, n.d.).

Case 1. Triangle ABC is acute.

Side BC is opposite angle A and has length a, side AC is opposite angle B and has length b, and side AB is opposite angle C and has length c. Again, construct the circles with diameters BC and AC, calling them circles BC and AC, respectively (see Figure 39.8). One intersection of circle AC and line BC is C; call the other intersection D. One intersection of circle BC and line AC is C; call the other intersection E. One intersection of circle BC and circle AC is C; call the other intersection F. It is clear that F lies on segment AB because $\angle$CFB and $\angle$CFA are right angles (they are inscribed in semicircles). Let x = BF and y = FA.

Because $\triangle$CDA and $\triangle$CEB are right triangles, CE = $a \cos$ C, and DC = $b \cos$ C. Applying the two-secants case of the Power of the Point theorem[3] to point B with respect to the circle AC yields $a (a - b \cos C) = xc$. Similarly, applying the two-

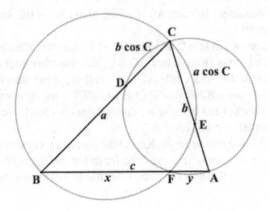

FIGURE 39.8. Diagram for Case 1 of Focus 5. ΔABC is acute.

secants case of the Power of the Point theorem to point A with respect to the circle BC results in $b\,(b - a\cos C) = yc$. Adding the corresponding members of the two equations yields

$$a\,(a - b\cos C) + b\,(b - a\cos C) = xc + yc$$

$$a^2 - ab\cos C + b^2 - ab\cos C = c\,(x + y)$$

$$a^2 + b^2 - 2ab\cos C = c^2.$$

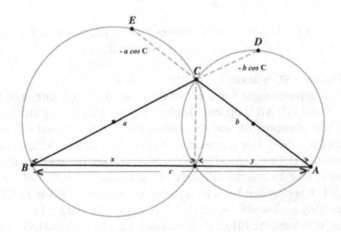

FIGURE 39.9. Diagram for Case 2 of Focus 5. ΔABC is obtuse and angle C is obtuse.

Case 2. Triangle ABC is obtuse and angle C is obtuse.

Side BC is opposite angle A and has length a, side AC is opposite angle B and has length b, and side AB is opposite angle C and has length c. Again, construct the circles with diameters BC and AC, calling them circles BC and AC, respectively (see Figure 39.9). One intersection of circle AC and line BC is C; call the other intersection D. One intersection of circle BC and line AC is C; call the other intersection E. One intersection of circle BC and circle AC is C; call the other intersection F. It is clear that F lies on segment AB because ∠CFB and ∠CFA are right angles (they are inscribed in semicircles). Let x = BF and y = FA.

Because △CDA and △CEB are right triangles, the lengths of CE and CD are $a \cos(180 - C)$ and $b \cos(180 - C)$, respectively. So CE = $-a \cos C$ and CD = $-b \cos C$. Following through with the same argument as previously presented yields

$$a(a - b \cos C) + b(b - a \cos C) = xc + yc$$

$$a^2 - ab \cos C + b^2 - ab \cos C = c(x + y)$$

$$a^2 + b^2 - 2ab \cos C = c^2.$$

Case 3. Triangle ABC is obtuse and angle C is acute.

Side BC is opposite angle A and has length a, side AC is opposite angle B and has length b, and side AB is opposite angle C and has length c. Again, construct the circles with diameters BC and AC, calling them circles BC and AC, respectively (see Figure 39.10). One intersection of circle AC and line BC is C; call the

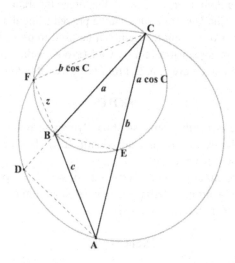

FIGURE 39.10.   Diagram for Case 3 of Focus 5. △ABC is obtuse and angle C is acute.

other intersection D. One intersection of circle BC and line AC is C; call the other intersection E. One intersection of circle BC and circle AC is C; call the other intersection F. It is clear that F lies on line AB because $\angle CFB$ and $\angle CFA$ are right angles (they are inscribed in semicircles) and because through a point on a line there is exactly one perpendicular.

Because angle C is acute and angle B is obtuse, the altitude from B cuts side AC. Call the point of intersection E. The altitudes from A and C, however, lie outside the triangle, and meet the lines in which sides CB and AB lie at the points D and F, respectively. Let the length of BF be represented by z. Because the length of DC is $b \times \cos C$ and the length of EC is $a \times \cos C$, the power of A with respect to circle BC applies, resulting in $b (b - a \cos C) = (c + z) c$. Similarly, using the two-chords case of the Power of a Point, the power of B with respect to circle AC yields $a (b \cos C - a) = zc$. Subtracting the expressions in the corresponding members of the two equations yields:

$$b (b - a \cos C) - a (b \cos C - a) = (c + z) c - zc$$

$$b^2 - ab \cos C - ab \cos C + a^2 = c^2 + zc - zc$$

$$a^2 + b^2 - 2ab \cos C = c^2.$$

POSTCOMMENTARY

The Foci in this situation serve to point out the connections between three important theorems in plane geometry: the Pythagorean theorem, the law of cosines, and the Power of a Point theorem. The Pythagorean theorem can be proved using the law of cosines, the law of cosines can be proved using the Pythagorean theorem, and the Power of a Point theorem can be used to prove the law of cosines and, thus, the Pythagorean theorem as well (Proof of the law of cosines, n.d.; Proofs of Pythagorean theorem, n.d.; Proof of Power of a Point theorem, n.d.).

NOTES

1. Note that this is the converse of the Pythagorean theorem.
2. Although circles typically are identified by their centers, in this case, circles are referenced by their diameters.
3. If two secants are drawn to a circle from an exterior point, the product of one secant and its external segment is equal to the product of the other secant and its external segment.

REFERENCES

Loomis, E. S. (1968). *The Pythagorean proposition* (2nd ed). Washington, DC: National Council of Teachers of Mathematics.

Proofs of Pythagorean theorem. (n.d.). *Pythagorean theorem* and *Law of cosines*. Retrieved from www.cut-the-knot.org/pythagoras/index.html

Proof of the law of cosines. (n.d.). *The law of cosines (cosine rule).* Retrieved from http://www.cut-the-knot.org/pythagoras/cosine.shtml

Proof of Power of a Point theorem. (n.d.). *Power of a Point theorem.* Retrieved from http://www.cut-the-knot.org/pythagoras/PPower.shtml

CHAPTER 40

CIRCUMSCRIBING POLYGONS

Situation 34 From the MACMTL-CPTM Situations Project

Shari Reed, AnnaMarie Conner, Heather Johnson,
M. Kathleen Heid, Bob Allen, Shiv Karunakaran,
Sarah Donaldson, and Brian Gleason

PROMPT

In a geometry class, after a discussion about circumscribing circles about triangles, a student asked, "Can you circumscribe a circle about any polygon?"

COMMENTARY

A polygon that can be circumscribed by a circle is called a *cyclic polygon*.[1] Not every polygon is cyclic, but there are infinitely many different cyclic polygons. This can be understood by considering a given circle and all the possibilities of how many points can be placed on the circle, and then connected to form a polygon. However, there are certain classes of polygons that are noteworthy because they are always cyclic. The conditions under which a circle circumscribes a given polygon are dependent upon the relationships among the angles, the sides, and the

Mathematical Understanding for Secondary Teaching: A Framework
and Classroom-Based Situations, pages 365–375.
Copyright © 2015 by Information Age Publishing
All rights of reproduction in any form reserved.

perpendicular bisectors of the sides of the polygon. The following Foci describe classes of cyclic polygons in order of the number of their sides: triangles, certain quadrilaterals, and regular polygons. Focus 3 provides one way to check whether a given polygon is cyclic: A polygon is cyclic if and only if the perpendicular bisectors of all its sides are concurrent. Although the inclusion of various geom- etries would provide interesting discussion, the Foci in this Situation are limited to Euclidean geometry in a plane.

MATHEMATICAL FOCI

Mathematical Focus 1

> *Every triangle is cyclic. This generalization is central to establishing a condition for other polygons to be cyclic.*

Because the center of a circle is equidistant from all points on the circle (this distance is the radius), an *inscribed triangle* is one in which the three vertices of the triangle lie on the circumscribed circle. Conversely, the circle circumscribed about a particular triangle must have as its center the point that is equidistant from the vertices of the triangle. Circumscribing a circle about a triangle, then, requires finding a point that is equidistant from the three vertices of the triangle. This point is called the *circumcenter* of the triangle.

A point in a plane is equidistant from points A and B in that plane if and only if it lies on $\overline{AB}$'s perpendicular bisector that lies in that plane. A proof of this theo- rem is included in the Postcommentary. Because of this, consider perpendicular bisectors of the sides of a triangle to find the circumcenter. Given $\triangle ABC$ (see Figure 40.1), the perpendicular bisectors (in the plane of $\triangle ABC$) of segments AB and BC intersect the sides at D and E, respectively. $\overline{AB}$ and $\overline{BC}$ are not parallel, so lines that are perpendicular to them are not parallel. Therefore, the perpendicular

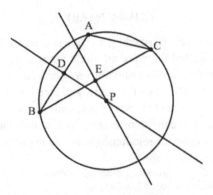

FIGURE 40.1. Location of the circumcenter, P, for triangle ABC.

bisectors of $\overline{AB}$ and $\overline{BC}$ must intersect at some point, call it P. P is equidistant from A and B because it lies on the perpendicular bisector of $\overline{AB}$, and P is equidistant from B and C because it lies on the perpendicular bisector of $\overline{BC}$. So P is equidistant from A, B, and C. That is, P is the circumcenter of $\triangle ABC$. So, given any set of three noncollinear points, it is possible to find the circumcenter of the triangle defined by those three points.

Another way to think about triangles being cyclic is to consider the equation of a circle in the coordinate plane: $(x - h)^2 + (y - k)^2 = r^2$ where (h, k) is the center of the circle, r is its radius, and every ordered pair (x, y) that satisfies the equation lies on the circle. To find a particular circle (that is, to find the three unknowns h, k, and r), one would need three equations. That is, if one had three ordered pairs (x, y) (i.e., three noncollinear points), one could determine the circle. This is another way of showing that three noncollinear points determine a unique circle. Given those three points, one could find the circumcenter of the triangle defined by those points.

Mathematical Focus 2

A convex quadrilateral in a plane is cyclic if and only if its opposite angles are supplementary.

A convex quadrilateral is a quadrilateral in a plane such that no two points in the interior of the quadrilateral can be connected by a segment that intersects one of the sides.[2] Proving that two conditions (a quadrilateral being cyclic and its opposite angles being supplementary) are equivalent requires proving an implication and its converse. That is, to prove $A \Leftrightarrow B$ one must prove that $A \Rightarrow B$ and prove that $B \Rightarrow A$. This proof uses a logically equivalent construction of proving $A \Rightarrow B$ and then proving *not A* $\Rightarrow$ *not B*.

a. First, prove that given a convex, cyclic quadrilateral, its opposite angles are supplementary. In the cyclic quadrilateral ABCD in Figure 40.2, $\angle ABC$ is opposite $\angle CDA$. Because the measure of an inscribed angle is half the measure of the arc in which it is inscribed,

$\text{m}\angle ABC = (1/2) \, \text{m(arcCDA)}$ and $\text{m}\angle CDA = (1/2) \, \text{m(arcABC)}$ [3]

$\text{m}\angle ABC + \text{m}\angle CDA = (1/2) \, \text{m(arcCDA)} + (1/2) \, \text{m(arcABC)}$

$= (1/2) \, [\text{m(arcCDA)} + \text{m(arcABC)}].$

Because the union of arcs CDA and ABC is a circle, $\text{m(arcCDA)} + \text{m(arcABC)} = 360°$. By substitution, $\text{m}\angle ABC + \text{m}\angle CDA = (1/2)(360) = 180°$. Therefore $\angle ABC$ and $\angle CDA$ are supplementary. (Angles BAD and DCB could be handled similarly.)

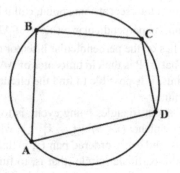

FIGURE 40.2. Cyclic quadrilateral ABCD.

b. Next prove that if the opposite angles of a quadrilateral are supplementary, then the quadrilateral is cyclic. Begin with convex quadrilateral ABCD such that angles BAD and DCB are supplementary and angles ABC and CDA are supplementary. Draw the circle defined by points A, B, and C. (This circle can be constructed because three points determine a circle—see Focus 1). Suppose D is located in the interior of the circle (see Figure 40.3). Then extend segments AD and CD until they each intersect the circle. By the inscribed angle theorem, the sum of angles BAD and DCB is less than 180 degrees (because together they subtend less than a whole circle[4]). This is a contradiction, therefore D cannot be inside the circle.

Suppose instead that D is outside the circle (see Figure 40.4). In this case, the sum of the measures of angles BAD and DCB will be greater than 180° because

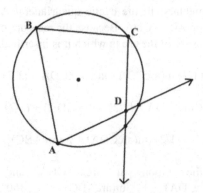

FIGURE 40.3. Convex quadrilateral ABCD; the circle defined by points A, B, and C; and point D in the interior of that circle.

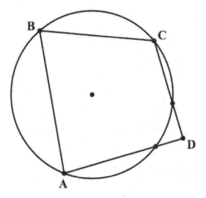

FIGURE 40.4.   Convex quadrilateral ABCD; the circle defined by points A, B, and C; and point D in the exterior of that circle.

together they subtend more than a whole circle (they subtend more than the whole circle, accounting for a portion of the circle twice). This is a contradiction, so D cannot be outside the circle. Because D cannot be inside or outside the circle, D must be on the circle. So, quadrilateral ABCD is cyclic.[5]

A corollary that is implied by the preceding result is that every rectangle is cyclic. Also, no parallelograms other than rectangles are cyclic. Moreover, every isosceles trapezoid is cyclic. [Note: A commonly used definition of *trapezoid* is that it is a quadrilateral with exactly one pair of parallel sides. However, trapezoids are defined by some sources as a quadrilateral with at least one set of parallel sides. If trapezoids are defined as in the latter definition, then every rectangle is an isosceles trapezoid.]

Mathematical Focus 3

There are four-sided figures in the plane that behave differently from convex quadrilaterals. Concave quadrilaterals are never cyclic, and a four-sided figure with nonsequential vertices is cyclic if and only if its "opposite" angles are congruent.

Cyclic quadrilaterals have opposite angles that are supplementary (see Focus 2). Consider a concave polygon such as quadrilateral EFGH in Figure 40.5. To show that quadrilateral EFGH is noncyclic, construct the circle passing through the points EFG (in general, through the 3 points that occur at the polygon's interior angles with measures that are not greater than 180°). Constructing this circle is possible because a concave quadrilateral will have exactly one interior angle greater than 180 degrees; in this case, it is angle EHG. Suppose that point H lies on this circle. But then the measure of (interior) angle EHG is less than 180 degrees because it cannot subtend the whole circle (much less even more than the whole circle), producing the contradiction that m∠EHG > 180° and m∠EHG < 180°. So,

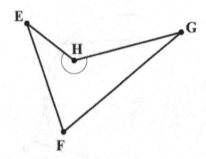

FIGURE 40.5. Concave quadrilateral EFGH.

point H cannot lie on the circle passing through E, F, and G, so concave quadrilateral EFGH is noncyclic.

A quadrilateral is commonly defined as a polygon with four sides. Therefore it is important that the definition of polygon be clear. If the definition of polygon requires that it be a simple, closed figure (as it is in many high school mathematics textbooks), then the figures in the following discussion are not polygons, and therefore not quadrilaterals. However, if the definition requires only that a quadrilateral be a closed figure in a plane with four straight sides, then a quadrilateral with nonsequential vertices can be considered in this Situation.

A quadrilateral with nonsequential vertices is cyclic if and only if its "opposite" angles are congruent. Here, the vertices of a quadrilateral ABCD are called *nonsequential* if side AB intersects side CD. Also, the term *opposite angles* of a quadrilateral is meant to convey two interior angles of the quadrilateral that do not share a common side. Such a quadrilateral has sides that "cross," as seen in Figure 40.6. "Opposite" angles in Figure 40.6, for example, are angles NMP and PON or angles ONM and MPO.

In this case, it must first be proved that if a quadrilateral with nonsequential vertices is cyclic, then its "opposite" angles are congruent. Consider quadrilateral MNOP in Figure 40.6. Because angles NMP and PON lie on the circle and intercept the same arc (arc NP), they are congruent. In the same way, angles ONM and MPO both intercept arc MO, so they are congruent. Note that the sum of the measures of the interior angles of this type of quadrilateral is not 360°.

Next, prove that if "opposite" angles of a quadrilateral with nonsequential vertices are congruent, then the quadrilateral is cyclic. The same strategy that was used for the converse in Part a of Mathematical Focus 2 can be used here: Begin with quadrilateral MNOP such that angles NMP and PON are congruent, angles ONM and MPO are congruent, and all the vertices except P lie on circle r (Figure 40.7). If P is inside circle r, then extend sides MP and OP to intersect circle r. Then angles NMP and PON subtend different size arcs, so their measures are not equal, which contradicts the statement that angles NMP and PON are congruent. If P is

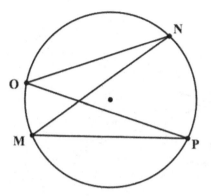

FIGURE 40.6.   Quadrilateral MNOP with nonsequential vertices.

outside circle r, then side MP intersects circle *r* in a different point than does side OP, so angles NMP and PON are again noncongruent, which again is a contradiction. Therefore, P must lie on circle *r*.

Mathematical Focus 4

Every planar regular polygon is cyclic. However, not every cyclic polygon is regular.

As was discussed in Focus 1, a point is equidistant from two points, A and B, if and only if it lies on the perpendicular bisector of the segment whose endpoints are A and B. For a polygon to be cyclic, there must be a circle that passes through all of its vertices. In other words, there must be a single point that is equidistant from all the vertices. This point must lie on the perpendicular bisectors of all the sides of the polygon. For a point to lie on all these bisectors, the bisectors must be concurrent, and the point of concurrency will be the circumcenter of the polygon. Because the statement about equidistance and the perpendicular bisector is biconditional, a biconditional statement can be made about the concurrency of the perpendicular bisectors of the sides of a polygon. That is, the perpendicular bisectors of the sides of a polygon are concurrent if and only if the polygon is cyclic.

Every triangle is a cyclic polygon, as was established in Focus 2. The question remains as to which other polygons are cyclic. By examining the perpendicular bisectors of the sides of a polygon, one can determine a set of conditions on a polygon that is sufficient to conclude whether a circle can circumscribe that polygon. In particular, one can show that every regular polygon is cyclic.

Having established that if the perpendicular bisectors of the sides of a polygon are concurrent, the polygon is cyclic, it remains to show that the perpendicular bisectors of the sides of a regular polygon are concurrent and conclude that every regular polygon is cyclic.

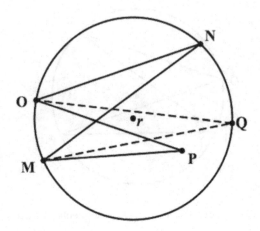

FIGURE 40.7. Quadrilateral MNOP with all vertices except P on circle r.

By definition, a *regular polygon* is an equilateral and equiangular n-sided polygon. Consider a regular polygon with adjacent vertices, A, B, C, and D. Let P be the point of intersection of the perpendicular bisectors ($\overline{FP}$ and $\overline{GP}$, respectively) of $\overline{AB}$ and $\overline{BC}$. It can be shown that $\triangle AFP \cong \triangle BFP \cong \triangle BGP \cong \triangle CGP$ using the fact that P is equidistant from A, B, and C, and using the HL (hypotenuse–leg) congruence theorem. Construct $\overline{PH}$ perpendicular to $\overline{DC}$ and consider $\triangle HCP$. $\angle FBG \cong \angle GCH$ because the polygon is equiangular. $\angle FBP \cong \angle GCP$ because $\triangle BFP \cong \triangle CGP$. By angle subtraction, $\angle GBP \cong \angle HCP$, so $\triangle FBP \cong \triangle HCP$ by AAS (angle–angle–side triangle congruence). Congruent triangles establish that HC = FB, and because the polygon is equilateral, AB = CD. Also, FB = (½)AB because F is the midpoint of AB. So HC = (½)CD by substitution. Because C, H, and D are collinear, HC = HD. Thus, $\overline{PH}$ is the perpendicular bisector of $\overline{DC}$. The argument can be extended to successive vertices of the polygon, resulting in establishing that each of the perpendicular bisectors of the sides contains the point P (Figure 40.8). That is, the perpendicular bisectors are concurrent. Note that this argument does break down if the polygon in question is not regular. If the polygon is not regular (specifically if the sides are not equal to each other) one cannot prove that all the triangles listed previously are congruent.

Therefore, every regular polygon has concurrent perpendicular bisectors and therefore is cyclic. Although every regular polygon is cyclic, it is not true that every cyclic polygon is regular. For example, every triangle is cyclic, but every triangle is not an equilateral triangle and thus not regular. So it is possible to have polygons that are not regular but are cyclic (see Figure 40.9).

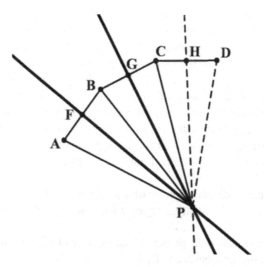

FIGURE 40.8.   Regular polygon ABCD   with point P the intersection of the perpendicular bisectors of its sides.

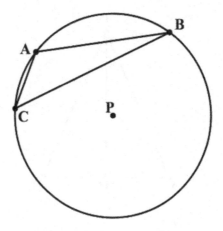

FIGURE 40.9.   A cyclic polygon, ΔABC, that is not regular.

POSTCOMMENTARY

In making mathematical statements, it is important to recognize which are biconditional and which are not. In Focus 2, the proof of a statement and its converse constituted a proof of the biconditional. In Focus 4, the converse of the statement proved is not true, and so the biconditional is not true. In Focus 1, the use of the biconditional was not as overt. That Focus drew upon the property that each point on a perpendicular bisector of a segment is equidistant from the endpoints of the segment. Later in the Focus, establishing uniqueness was based upon the result that, in a plane, points that are equidistant from the endpoints of a segment lie on the perpendicular bisector of the segment. Because this statement was the converse of a previously established property, it requires a proof. That proof follows here:

A line is a perpendicular bisector of a segment AB if and only if it is the set of all points in a plane containing segment AB that are equidistant from A and B.

a. If a point lies on the perpendicular bisector of a line segment AB, then it is equidistant from A and B.

Let M be the midpoint of segment AB, and let P be a point that lies on the perpendicular bisector of segment AB such that P does not lie on segment AB (see Figure 40.10). By the definition of perpendicular bisector, AM = BM and m∠BMP = m∠AMP = 90°. Now consider ΔAMP and ΔBMP. These triangles share side PM, so they are congruent by SAS (side–angle–side triangle congruence). Therefore PA = PB.

b. If a point is equidistant from A and B then it lies on the perpendicular bisector of the line segment AB.

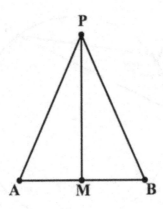

FIGURE 40.10.   Segment AB with midpoint M and point P (≠ M) on its perpendicular bisector.

Let P be a point not on segment AB such that PA = PB. This means ΔPAB is isosceles and its base angles are congruent, so m∠BAP = m∠PBA. Let M be the midpoint of segment AB. Because triangles PAM and PBM share side PM, they are congruent by SAS. Because ∠BMA is a straight angle, m∠PMA + m∠BMP = 180°. Also, m∠PMA = m∠BMP because ΔPAM ≅ ΔPBM. Thus,

$$m\angle PMA + m\angle BMP = 180°$$

$$\Rightarrow 2(m\angle PMA) = 180°$$

$$\Rightarrow m\angle PMA = 90°$$

Therefore segment PM is a perpendicular bisector of segment AB.

NOTES

1. See Leung and Lopez-Real (2002) for additional discussion of conditions under which a polygon may or may not be cyclic.
2. This definition can be generalized to that of a convex object: An object O in a vector space is said to be *convex* if for any two points p and q in O, the line segment from p to q lies entirely inside O (meaning in its interior and/or on its boundary). The line segment can be described as the set of points $(1 - t) \cdot p + t \cdot q$, for t in [0, 1].
3. Arc measures and angle measures are assumed to be in degrees.
4. The claim that these angles subtend less than the whole circle relies on the figure, which shows that part of the circle is unaccounted for in the two subtended arcs.
5. One can give a nonpicture-based proof of the fact that if D is exterior to the circle, then the corresponding subtended arcs can be shown to have an arc of the circle as intersection, and cover more than the entire circle. And similarly, if D is interior to the circle, then the union of the corresponding subtended arcs can be shown to omit an arc of the circle.

REFERENCE

Leung, A., & Lopez-Real, G. (2002). Theorem justification and acquisition in dynamic geometry: A case of proof by contradiction. *International Journal of Computers for Mathematical Learning, 7,* 145–165.

CHAPTER 41

CALCULATION OF SINE

Situation 35 From the MACMTL–CPTM
Situations Project

Patricia S. Wilson, Heather Johnson, Jeanne Shimizu, Evan McClintock,
Rose Mary Zbiek, M. Kathleen Heid, Maureen Grady, and Svetlana Konnova

PROMPT

After completing a discussion on special right triangles (30°–60°–90° and 45°–
45°–90°), the teacher showed students how to calculate the sine of various angles
using a calculator.

A student then asked, "How could I calculate sin(32°) if I do not have a calculator?"

COMMENTARY

The set of Foci provide interpretations of sine as a ratio and sine as a function, using graphical and geometric representations. The first three Foci highlight sin(θ)
as a ratio, appealing to the law of sines, right-triangle trigonometry, and unit-
circle trigonometry. The next three Foci highlight sin(x) as a function and use
tangent and secant lines as well as polynomials to approximate sin(x).

*Mathematical Understanding for Secondary Teaching: A Framework
and Classroom-Based Situations*, pages 377–384.

The question of "how good" an approximation one gets using secants, tangents, or Taylor polynomials depends on the size of the x-interval, the order of the highest derivative, and the function(s) in question. These challenging problems are usually taken up in courses on mathematical analysis and numerical analysis.

MATHEMATICAL FOCI

Mathematical Focus 1

Ratios of lengths of sides of right triangles can be used to compute and approximate trigonometric function values.

A ratio of measures of sides of a right triangle with an acute angle of measure $x°$ can be used to approximate $\sin(x)$. $\sin(x)$ can be approximated by sketching a 32°–58°–90° right triangle with a protractor or with dynamical geometry software, measuring the length of the hypotenuse and leg opposite the 32° angle, and computing the sine ratio (see Figure 41.1).

Hence, $\sin(32°) \approx 0.53$.

Mathematical Focus 2

Coordinates of points on the unit circle represent ordered pairs of the form $(\cos(\theta), \sin(\theta))$ that can be used to approximate trigonometric values.

The unit circle is the locus of all points one unit from the origin (0, 0). The equation for a circle with radius 1 centered at the origin is $x^2 + y^2 = 1$. Consider the angle θ in standard position formed by the x-axis and a ray from the origin through a point A on the unit circle. Then, $\cos(\theta) = \dfrac{x}{1}$ and $\sin(\theta) = \dfrac{y}{1}$. Hence, the coordinates of A are $(\cos(\theta), \sin(\theta))$, and another equation for a circle with radius 1 centered at the origin is $(\cos(\theta))^2 + (\sin(\theta))^2 = 1$.

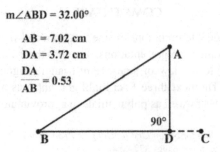

$m\angle ABD = 32.00°$

$AB = 7.02$ cm

$DA = 3.72$ cm

$\dfrac{DA}{AB} = 0.53$

FIGURE 41.1. Right triangle ABD with a 32° angle.

Let A be positioned on the unit circle so that ∠ABD has degree-measure 32° (see Figure 41.2). Then, the signed length of segment AD is equal to sin(32°). The signed length of segment AD is approximately 0.53, and so sin(32°) ≈ 0.53.

Mathematical Focus 3

The law of sines can be used to compute and approximate the sine function value through the measurement of geometric constructions.

The law of sines applies to any triangle in a plane. Consider triangle ABC, with sidelengths a, b, and c for $\overline{BC}$, $\overline{AC}$, and $\overline{AB}$ respectively. The law of sines states:

$$\frac{a}{\sin(A)} = \frac{b}{\sin(B)} = \frac{c}{\sin(C)}.$$

Sin(32°) can be approximated by sketching any triangle the degree-measure of one of whose angles is 32° and the degree-measure of another of whose angles has a known sine value (e.g., 30°, 45°, 60°, or 90°).

For example, a triangle can be sketched (with dynamical geometry software) with m∠A = 32° and m∠B = 90° (see Figure 41.3). Using the measure a and the measure b (the length of the side opposite the 90° angle), sin(32°) can be calculated using the law of sines.

$$\frac{a}{\sin(32°)} = \frac{b}{\sin(90°)}$$

Because sin(90°) = 1, $\sin(32°) = \frac{a}{b}$.

Hence, sin(32°) ≈ 0.53.

For another example, a triangle can be sketched (with dynamical geometry software) with m∠A = 32° and m∠B = 30° (see Figure 41.4). Using the measure a and the measure b (the length of the side opposite the 30° angle), sin(32°) can be calculated using the law of sines.

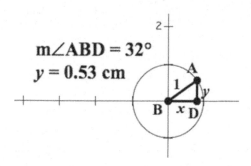

m∠ABD = 32°
y = 0.53 cm

FIGURE 41.2.   Right triangle ABD with a 32° angle on a unit circle.

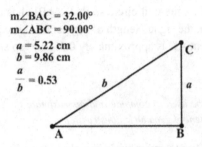

$m\angle BAC = 32.00°$
$m\angle ABC = 90.00°$
$a = 5.22$ cm
$b = 9.86$ cm
$\dfrac{a}{b} = 0.53$

FIGURE 41.3. Using the law of sines and sin(90°) to calculate sin(32°).

$$\frac{a}{\sin(32°)} = \frac{b}{\sin(30°)}$$

Because $\sin(30°) = \dfrac{1}{2}$, $\sin(32°) = \dfrac{a}{2b}$. Hence, sin(32°) $\approx$ 0.53.

Mathematical Focus 4

A continuous function, such as f(x) = sin(x), can be represented locally by a linear function and that linear function can be used to approximate local values of the original function.

The function $f(x) = \sin(x)$ is not a linear function; however, linear functions can be used to approximate nonlinear functions over sufficiently small intervals. Measuring angles in radians, 180° is equivalent to π radians. Therefore:

$$30° \text{ is equivalent to } \frac{30\pi}{180} = \frac{\pi}{6}, \text{ or } 0.5236 \text{ radians,}$$

$$32° \text{ is equivalent to } \frac{32\pi}{180} = \frac{8\pi}{45}, \text{ or } 0.5585 \text{ radians, and}$$

$$45° \text{ is equivalent to } \frac{45\pi}{180} = \frac{\pi}{4}, \text{ or } 0.7854 \text{ radians.}$$

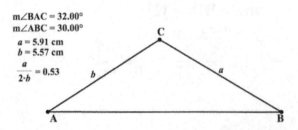

$m\angle BAC = 32.00°$
$m\angle ABC = 30.00°$
$a = 5.91$ cm
$b = 5.57$ cm
$\dfrac{a}{2\cdot b} = 0.53$

FIGURE 41.4. Using the law of sines and sin(90°) to calculate sin(32°).

Figure 41.5 shows the graph of the function $f(x) = \sin(x)$ and the graph of the secant line $\overleftrightarrow{AB}$, where the coordinates of A are $\left(\dfrac{\pi}{6},\ \sin\left(\dfrac{\pi}{6}\right)\right) = \left(\dfrac{\pi}{6},\ \dfrac{1}{2}\right) = \left(\dfrac{\pi}{6},\ 0.5\right)$

and the coordinates of B are $\left(\dfrac{\pi}{4},\ \sin\left(\dfrac{\pi}{4}\right)\right) = \left(\dfrac{\pi}{4},\ \dfrac{\sqrt{2}}{2}\right) = \left(\dfrac{\pi}{4},\ 0.7071\right)$. Because the

function $f(x) = \sin(x)$ is approximately linear between points A and B, the values of the points on the secant line, $\overleftrightarrow{AB}$, provide reasonable approximations for the values of $f(x) = \sin(x)$ between points A and B (see Figure 41.5). Because $\sin(x)$ is concave down in the x-interval $\left(\dfrac{\pi}{6},\ \dfrac{\pi}{4}\right)$, the estimate for $\sin(32°)$ will be an underestimate.

In Figure 41.6, point D on secant line $\overleftrightarrow{AB}$ with coordinates (0.5585, 0.5276) provides a reasonable approximation for the location of point C on $f(x) = \sin(x)$ with coordinates (0.5585, sin(0.5585)). Therefore, $\sin(32°) \approx 0.5276$.

An approximation for $\sin(32°)$ can also be found by using the equation for secant line $\overleftrightarrow{AB}$. Because secant line $\overleftrightarrow{AB}$ passes through the points $\left(\dfrac{\pi}{6},\ \sin\left(\dfrac{\pi}{6}\right)\right) \approx (0.5236,\ 0.5)$ and $\left(\dfrac{\pi}{4},\ \sin\left(\dfrac{\pi}{4}\right)\right) \approx (0.7854,\ 0.7071)$, its equation can be approximated as follows:

$$y - 0.5 = \frac{0.7071\ \ 0.5}{0.7854\ \ 0.5236}(x - 0.5236)$$
$$y = 0.7911(x - 0.5236) + 0.5$$

When $x = 0.5585$, $y = 0.5276$. Therefore, $\sin(32°) \approx 0.5276$.

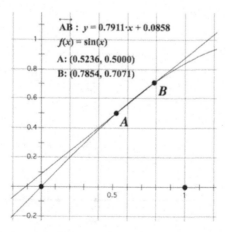

FIGURE 41.5.   Using a secant line to estimate sin(32°).

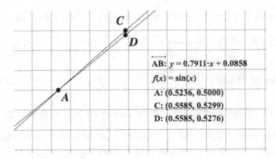

AB: $y = 0.7911 \cdot x + 0.0858$
$f(x) = \sin(x)$
A: (0.5236, 0.5000)
C: (0.5585, 0.5299)
D: (0.5585, 0.5276)

FIGURE 41.6. Zooming in on the estimate of sin(32°) using a secant line.

Mathematical Focus 5

> *Given a differentiable function and a line tangent to the function at a point, values of the tangent line will approximate values of the function near the point of tangency.*

Because the function $f(x) = \sin(x)$ is differentiable, given a point $(a, \sin(a))$ on $f(x) = \sin(x)$, the line tangent to $f(x) = \sin(x)$ at $(a, \sin(a))$ can be used to approximate $(a + dx, \sin(a + dx))$ at a nearby point with x-coordinate $a + dx$. When dx is small, the y-value of the tangent line at the point with x-coordinate $a + dx$ will be very close to the value of $\sin(a + dx)$. Using radian measure, $32°$ is equivalent to $\dfrac{32\pi}{180} = \dfrac{8\pi}{45}$, or 0.5585 radians.

Consider a geometric interpretation of differentials dx and dy and their relation to Δx and Δy, where a tangent line is used to approximate $f(x)$ near a given value (see Figure 41.7).

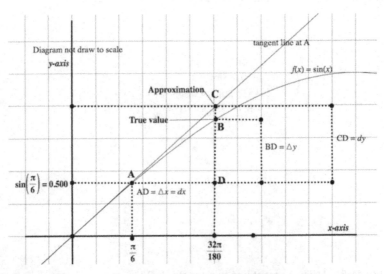

FIGURE 41.7. A geometric interpretation of differentials to estimate sin(32°).

$$f'(x) \approx \frac{\Delta y}{\Delta x} \rightarrow \Delta y \approx (\Delta x) f'(x)$$

Because $\left(a,\ f(a)\right) = \left(\frac{\pi}{6},\ \sin\left(\frac{\pi}{6}\right)\right)$ and $f'(x) = \cos(x)$,

$$\Delta y \approx (\Delta x) f'(x)$$

$$\Rightarrow \sin\left(\frac{32\pi}{180}\right) - \sin\left(\frac{\pi}{6}\right) \approx \left(\frac{32\pi}{180} - \frac{\pi}{6}\right)\cos\left(\frac{\pi}{6}\right)$$

$$\Rightarrow \sin\left(\frac{32\pi}{180}\right) - \sin\left(\frac{\pi}{6}\right) \approx 0.0302$$

$$\Rightarrow \sin\left(\frac{32\pi}{180}\right) \approx 0.0302 + \sin\left(\frac{\pi}{6}\right)$$

$$\Rightarrow \sin\left(\frac{32\pi}{180}\right) \approx 0.5302 .$$

Mathematical Focus 6

Mathematical theory involving Taylor series provides a definition of the sine function based on the foundations of the real number system, independent of any geometric considerations.

The sine function could be defined using an infinite series. The following identity holds for all real numbers x, with angles measured in radians:

$$\sin x = x - \frac{x^3}{3!} + \frac{x^5}{5!} - \frac{x^7}{7!} + \dots = \sum_{n=0}^{\infty} \frac{(-1)^n x^{2n+1}}{(2n+1)!}$$

The sine function is closely approximated by its Taylor polynomial of degree 7 for a full cycle centered on the origin, $-\pi \le x \le \pi$ (see Figure 41.8).

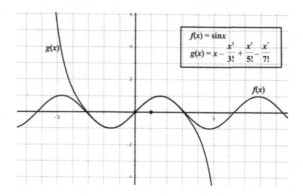

FIGURE 41.8.   A Taylor polynomial approximating the sine function.

POSTCOMMENTARY

Although the methods shown in the Foci differ in the use of ratios versus the use of lines as approximation tools for the sine function, each of the methods involves approximations. The ratio methods depend on a definition of the trigonometric functions and therefore are not generalizable to other types of functions, whereas the line methods depend on characteristics of continuous functions and therefore can be used for a wider range of functions.

CHAPTER 42

GRAPHING SIN(2X)

Situation 36 From the MACMTL-CPTM Situations Project

Heather Johnson, Evan McClintock, Rose Mary Zbiek,
Brian Gleason, Shawn Broderick, and James Wilson

PROMPT

During a lesson on transformations of the sine function, a student asked, "Why is the graph of $y = \sin(2x)$ a horizontal shrink of the graph of $y = \sin(x)$ instead of a horizontal stretch?"

COMMENTARY

The period of the function $\sin(kx)$, $k \neq 0$ is $\dfrac{2\pi}{|k|}$. Thus, the case of $k = 2$ represents a horizontal shrink of the function $\sin(x)$, since the period is smaller. Focus 1 and Focus 2 emphasize the periodicity of the sine function, appealing to composition of functions and to the unit circle. In contrast, Focus 3 emphasizes the first derivative of $y = \sin(2x)$ as the instantaneous rate of change of the function.

Mathematical Understanding for Secondary Teaching: A Framework
and Classroom-Based Situations, pages 385–389.

MATHEMATICAL FOCI

Mathematical Focus 1

The periodic function g(x) = sin(2x) is a composition of the periodic function f(x) = sin(x) and the linear function h(x) = 2x. The period of g is smaller than the period of f, which is represented graphically by a horizontal "shrink."

TABLE 42.1   Values of sin(x) and sin(2x) for Selected Values of x Between 0 and 2π.

x	2x	sin(x)	sin(2x)
0	0	0	0
$\pi/12$	$\pi/6$	0.259	1/2
$2\pi/12 = \pi/6$	$\pi/3$	1/2	$\sqrt{3}/2$
$3\pi/12 = \pi/4$	$\pi/2$	$\sqrt{2}/2$	1
$4\pi/12 = \pi/3$	$2\pi/3$	$\sqrt{3}/2$	$\sqrt{3}/2$
$5\pi/12$	$5\pi/6$	0.966	1/2
$6\pi/12 = \pi/2$	π	1	0
$7\pi/12$	$7\pi/6$	0.966	-1/2
$8\pi/12 = 2\pi/3$	$4\pi/3$	$\sqrt{3}/2$	$-\sqrt{3}/2$
$9\pi/12 = 3\pi/4$	$3\pi/2$	$\sqrt{2}/2$	-1
$10\pi/12 = 5\pi/6$	$5\pi/3$	1/2	$-\sqrt{3}/2$
$11\pi/12$	$11\pi/6$	0.259	-1/2
$12\pi/12 = \pi$	2π	0	0
$13\pi/12$	$13\pi/6$	-0.259	1/2
$14\pi/12 = 7\pi/6$	$7\pi/3$	-1/2	$\sqrt{3}/2$
$15\pi/12 = 5\pi/4$	$5\pi/2$	$-\sqrt{2}/2$	1
$16\pi/12 = 4\pi/3$	$8\pi/3$	$-\sqrt{3}/2$	$\sqrt{3}/2$
$17\pi/12$	$17\pi/6$	-0.966	1/2
$18\pi/12 = 3\pi/2$	3π	-1	0
$19\pi/12$	$19\pi/6$	-0.966	-1/2
$20\pi/12 = 5\pi/3$	$10\pi/3$	$-\sqrt{3}/2$	$-\sqrt{3}/2$
$21\pi/12 = 7\pi/4$	$7\pi/2$	$-\sqrt{2}/2$	-1
$22\pi/12 = 11\pi/6$	$11\pi/3$	-1/2	$-\sqrt{3}/2$
$23\pi/12$	$23\pi/6$	-0.259	-1/2
$24\pi/12 = 2\pi$	4π	0	0

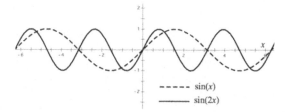

FIGURE 42.1. Graphs of sin(x) and sin(2x) for values of *x* between -2π and 2π.

The composition uses the outputs of $h(x) = 2x$ as inputs of $f(x) = \sin(x)$. Table 42.1 lists these inputs and outputs for selected values of *x* from 0 to 2π, and it should be apparent from this table that sin(2x) "progresses" through the cycle of output values twice as fast as sin(x) does, because 2x "progresses" through its output values twice as fast as *x* does (see the entries in boldface type). Figure 42.1 allows one to compare the graphs of these periodic functions, from which it is also apparent that sin(x) has a larger period, 2π, than does sin(2x), which has period π. Therefore, the graph of sin(2x) is a horizontal shrink of the graph of sin(x).

Mathematical Focus 2

The unit circle is the locus of all points with coordinates (cosθ, sinθ), which illustrates that the period of sinθ is 2π, whereas the period of sin(2θ) is π.

Consider the unit circle with embedded triangles ABC and A'B'C' as shown in Figure 42.2. If θ is allowed to vary from 0 to 2π, then it is apparent that A' will traverse the unit circle twice while A traverses the unit circle only once. Because

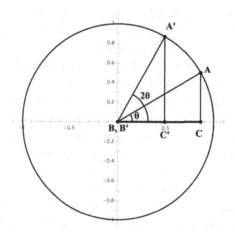

FIGURE 42.2. Unit circle illustrating sin(θ) and sin(2θ).

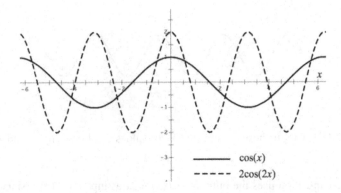

FIGURE 42.3. Graphs of cos(x) and 2cos(2x) for values of x between -2π and 2π.

$|A'C'| = \sin(2\theta)$, this double traversing of the unit circle by A' represents two periods of sin(2θ). Similarly, $|AC|$ sin( ) implies that the single traversing of the unit circle by A represents one period of sin(θ). Thus, sin(2θ) has half the period of sin(θ), and so its graph will be a horizontal shrink (by a factor of $\frac{1}{2}$), of the graph of sin(θ).

Mathematical Focus 3

The first derivative, as the instantaneous rate of change of a function, can be used in particular cases to locate maximum and minimum values of a function, and in the case of a regularly oscillating function, determine the period.

Because sin(kx) is composed of regular oscillations, one may judge its period (and hence its stretch) by how quickly these oscillations occur. A rough measure of this quickness is given by the derivative, because a faster change in the sign of the slope (or changes in rate of change from positive to negative) implies a faster oscillation. The derivative of sin(x) is cos(x), and the derivative of sin(2x) is 2cos(2x). Although the symbolic expressions cos(x) and 2cos(2x) are not immediately comparable—as cos(x) and 2cos(x) would be—because of the factor of 2 in 2x, the factor of 2 in 2cos(<some value>) may suggest that 2cos(2x) in general represents a larger value than cos(x). Examination of the graphs of cos(x) and 2cos(2x) challenges that conjecture (see Figure 42.3). Nevertheless, the zeros of 2cos(2x) are closer together than the zeros of cos(x), and because these zeros represent the relative extrema of the respective sin(<some value>) functions, the extrema of sin(2x) are closer together than the extrema of sin(x). This implies a faster oscillation and hence a horizontal "shrink."

POSTCOMMENTARY

The three Foci attempt to provide some intuitive understanding of why the graph of $y = \sin(2x)$ is a horizontal shrink—instead of a horizontal stretch—of the graph of $y = \sin(x)$. This can be counterintuitive to students because the graph of $y = 2\sin(x)$ is a vertical stretch of the graph of $y = \sin(x)$ and because students often associate multiplication with "making things larger," especially when the multiplicative factor is greater than 1.

CHAPTER 43

TRIGONOMETRIC IDENTITIES

Situation 37 From the MACMTL-CPTM Situations Project

Bob Allen, Sharon K. O'Kelley, and Erik Jacobson

PROMPT

While proving a trigonometric identity a student produced the following sequence of equations.

$$\sin x \cdot \cos x \cdot \tan x = \frac{1}{\csc^2 x}$$

$$\csc^2 x \cdot \sin x \cdot \cos x \cdot \tan x = 1$$

$$\frac{1}{\sin^2 x} \cdot \sin x \cdot \cos x \cdot \frac{\sin x}{\cos x} = 1$$

$$\cos x \cdot \frac{1}{\cos x} = 1$$

$$1 = 1$$

When asked about her reasoning, the student replied, "I just treated the equation like any algebra equation. You know, what you do to one side, you have to do

Mathematical Understanding for Secondary Teaching: A Framework and Classroom-Based Situations, pages 391–396.
Copyright © 2015 by Information Age Publishing

to the other, and then I showed it was the same as $1 = 1$. I know $1 = 1$ is true, so the identity must be true."

COMMENTARY

Proof in trigonometry is like proof in other areas of mathematics: One cannot assume a result in order to prove it but must progress by logical steps from a given statement or established result to the conclusion. When reasoning using a series of equations, it is important to show that each step necessarily and sufficiently follows from the previous step. The facts that only some manipulations of equations are reversible and that not all algebraic operations have inverses that preserve the domain of the original variable can introduce logical errors in trigonometric proofs.

MATHEMATICAL FOCI

Mathematical Focus 1

Assuming that a result is true when trying to prove its validity is invalid reasoning.

In the equation

$$\sin x \cdot \cos x \cdot \tan x = \frac{1}{\csc^2 x}$$

the student is asked to prove whether the equality of the identities is true. By multiplying both sides by $\csc^2 x$, however, the student is acting under the assumption that the equality is indeed true. In other words, multiplying both members of the equation by some factor presupposes that the equation is true.

This issue may be clarified by notation. For example, a question mark might be written above the equation as a reminder that the truth of the identity is conditional on the reversibility of the sequence of equations and the truth of the final step.

$$\sin x \cdot \cos x \cdot \tan x \overset{?}{=} \frac{1}{\csc^2 x} \ .$$

A better solution might be to use a different structure in the proof, as described in Focus 2 and Focus 3.

Mathematical Focus 2

A valid argument progresses by logical steps from a given statement or known result to the conclusion.

When proving trigonometric identities, it is most direct to set one side of the equation equal to itself (the known result) and by successive manipulations (logical steps) produce the desired result (conclusion).

Applying this method to the example in the Prompt yields:

$$\sin x \cdot \cos x \cdot \tan x = \sin x \cdot \cos x \cdot \tan x$$

$$\textit{iff } \sin x \cdot \cos x \cdot \tan x = \sin x \cdot \cos x \cdot \frac{\sin x}{\cos x}$$

$$\textit{iff } \sin x \cdot \cos x \cdot \tan x = \sin^2 x$$

$$\textit{iff } \sin x \cdot \cos x \cdot \tan x = \frac{1}{\csc^2 x}$$

In many textbooks, students are instructed to prove trigonometric identities by manipulating only one side of the identity using substitution until it is identical to the other side. The sequence of statements recorded for this proof method can be misleading, as discussed in Focus 1. Although it appears that one assumes the identity that is to be proved, in fact, because only one side of the equation is manipulated, the sequence of equations can be written as a chain of equivalent statements with one member of the identity's equation at each end. Thus, to prove the equality of trigonometric identities, substitutions are made on only one member of the equation while the other member is held constant.

$$\sin x \cdot \cos x \cdot \tan x = \frac{1}{\csc^2 x}$$

$$\sin x \cdot \cos x \cdot \frac{\sin x}{\cos x} = \frac{1}{\csc^2 x}$$

$$\sin x \cdot \sin x = \frac{1}{\csc^2 x}$$

$$\sin^2 x = \frac{1}{\csc^2 x}$$

$$\frac{1}{\csc^2 x} = \frac{1}{\csc^2 x}$$

As illustrated here, note that the proof of an identity is possible by producing successive equivalent expressions in only the member on one side (say the left) of the purported equality. During this manipulation, no reference is made to equating the left-hand member to the right-hand member until the expression in the left-hand member appears identical to the expression in the right-hand member.

Mathematical Focus 3

Another valid style of argument shows that the identity to be proven is necessary and sufficient for a known identity.

In the Prompt, the student shows that if the identity to be proved is true, then $1 = 1$. In so doing, the student implies that the converse is also true. Yet a statement and its converse are not logically equivalent.

The student's proof is not valid as it stands, but it might be made valid by reversing the steps. In other words, the identity is shown to be true only if one can start with a known identity (such as $1 = 1$) and, by working backwards, using algebraic operations that are inverses of the ones used in proceeding in the other direction, one can demonstrate the equality of the identity to be proved.

$$1 = 1$$

$$\cos x \cdot \frac{1}{\cos x} = 1$$

$$\frac{1}{\sin^2 x} \cdot \sin x \cdot \cos x \cdot \frac{\sin x}{\cos x} = 1$$

$$\csc^2 x \cdot \sin x \cdot \cos x \cdot \tan x = 1$$

$$\sin x \cdot \cos x \cdot \tan x = \frac{1}{\csc^2 x}$$

Because each step follows sufficiently from the previous one, this is a valid argument that the identity to be proved is true. This procedure does not always work, however, as described in Focus 4.

Mathematical Focus 4

Only some algebraic operations have inverses that preserve domain; only some manipulations of equations are reversible.

In general, algebraic manipulations of an equation show that the original equation is sufficient for the new equation, but not necessary. In the following argument (Example 1), each member of an equation is squared, introducing extraneous roots. This example demonstrates that the statement $\sin \phi = -1$ is sufficient but not necessary for the statement $(\sin \phi)^2 = 1$.

Example 1

The solution of $\sin \phi = -1$ is $\phi = \frac{3\pi}{2} \pm 2n\pi$, $n = 0, 1, 2, \dots$.

But, squaring both members of the equation $\sin \phi = -1$ results in $(\sin \phi)^2 = 1$.

$$(\sin \phi)^2 = 1 \Rightarrow (\sin \phi)^2 - 1 = 0$$

$$\Rightarrow (\sin \phi - 1)(\sin \phi + 1) = 0$$

The solutions of $(\sin \phi - 1)(\sin \phi + 1) = 0$ are $\frac{\pi}{2} \pm 2n\pi$, $n = 0, 1, 2, \dots$, relating to the first factor $(\sin \phi - 1)$, and $\phi = \frac{3\pi}{2} \pm 2n\pi$, $n = 0, 1, 2, \dots$, relating to the second factor $(\sin \phi + 1)$.

So the solutions of $\sin \phi = 1$ and $(\sin \phi)^2 = 1$ are not the same.

The trouble is that the inverse operation of squaring is not a function—each positive real number has two square roots. Another common problem of this type occurs when one multiplies both sides of an equation by an expression that might take on the value of 0. The inverse operation would require one to divide by that expression, but this is undefined whenever the expression has a value of 0.

Consider Example 2. Beginning with the equation $x = -3$, first multiply both sides of the equation by $\sin x$. Reversing the operation involves dividing by $\sin x$, but this does not make sense whenever x is a multiple of π because $\sin n\pi = 0$, for all integers n. Rather than regaining $x = -3$, the nonequivalent statement "$x = -3$ whenever x is not a multiple of π" is produced. Here, the trouble is that the inverse operation of multiplying by $\sin x$ is not defined for some of the values of x to which the original equation applied. The equation $x = -3$ is sufficient to show $x \sin x = -3 \sin x$, but it is not necessary.

Example 2

$x = -3$ implies $x \sin x = -3 \sin x$, for all real values of x but $x \sin x = -3 \sin x$ does not imply $x = -3$ because the principle that $ac = bc \; a = b$ does not apply when $c = 0$ (i.e., $a(0) = b(0)$ does not imply that $a = b$).

The problem of the domain of an expression is particularly applicable when proving trigonometric identities. Consider the following possible identity: $\csc x - \cos x \cdot \cot x \overset{?}{=} \sin x$. Does the following argument show that the identity is true?

$$\csc x - \cos x \cdot \cot x = \sin x$$

$$\sin x \left(\frac{1}{\sin x} - \cos x \cdot \frac{\cos x}{\sin x} \right) = (\sin x) \sin x$$

$$1 - \cos^2 x = \sin^2 x$$

No, it merely shows that $\csc x - \cos x \cdot \cot x = \sin x$ is sufficient for $1 - \cos^2 x = \sin^2 x$. In fact, it is not necessary; whenever $x = 0$, $\sin x = 0$ but $\csc x - \cos x \cdot \cot x$ is undefined. To see why this is the case, one can try to reverse the argument. Certainly, $1 - \cos^2 x = \sin^2 x$ is true, and then using the inverse operations of those used above:

$$1 - \cos^2 x = \sin^2 x$$

$$\frac{1}{\sin x}(1 - \cos x \cdot \cos x) = (\sin^2 x)\frac{1}{\sin x}; \; x \neq n\pi, \; n \in \mathbb{Z}$$

$$\csc x - \cos x \cdot \cot x = \sin x; \; x \neq n\pi, \; n \in \mathbb{Z}.$$

Just as in the second example, dividing by $\sin x$ is undefined whenever x is a multiple of π. The order of proof used by the student in the Prompt (assuming the identity to be proved and then showing it sufficiently implies a known identity) works only when each algebraic manipulation of the equation is reversible and the domains of relevant functions are preserved.

A simple example serves to illustrate the point. Consider this four-step deduction:

Step 1: -1 = 1

Step 2: $(-1)^2 = 1^2$

Step 3: Because $(-1)^2 = 1$ and $1^2 = 1$, step 2 gives 1 = 1.

Step 4: Because 1 = 1 is true, step 1 (-1 = 1) must be true.

Because Step 1 clearly is not true, the lesson is that a correct resulting equation does not mean the "equation" one started with was correct. Starting with a statement and producing a sequence of correct implications from that statement and arriving at a statement known to be true does not guarantee that the original statement is true. Unless the steps in the deductive sequence are reversible, the original statement might not be true, as illustrated in this four-step example. In this case, Step 2 follows from Step 1, but this implication is not reversible (Step 1 does not follow from Step 2).

POSTCOMMENTARY

Students who are familiar with proofs by contradiction may be confused by arguments that appear to assume the conclusion (instead of its negation) in order to prove it. It may be helpful to think of this assumption as an exploration separate from the proof. The question during the phase of exploration might be, "If this equation were an identity, what other (known) identities could be confirmed?" As soon as the student has found such an identity (in the Prompt, the student used 1 = 1), the "real" proof begins. Now the question is, "Can each step be reversed?" This way of thinking highlights the fact that a proof by contradiction works backwards, starting with the negated conclusion and then contradicting a premise or a known fact, whereas proofs of identities are usually direct arguments. They start with a known premise and proceed logically to the conclusion.

CHAPTER 44

MEAN AND MEDIAN

Situation 38 From the MACMTL–CPTM Situations Project

Susan Peters, Evan McClintock, Donna Kinol, Shiv Karunakaran,
Rose Mary Zbiek, M. Kathleen Heid,
Laura Singletary, and Sarah Donaldson

PROMPT

The following task was given to students at the end of the year in an AP Statistics class.

Consider the box plots and five-number summaries[1] for two distributions, each of which is comprised of a finite number of data values (see Figure 44.1 and Figure 44.2). Which of the distributions (Data Set 1 or Data Set 2) has the greater mean?

One student's approach to this problem was to construct what he thought were probability distributions for each data set and to compare the corresponding expected values to determine which data set had the greater mean. The student formed four intervals using the five-number summaries and calculated the midpoint of each interval (i.e., he defined the intervals as the four quarters of the distributions, with each quarter containing 25% of the values for the distribu-

Mathematical Understanding for Secondary Teaching: A Framework and Classroom-Based Situations, pages 397–404.
Copyright © 2015 by Information Age Publishing

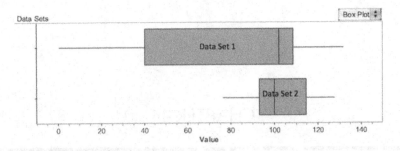

FIGURE 44.1. Box plots for Data Set 1 and Data Set 2.

Data Sets

	Group	
	one	two
Value	0	76
	40	93
	102	100
	109	115
	132	128

S1 = min ( )
S2 = Q1 ( )
S3 = median ( )
S4 = Q3 ( )
S5 = max ( )

FIGURE 44.2. Five-number summaries for Data Set 1 and Data Set 2.

Data Set 1:

E(X) = 79.25

	0 – 40	40 - 102	102 - 109	109 - 132
X	20	71	105.5	120.5
P(X)	0.25	0.25	0.25	0.25

Data Set 2:

E(X) = 102.75

	76-93	93-100	100-115	115-128
X	84.5	96.5	107.5	121.5
P(X)	0.25	0.25	0.25	0.25

FIGURE 44.3. The student's calculations for the two data sets.

tion). Using the midpoint of each interval as the X-value for that interval, he then calculated the weighted mean for each probability distribution (see Figure 44.3). After completing his calculations, the student responded that the second data set had the larger mean.

COMMENTARY

This Prompt deals with the differences and similarities between the mean and median of a particular data set when the data set is displayed as a box plot using its five-number summary. It is likely that the intent of the question was not to encourage mathematical calculations, but rather to ask students to predict which distribution would have a greater mean based on what is expected from the visual display of the box plots.

Another important aspect to consider is that the problem given in the Prompt is given without a context. Without knowing the context of the data given, one does not know whether the data are continuous or discrete. Also, one does not know the statistical question being considered, namely, why the data were collected.

MATHEMATICAL FOCI

Mathematical Focus 1

The skewness of a data distribution affects the relationship between the mean and median of that set of data.

The box plots in Figure 44.1 display information about the distributions of the two data sets. Data Set 2 appears to be roughly symmetric with possible right skewness (note that the median is pulled to the left of central box). For this case, one would expect the mean and the median to be approximately equal, or the mean to be slightly greater than the median. Data Set 1 appears skewed left. If the distribution for the first data set is skewed to the left, then the smaller values have a stronger impact on the mean than the larger values. Therefore, one expects the mean to be less than the median. On the other hand, the median is the "middle" value of the data set after the data set is arranged in increasing order, and it is resistant to the larger spread in the smaller values. Because the medians are similar in each distribution, one expects Data Set 2 to have the larger mean. Although reasoning via the shape of the distribution is an approach that typically works when making comparisons about distributions, it is not always possible to make definitive statements about the relative locations of the means for some pairs of box plots.

Mathematical Focus 2

A box plot display of data does not necessarily give the data values or information about the "distribution" of the data within each quarter.[2]

How the data are distributed within each interval determined by the five-number summary is not represented in a box plot. The mean of a particular interval represented by the midpoint would be representative of the data points in that interval only when the data are distributed normally, uniformly, or symmetrically within that interval. The information given in the Prompt does not allow one to make such an assumption.

In the Prompt the student assumed that each interval contains exactly 25% of the data points. However, this is true only when the number of data points is divisible by 4. Also, some of the numbers in the five-number summary may not be members of the data set. The only values in the data set that are known for certain from the box plot are the minimum and maximum values. The median will be a member of the data set only when the number of data points is odd.[3] Q_1 and Q_3 are members of the data set only when the size of the data set has a remainder of 2 or 3 when divided by 4.[4]

Mathematical Focus 3

When exact values of two data sets are not known, comparisons between the two data sets can sometimes be made by comparing the ranges of their possible values.

For the given box plots, one can calculate an upper bound and a lower bound for the means of the data sets. Because of the apparent skewness discussed in Focus 1, one can assume that Data Set 2 has a greater mean than Data Set 1. To investigate this, consider the upper bound for the mean of Data Set 1 and compare it to the lower bound for the mean of Data Set 2. The greatest possible value of the mean of Data Set 1 is strictly less than the least possible value of the mean of Data Set 2. Therefore, it can be concluded that the mean of Data Set 2 is greater than the mean of Data Set 1.

To find an upper bound for the mean of Data Set 1, assume that in each interval the data points are located at the greatest possible value within the interval. (Of course, strictly speaking, since 0 is the minimum value of the data set, 0 is also a data point in the first interval; however, because an upper bound is being investigated and n is not known, 0 is not included in the calculation.) Let 25% of the values in Data Set 1 be located at the maximum value of each interval, that is, at Q1, at the median, at Q3, and at the maximum. In this extreme case, the mean of the data set is given by:

$$E(\text{Data Set 1}) = 0.25(40) + 0.25(102) + 0.25(109) + 0.25(132) = 95.75$$

Similarly, to find a lower bound of the mean for Data Set 2, assume that in each interval the data points are located at the least possible value within the interval. (Using this method, do not include the maximum value, 128, in the calculation.) Let 25% of the values in Data Set 2 be located at the minimum value of each interval, that is, at the minimum, at Q1, at the median, and at Q3. In this extreme case, the mean of the data set is given by:

$$E(\text{Data Set 2}) = 0.25(76) + 0.25(93) + 0.25(100) + 0.25(115) = 96.00$$

Because the lower bound of the mean of Data Set 2 is greater than the upper bound of the mean of Data Set 1, one can conclude that the mean of Data Set 2 is greater than the mean of Data Set 1.

The previous example assumed that each interval contained exactly 25% of the data set. This assumption may often be wrong, because this situation occurs only when the size of the data set is equivalent to $4n$ (i.e., is divisible by 4). The Postcommentary contains further investigation of Data Set 1, taking into consideration the possibility that the data set is of size $4n + 1$, $4n + 2$, or $4n + 3$. Even in these cases, however, the mean of Data Set 2 can be shown to be greater than the mean of Data Set 1 (see Appendix to Focus 3).

Mathematical Focus 4

Stating a definitive conclusion about a comparison of the means using the five-number summaries and box plots is not always possible because the size of the data set may influence the relationship between the means for these distributions.

The following example portrays the importance of sample size and its effect on the relationships between the means and the five-number summaries for the given distributions.

For this example, consider the possibility that each data set contained 12 values (the box plots and five-number summary of which appear in Figure 44.4 and Figure 44.5). If the values in Data Set A were 14, 14, 17, 17, 17, 21, 21, 21, 24, 24, 24, and 30, then the mean of Data Set A would be 20.333. If Data Set B contained the values of 0, 12, 12, 12, 22, 22, 22, 26, 26, 26, 31, and 31, then the mean of

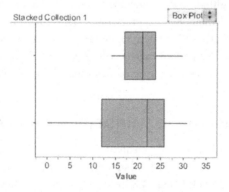

FIGURE 44.4.   Box plot for Data Set A and Data Set B.

	Group	
	One	Two
	14	0
	17	12
Value	21	22
	24	26
	30	31

S1 = min( )
S2 = Q1( )
S3 = median( )
S4 = Q3( )
S5 = max( )

FIGURE 44.5.   Five-number summary for Data Set A and Data Set B.

Data Set B would be 20.167. Thus, the mean of Data Set A would be larger than the mean of Data Set B.

However, what if each set of data contained 100 values and the five-number summaries were maintained? If Data Set A were distributed with 24 values at 14, 25 values at 17, 25 values at 21, 25 values at 24, and 1 value at 30, then the mean of Data Set A would be 19.16. If Data Set B were distributed with 24 values at 31, 25 values at 26, 25 values at 22, 25 values at 12, and 1 value at 0, the mean of Data Set B would be 22.44. Thus, the mean of Data Set B would be larger than the mean of Data Set A.

Because the sizes of the data sets may influence the relationship between their means, stating definitive conclusions about a comparison of means is not always possible for a pair of box plots.

POSTCOMMENTARY

Each of the Foci highlight a difference between information that allows conclusions to be made with mathematical precision and information that allows only for general claims to be made. For example, Focus 3 contains a conclusive argument for the relative sizes of the two means, whereas Focus 1 describes how claims can be made based on what is expected using reasoning about the shape of the distribution.

APPENDIX TO MATHEMATICAL FOCUS 3

In Focus 3, the upper bound of the mean of Data Set 1 and the lower bound of the mean of Data Set 2 were calculated assuming the size of each data set was divisible by 4. Figures 44.6, 44.7, and 44.8 illustrate the possibilities not considered in Focus 3 (i.e., that the size of a data set might be $4n + 1$, $4n + 2$, or $4n + 3$).

Consider the possibility that Data Set 1 contains $4n + 1$ data values. To maintain the same five-number summary, the extra data value would be located at the median, 102. The mean value of 95.75 previously calculated for Data Set 1 assumed 25% of the data values would lie in each segment; however, the lower extreme value of 0 was not accounted for in the calculation. Note that accounting for the lower extreme of 0 (by effectively removing a value of 40) will lower the mean more than the addition of a value at 102 will increase the mean. Thus, the mean of Data Set 2 is still greater than the mean of Data Set 1.

If Data Set 1 contains $4n + 2$ values, then one additional value will be located at Q1 and one additional value will be located at Q3. Accounting for the minimum of 0, this situation nets the addition of a value at 0 and a value at 109. The value added at 0 lowers the mean more than the value of 109 increases the mean, and the mean of Data Set 2 is still greater than the mean of Data Set 1.

Finally, if Data Set 1 contains $4n + 3$ values, then a value is added at each of Q1, the median, and Q3. Accounting for the minimum of 0, this situation nets the addition of a value at 0, a value at 102, and a value at 109. The value added at 0 lowers the mean more than the addition of values both at 102 and 109. Therefore, the mean of Data Set 2 is still greater than the mean of Data Set 1.

In all cases, the mean of Data Set 2 is greater than the mean of Data Set 1. Similar arguments can be made for different sample sizes and their effects on the mean of Data Set 2.

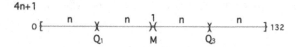

FIGURE 44.6. Data Set 1 contains $4n + 1$ data values.

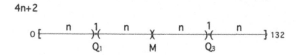

FIGURE 44.7. Data Set 1 contains $4n + 2$ data values.

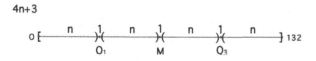

FIGURE 44.8. Data Set 1 contains $4n + 3$ data values.

NOTES

1.   Box plots are sometimes referred to as boxplots or box-and-whisker plots. The box plot is a visual display of the five statistics values that comprise the five-number summary.
2.   A *quartile* is the boundary point for quarters of the data.
3.   Note that if there are multiple data values that all have the same value as the median, the median may not be a specific member of the data set but will have the same value as members of the data set. In that case, the average of the middlemost two values would equal the two values and hence the median would be equal in value to members of the data set. For example, if the data set consists of an even number of data points, say 5, 5, 5, and 5, the median is 5, which is not one of the specific members of the data set but which has the same value as each of the members of the data set.
4.   The first and third quartiles were calculated using one particular convention. As noted by Kadar and Jacobbe (2013), "There are several different methods for determining quartiles. For example, when *n* is odd, the ordered data cannot be evenly divided in half. One method excludes the median from the lower and upper 'halves' when determining the quartiles, while another method includes the median in both 'halves'" (p. 35). The convention assumed in this Situation excluded the median from the lower and upper "halves," which is a method typically used in schools.

REFERENCE

Kadar, G., & Jacobbe, T. (2013). *Developing essential understanding of statistics for teaching mathematics in Grades 6–8*. Essential understanding series. (P. Wilson, Vol. Ed.; R. M. Zbiek, Series Ed.). Reston, VA: National Council of Teachers of Mathematics.

CHAPTER 45

REPRESENTING STANDARD DEVIATION

Situation 39 From the MACMTL–CPTM Situations Project

Rose Mary Zbiek, M. Kathleen Heid, Shiv Karunakaran,
Ryan Fox, Eric Gold, Sarah Donaldson, and Laura Singletary

PROMPT

In prior lessons, students learned to compute mean, mode, and median. The teacher presented the formula for standard deviation and had students work through an example of computing the standard deviation with data from a summer-job context. The written work in Figure 45.1 developed during the example.

The teacher then said, "Standard deviation is a measure of the consistency of our data set. Do you know what *consistency* means?" To explain consistency the teacher used the idea of throwing darts. One student pursued the analogy, "If you hit the bull's eye, your standard deviation would be lower. But if you're all over the board, your standard deviation would be higher." The student drew the picture in Figure 45.2 to illustrate his idea:

*Mathematical Understanding for Secondary Teaching: A Framework
and Classroom-Based Situations,* pages 405–413.

x	$\bar{x}$	$(x-\bar{x})$	$(x-\bar{x})^2$
140	200	−60	3600
190	200	−10	100
210	200	10	100
260	200	60	3600

$$\sum(x-\bar{x})^2 = 7400$$

$$\sigma = \sqrt{\frac{\sum(x-\bar{x})^2}{n}}$$

$$\sigma = \sqrt{\frac{7400}{4}}$$

$$\sigma = \sqrt{1850}$$

$$\sigma = 43.01$$

FIGURE 45.1.   Written work accompanying the summer-job example.

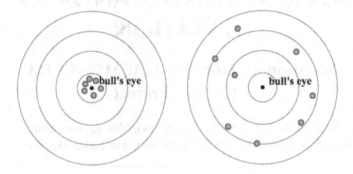

FIGURE 45.2.   Student's drawing illustrating "low" and "high" standard deviations.

A student raised her hand and asked, "But what does this tell us about what we are trying to find?"

COMMENTARY

This Prompt seems to deal with the analogy between the data points, mean, and the standard deviation of a data set and the places on a dartboard target where the darts land. The main issue raised by the student in the Prompt is the representation of standard deviation as either "hit[ting] the bull's eye" or being "all over the board." In the student's analogy, the bull's eye is treated as the mean of the data points. Standard deviation is a measure of the spread of a data set, so there are aspects of the target analogy that make it a viable representation because it addresses two key aspects of a visual representation: the objects and the relative positions of those objects with respect to the mean. However, the robustness of a representation depends upon how well these two aspects collectively represent the key mathematical objects being represented (e.g., data points, mean, standard deviation) and the relationships among those objects. The target representation contains severe limitations in this regard. An alternative representation (using a one-dimensional number line rather than the two-dimensional target) is suggested in Focus 5.

Note that within the Prompt, there is not a consistent use of notation. To clarify, it is necessary to distinguish between *parameters*, which refer to the population, and *statistics*, which refer to a sample. This distinction is seen in the two formulas for standard deviation. For the population, the standard deviation is defined to be $\sigma = \sqrt{\dfrac{\Sigma(x-\mu)^2}{N}}$, where μ is the population mean and N is the population size. For a sample, the standard deviation is defined to be $s = \sqrt{\dfrac{\Sigma(x-\bar{x})^2}{n-1}}$, where $\bar{x}$ is the sample mean and n is the sample size. The intent in the Prompt is to calculate a statistic, but a mixture of parameter and statistic notation is used, resulting in an incorrect formula for standard deviation of the sample. From this point forward, this Situation uses parameter notation when referring to the population and statistic notation when referring to a sample.

MATHEMATICAL FOCI

Mathematical Focus 1

Standard deviation is a measure of the spread of the data with respect to the mean.

Standard deviation is one way of indicating the spread of the data with respect to the mean. The more spread out the data are from the mean, the larger the standard deviation for that particular data set. A smaller standard deviation indicates that the data are more tightly grouped about the mean. The standard deviation for a data set is somewhat like the average distance that the data points are from the mean. However, this is not a strict average, as is seen in the formula for standard deviation: $s = \sqrt{\dfrac{\Sigma(x-\bar{x})^2}{n-1}}$. What some[1] see as a true average of the spread of the data with respect to the mean is known as the Mean Absolute Deviation (*MAD*), and is calculated by the following formula: $MAD = \dfrac{\Sigma|x-\bar{x}|}{n}$. Both of these formulas provide information about the spread of the data with respect to the mean, and it is desirable that they yield nonnegative values. In the formula for standard deviation, the values are nonnegative because each difference, $(x-\bar{x})$, is squared. In the formula for the *MAD*, the values are nonnegative because the absolute value is calculated for each difference, $(x-\bar{x})$. The *MAD* is a good introduction to the concept of measuring the spread of the data distribution, and therefore is a nice transitional way to help students understand the statistical concept of standard deviation. In theoretical statistics however, the formula for standard deviation is used rather than the *MAD*, because, among other things, $(x-\bar{x})^2$ is easier to work with mathematically than $|x-\bar{x}|$. Further rationale for the formula for standard deviation of a sample (in particular the denominator $n - 1$) is given in Situation 40 (see Chapter 46).

Mathematical Focus 2

If elements of a visual representation of a formula represent symbolic elements of the formula, the relative positions of those elements should be consistent with the meaning of the symbols.

The formula for the sample standard deviation contains the symbols, n, x, $\bar{x}$ and s. The number of data points, n, is represented by the number of darts on the target. Every data point (that is, each x-value) is represented by the place a dart lands. For the two targets shown in the Prompt, the bull's eye (center of the target) is treated as the sample mean, $\bar{x}$. Because distance is nonnegative, each distance between a dart and the bull's eye is associated with a value of $|x - \bar{x}|$. The standard deviation, s, is indicated by how spread out the darts are. For example, if the darts are tightly clustered around the bull's eye, this represents a small standard deviation, and if the darts are "all over the board," the data set has a larger value for s. If, on the other hand, the darts are clustered in an area of the board that is not near the bull's eye, the bull's eye is not an appropriate representation for the mean.

Mathematical Focus 3

Data points may be compared either to the sample mean or the population mean. In either case, placement of these points in a visual representation may indicate something about the standard deviation of the data set.

Targets A and B, shown in Figure 45.3, represent the distribution of two different sets of data. For the targets in each case, the exact center of the circle,

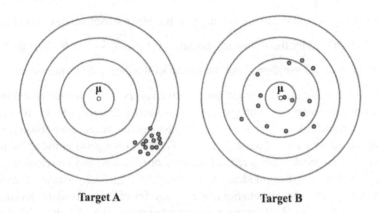

Target A   Target B

FIGURE 45.3.   Data sets with different relationships to μ.

indicated by the open circle on the target, represents the population mean, μ. The center of the data points is the sample mean, $\bar{x}$. The exact center of Target A is not the same as the center of the data points, indicating that the mean calculated from this sample is not a desirable estimate of the population mean. It may be said that the sample mean in Target A is a *biased* estimate (either an overestimate or underestimate) of the population mean. The bias of the sample mean may have occurred because of the way the sample was selected. Target B, on the other hand, shows a sample whose mean is a closer estimate of the population mean, because the center of the data points is close to the center of the target. That is, in Target B, the sample mean, $\bar{x}$, is closer to an *unbiased* estimate of the population mean, μ.

Clearly, the spread of the distribution of the data points is different in each target representation. In Target A, the data points are more tightly clustered than the data points in Target B. This indicates a smaller sample standard deviation for the set of data points in Target A. That is, the data points in Target A have a greater consistency (i.e., smaller variability) than those in Target B.

Therefore, each target representation displays a characteristic of an ideal sample, which typically is achieved through random sampling and using large sample sizes. Random sampling, in the long run, would yield an unbiased estimate of the population mean (as in Target B). Using a large sample size yields more consistency in the distribution of data points (as in Target A). Therefore random sampling reduces bias and large sample size reduces variability.

Mathematical Focus 4

The position of data points relative to the mean of the data set is best represented using a one-dimensional analogy because each data point is either greater than, equal to, or less than the mean.

A one-dimensional representation of the data can be created from the two-dimensional target representation. In what follows, the small open circles in Figure 45.4 represent the sample mean, $\bar{x}$. To construct this representation, begin by calculating and placing the sample mean on the target. Then place each data point so that its distance from the sample center represents the absolute value of the difference between the data value and the sample mean, $|x-\bar{x}|$. The point may be anywhere on the target such that this criterion is met (i.e., the point may be anywhere on a circle with radius $|x-\bar{x}|$). For example, if the sample mean is 200, the data points 190 and 210 will lie on the same circle, because in each case the value of $|x-\bar{x}|$ is 10. However, it may seem counterintuitive to randomly place these two data points on this circle; rather, one may prefer to place them on different sides of the sample mean based on whether the data values are greater than or less than the sample mean. In Figure 45.4, the random placement of the data points may be confusing. In Figure 45.4, the data set from the Prompt is being used: The *x*-values from the sample are 140, 190, 210, and 260, and the sample mean is 200.

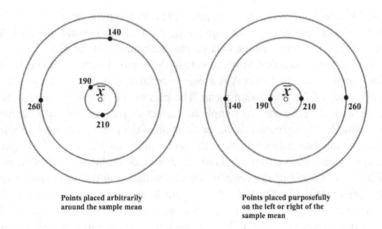

FIGURE 45.4. Placement of data values, x_i, so that $|x_i - \bar{x}|$ is represented by placement on circles with radii $|x_i - \bar{x}|$.

Several things are worth noting about the placement of data values in the two targets in Figure 45.4. In the target on the left, the data values are placed arbitrarily on circles with radii, $|x - \bar{x}|$; the only "rule" is that a point lies a certain distance from the center. In contrast, the target on the right displays an additional feature: The data points lie on a diameter in increasing order from left to right. Note that this more purposeful placement of the data points implies a linear representation. The target on the right has the limitation that there is no way to indicate repetition of data points. That is, the distribution of the data remains unclear, and therefore information about the standard deviation also is unclear.

An alternative to the two-dimensional target representation involves a linear placement of the data points. Such a representation shows where data points lie in relation to the mean, and this relationship is one-dimensional; that is, the points are either less than or greater than the mean (or are equal to the mean). The dot plot in Figure 45.5 is a linear representation that shows the spread of the same data points used in the Prompt.

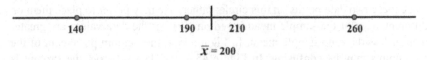

FIGURE 45.5. A linear representation that shows the spread of a set of data without duplicate data.

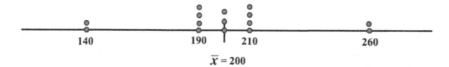

FIGURE 45.6.   A linear representation that shows the spread of a set of data with duplicate data.

A dot plot has several advantages over either target representation. First, if there is more than one data point at a particular value, this can be indicated as shown in Figure 45.6.

Second, the spread of the data about the sample mean provides a visual representation of the concept of standard deviation. It is possible to compare the standard deviations of two data sets based on their dot plots. For example, in the dot plot in Figure 45.6, data values are more tightly grouped about the mean than the data values in the dot plot in Figure 45.5. This implies that the data distribution in Figure 45.6 has a smaller standard deviation than the data distribution represented in Figure 45.5. The larger sample size in Figure 45.6 also plays a role in the standard deviation being smaller than the standard deviation for the data distribution in Figure 45.5. Specifically, the standard deviation of the data set in Figure 45.6 is approximately 32.95, which is less than the standard deviation of the data set in Figure 45.5, which is approximately 49.67.

Mathematical Focus 5

Different representations may be more or less effective in representing the spread of a data set.

A *box-and-whisker* plot[2] also conveys something about the spread of a data set. However, in this type of representation, the measure of central tendency is the *median* of the data set rather than the mean. The box-and-whisker plot provides a visual representation of what is called the *five-number summary* of the data set (minimum, Q_1, median, Q_3, and the maximum),[3] so that the data set is separated into four intervals. Whereas the box-and-whisker plot does not indicate the shape of the distribution of data values within a particular interval, the box-and-whisker plot as a whole shows something about the spread and possible skewness of the data set around the median.

There are limitations to using either the target representation or the box-and-whisker plot. Neither representation indicates the sample size, n. In the target representation, repeated data values are not necessarily apparent, and the box-and-whisker plot does not show particular data values other than the minimum and maximum of the data set. Because n is not necessarily known in either representation, there are limitations to the conclusions that can be made about the spread of

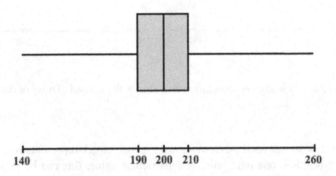

FIGURE 45.7. Box-and-whisker plot for the data values in Figure 45.5.

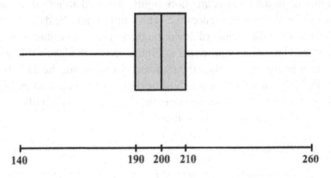

FIGURE 45.8. Box-and-whisker plot for the data values in Figure 45.6.

the data. For example, the data sets represented in Figures 45.5 and 45.6 have the same five-number summary, so their box-and-whisker plots are exactly the same (see Figures 45.7 and 45.8). Also, as discussed previously, the standard deviations of these two data sets are different. Lastly, because of the arbitrary placement of points on circles of radius $|x - \bar{x}|$ in the representation as shown in Figure 45.4, the possible skewness of the distribution of the data is not evident. The box-and-whisker plot, on the other hand, can illustrate possible skewness through the five-number summary. In the case of the data on which this discussion has focused, the box plots indicate that the data distribution appears to be symmetric.

In conclusion, there is value to each representation described in this Situation depending upon the intended purpose. The dot plot is useful in that it displays the actual data points and therefore illustrates the spread of these data values about the mean, possible bias, and consistency of the data distribution. The box-and-whisker plot is useful in that the five-number summary indicates the spread of the data in relation to measures of position (i.e., minimum, Q_1, median, Q_3, and the maximum). The box-and-whisker plot also allows for comparison of different data sets measured on a similar scale, such as scores on a standardized test with different groups represented with different box-and-whisker plots, aligned vertically (e.g., males/females, and other subgroup classifications). The target representation is useful for indicating possible bias and consistency of the data distribution.

NOTES

1. See http://www.leeds.ac.uk/educol/documents/00003759.htm for a description of and comparison of the standard deviation and the mean absolute deviation.
2. Box-and-whisker plots are also referred to as *box plots* or *boxplots*.
3. Q_1 and Q_3 refer to the first quartile and the third quartile, respectively.

CHAPTER 46

SAMPLE VARIANCE AND POPULATION VARIANCE

Situation 40 From the MACMTL-CPTM Situations Project

Ken Montgomery and Sarah Donaldson

PROMPT

A student in a statistics class had observed that the definition of sample variance (s^2) resembles the average of the squares of the deviations from the mean of the data set.

$$s^2 = \frac{\sum_{i=1}^{n}(x_i - \bar{x})^2}{n-1}$$

Yet, these squared deviations are summed and then divided by ($n-1$). "Why," asked the student, "do you not divide by n? Is this an actual mean or some sort of 'pseudomean'?"

Mathematical Understanding for Secondary Teaching: A Framework and Classroom-Based Situations, pages 415–419.

COMMENTARY

The concept in statistics known as *variance* is closely related to standard deviation: Both indicate the spread of the data distribution about the mean. In fact, the standard deviation is simply the principal square root of the variance. There are a number of reasons why the variance is calculated the way it is. As the formula for sample variance shows, the sum of the squared deviations is divided by $n-1$. The following Foci are investigations of reasons why division by $n-1$, rather than division by n, is necessary to calculate the sample variance. For discussion of division by $n-1$ and the concept of variability, see references such as Agresti and Franklin (2007); Freund (1992); Mendenhall, Beaver, and Beaver (2005); Peck, Gould, and Miller (2013); and Shaughnessy and Chance (2005).

MATHEMATICAL FOCI

Mathematical Focus 1

Because the deviations sum to 0, only n *– 1 of the data values may "vary freely."*

One way to think about a data set is from the perspective of variability. That is, one might consider how much the data points in a sample differ (or deviate) from the sample mean, which may be interpreted as the balance point of the distribution of the data in the sample. The deviation of a single data point, x, from the sample mean, $\bar{x}$, is simply the difference $x - \bar{x}$. The sum of all these differences must be 0 because the sample mean, $\bar{x}$, is the balance point. This can be seen algebraically. For example, consider a data set of three values, x_1, x_2, and x_3. The value of $\bar{x}$ is the arithmetic mean of these values: $\bar{x} = \dfrac{x_1 + x_2 + x_3}{3}$. An equation equivalent to this one is $x_1 + x_2 + x_3 = 3\bar{x}$. Next, consider the sum of the deviations: $(x_1 - \bar{x}) + (x_2 - \bar{x}) + (x_3 - \bar{x})$. This expression is equivalent to $x_1 + x_2 + x_3 - 3\bar{x}$. Because $x_1 + x_2 + x_3 = 3\bar{x}$, it follows that, $x_1 + x_2 + x_3 - 3\bar{x} = 0$, so the sum of the deviations is 0. That is, $(x_1 - \bar{x}) + (x_2 - \bar{x}) + (x_3 - \bar{x}) = 0$.

The fact that the deviations sum to 0 helps one to think about the role of $n-1$ in the formula for sample variance. Because the deviations sum to 0, if all but one of them are known (that is, if $n-1$ of them are known), the value of the last (i.e., the nth) deviation is determined. In other words, only $n-1$ of the data values may "vary freely." In statistics this is known as *degrees of freedom*: the number of data values that have "freedom" to vary such that the mean remains the same.

For example, suppose a data set of five values is given and it is known that $\bar{x} = 9$. There are many sets of five data values that have a mean of 9, but as soon as four of them are known, the fifth is determined. One possibility is that four of the data values are 5, 8, 12, and 13, as shown in Table 46.1.

TABLE 46.1   Four Members of a Set of Five Data Values That Has a Mean 9 and Their Deviations From the Mean

x_i	$\bar{x}$	$x_i - \bar{x}$
5	9	-4
8	9	-1
12	9	3
13	9	4
?	9	

Given these four data values and their mean, one can determine the fifth data value, d. Because the sum of the deviations is 0,

$$-4 + -1 + 3 + 4 + d = 0$$

$$d = -2$$

Therefore the fifth data value is 7 because $7 - 9 = -2$.

Why, then, can it be concluded that the sample variance should be calculated by dividing by $n - 1$? The reason is that to obtain an unbiased estimate of the population variance, it is necessary to divide by the number of data values that are independent (that can vary freely). That is, it is necessary to divide by the degrees of freedom. By dividing by the degrees of freedom, the sample variance is an unbiased estimate (see Focus 2 for an explanation of *unbiased estimate*) of the population variance.

Mathematical Focus 2

The sample variance is an unbiased estimator of the population variance.

If the expected value of the sample mean, $\bar{x}$, is the same as the value of the mean, μ, of the population from which the sample was taken, the sample mean is said to be an *unbiased estimate* of the population mean. It is also possible to refer to bias as it relates to the *variance* of a sample—that is, the bias of the sample variance, s^2, as an estimate of the population variance, σ^2. In calculating the sample variance, dividing the sum of the squared deviations by n may seem intuitive, but doing so will result in the sample variance being a biased estimate of the population variance. In order for the sample variance to be an *unbiased* estimate of the population variance, it is necessary to divide by $n - 1$ (called the degrees of freedom—see Focus 1 and the Postcommentary).

The following example illustrates what would happen if one were to divide by n when calculating the sample variance, rather than dividing by $n - 1$. Figure 46.1

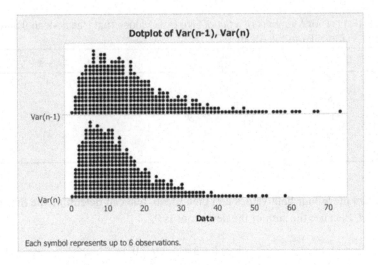

Each symbol represents up to 6 observations.

FIGURE 46.1.   The top dot plot (indicated by Var(n – 1)) is one in which n – 1 was used in the calculation. The bottom dot plot (indicated by Var(n)) is one in which n was used in the calculation.

shows two dot plots—one in which n – 1 was used in the calculation (the first plot, indicated by *Var(n-1)*) and one in which n was used (indicated by *Var(n)*).

In the example portrayed in Figure 46.1, a computer program was used to simulate taking 2000 samples, with the size of each sample being $n = 5$. Because 5 is a relatively low value for n, the difference between n and n – 1 is large enough in comparison to the sample size to illustrate the point. (For very large values of the sample size, n, the difference between n and n – 1 is minute in comparison to the sample size.) For the data in this example, it is known that the population variance, σ^2, is 16. So, to be an unbiased estimate of this population variance, the value of the sample variance, s^2, is expected to be 16. That is, the average value of the variance (i.e., the average of all the data points in the dot plot above) should be at or near 16.

The statistics in Table 46.2 indicate the actual average of these sample variances when the variance is calculated using n – 1 and when it is calculated using n.

The average sample variance when n – 1 is used is 16.126, but when n is used, the average sample variance, 12.901, is much too low. Clearly, the calculation us-

TABLE 46.2   Average Sample Variances Calculated Using n – 1 and n

Descriptive statistics	n	Mean
Var(n – 1)	2000	16.126
Var(n)	2000	12.901

ing $n - 1$ gives the better estimate of the true population variance, 16, whereas a calculation using n will, on average, underestimate the population variance.

POSTCOMMENTARY

One can investigate further the topic of an unbiased estimate by considering the concept *expected value*. This may shed light on why one divides by $n - 1$ in calculating the variance.

> **Definition:** A statistic $\hat{\Theta}$ is an unbiased estimator of the parameter θ if and only if $E(\hat{\Theta}) = \theta$ (in other words, if the expected value of the statistic equals the parameter).

> **Theorem:** If S^2 is the variance of a random sample from an infinite population with the finite variance σ^2, then $E(S^2) = \sigma^2$.

Proof:

$$E(S^2) = E\left(\frac{\sum_{i=1}^{n}(x_i - \bar{x})^2}{n-1}\right) = \frac{1}{n-1} \cdot E\left(\sum_{i=1}^{n}\left[(x_i - \mu) - (\bar{x} - \mu)\right]^2\right)$$

$$= \frac{1}{n-1} \cdot \left\{\sum_{i=1}^{n} E\left[(x_i - \mu)^2\right] - n \cdot E\left[(\bar{x} - \mu)^2\right]\right\}$$

Then, because $E\left[(x_i - \mu)^2\right] = \sigma^2$ and because $E\left[(\bar{x} - \mu)^2\right] = \dfrac{\sigma^2}{n}$,

$$E(S^2) = \frac{1}{n-1} \cdot \left\{\sum_{i=1}^{n}\sigma^2 - n \cdot \frac{\sigma^2}{n}\right\} = \sigma^2,$$

and thus, by the definition, the sample variance is an unbiased estimator of the population variance.

REFERENCES

Agresti, A., & Franklin, C. (2007). *Statistics: The art and science of learning from data.* Upper Saddle River, NJ: Pearson/Prentice Hall.

Freund, J. E. (1992). *Mathematical statistics* (5th ed.). Upper Saddle River, NJ: Prentice Hall.

Mendenhall, W., Beaver, R. J., & Beaver, B. M. (2005). *Introduction to probability and statistics* (12th ed.). Pacific Grove, CA: Thomson Brooks/Cole.

Peck, R., Gould, R., & Miller, S. J. (2013). *Developing essential understanding of statistics: Grades 9–12. Essential understanding series.* (P. Wilson, Vol. Ed.; R. M. Zbiek, Series Ed.). Reston, VA: National Council of Teachers of Mathematics.

Shaughnessy, J. M., & Chance, B. (2005). *Statistical questions from the classroom.* Reston, VA: National Council of Teachers of Mathematics.

CHAPTER 47

LEAST SQUARES REGRESSION

Situation 41 From the MACMTL-CPTM Situations Project

Susan Peters, Evan McClintock, Donna Kinol, Maureen Grady, Heather Johnson, Svetlana Konnova, and M. Kathleen Heid

PROMPT

During a discussion of lines of best fit, a student asked why the sum of the squared differences between predicted and actual values was used. She asked, "Why use squared differences to find the line of best fit? Why use differences rather than some other measure to find the line of best fit?"

COMMENTARY

The set of Foci provide reasons why the sum of the squared residuals, differences between the predicted and actual values, is used when determining lines of best fit. Focus 1 highlights why summing the residuals is not sufficient for determining a line of best fit. The remaining Foci highlight the computational advantages of the squared residuals over other alternatives. Focus 2 examines why one would sum the squared residuals as opposed to summing the absolute value of the residu-

Mathematical Understanding for Secondary Teaching: A Framework and Classroom-Based Situations, pages 421–424.

als. Focus 3 deals with why one would sum the squared residuals as opposed to summing the squared perpendicular distances.

MATHEMATICAL FOCI

Mathematical Focus 1

The line of best fit is not the only line having residuals sum to 0.

The sum of the residuals for the line of best fit is 0. However, the line of best fit is not unique in producing this sum. For any bivariate set of data, the sum of the residuals (signed lengths) of the vertical segments from each y_i to any line passing through the mean of the x's and the mean of the y's will also equal 0.

Consider the following data set: $\{(5, 20), (10, 30), (15, 33), (20, 39), (25, 48)\}$. The line of best fit is $y = 1.3x + 14.5$. The sum of the residuals is 0, and the sum of the squared residuals is 11.5 (Figure 47.1). For this data set, the line $y = -x + 49$ also yields residuals totaling 0, but the sum of the squared residuals is now 1334. The line $y = -x + 49$ has slope -1 and is not a good fit for the data (see Figure 47.2).

Mathematical Focus 2

Although summing the absolute value residuals seems reasonable, it presents more calculation challenges than summing the squared residuals.

The least squares regression line results from minimizing the sum of squared residuals. To examine the difference between summing the squared residuals and

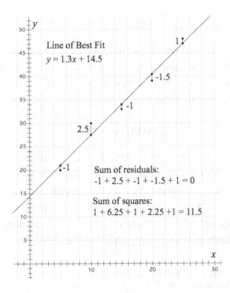

FIGURE 47.1. The line of best fit, given by $y = 1.3x + 14.5$.

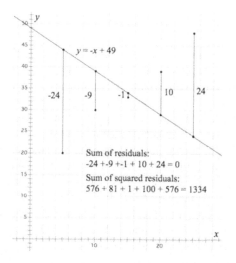

FIGURE 47.2. The line given by $y = -x + 49$.

summing the absolute value residuals, one needs to understand that the sum of quadratic expressions[1] is quadratic, whereas the sum of absolute value expressions cannot be written as a single absolute value expression. Further, to minimize the sum by using calculus (specifically the derivative), the calculations are straightforward in the case of a quadratic function but quite cumbersome for a function defined as a sum of absolute values. This argument is not as compelling if powerful technology is available to perform the calculations.[2]

Mathematical Focus 3

The sum of the squared perpendicular distances from data points to a potential line of best fit provides less helpful information regarding the goodness-of-fit of predictions than does the sum of the squared residuals.

The main purpose for using least squares regression is to make predictions for the response variable, y, based on a given value for the explanatory variable, x. Thus, statisticians are interested in determining the goodness-of-fit of their predictions, that is, they are interested in finding the prediction error, which is the residual, and minimizing this error. Additionally, the calculations for minimizing the sum of squared residuals for ordinary least squares regression are less cumbersome (minimizing $\sum_{i=1}^{n}\left[y_i-\left(a+bx_i\right)\right]^2$) than the calculations for minimizing the sum of squared perpendicular distances (minimizing $\sum_{i=1}^{n}\dfrac{\left[y_i-\left(a+bx_i\right)\right]^2}{1+b^2}$), which requires using a formula for the distance from a point to a line.[3]

NOTES

1. In this case, the function to be minimized is quadratic in the two variables m and b (the slope and y-intercept of the unknown least squares line). In general it involves sums of the terms m^2, $m \cdot b$, and b^2, all second-degree expressions in the relevant variables, and lower degree terms as well. Thinking of *quadratic* as meaning second-degree polynomial in x, is not the relevant mathematical idea in this case.

2. For discussion of and comparison of using the squared difference rather than the absolute value of the differences, see http://stats.stackexchange. com/questions/118/why-square-the-difference-instead-of-taking-the-absolute-value-in-standard-devia

3. The National Council of Teachers of Mathematics applet, *Understanding Best-Fit With a Visual Model: Measuring Error in a Linear Model*, provides a visual comparison of absolute differences, least squares differences, and perpendicular distances as measures of goodness of fit. This Java applet for *Principles and Standards for School Mathematics* (NCTM, 2000) is available at http://www.nctm.org/standards/content. aspx?id=33199.

REFERENCE

National Council of Teachers of Mathematics. (2000). *Principles and standards for school mathematics*. Reston, VA: Author.

CHAPTER 48

THE PRODUCT RULE FOR DIFFERENTIATION

Situation 42 From the MACMTL–CPTM Situations Project

**Heather Johnson, Shari Reed, Evan McClintock,
Erik Jacobson, and Kelly Edenfield**

PROMPT

In an introductory calculus classroom, a student asked the teacher the following question:

Why isn't the derivative of $y = x^2 \sin x$ just $y' = 2x \cos x$?

COMMENTARY

This Situation deals with a common student misconception that the derivative of the product of functions is the product of the derivatives of the functions. The Foci describe different ways of challenging this misconception and substantiating the true derivative of a product of functions. Focus 1 presents the slope of the tangent

Mathematical Understanding for Secondary Teaching: A Framework and Classroom-Based Situations, pages 425–431.

line as one way to think about the value of the derivative. In Focus 2, the definition of the derivative as a limit of a difference quotient is used to motivate the product rule for differentiation. In Focus 3, a property of even and odd functions and their derivatives contradicts the misconception.

MATHEMATICAL FOCI

Mathematical Focus 1

The slope of a secant line drawn through two points that are near each other on the graph of a function can approximate the derivative of the function between those points.

The derivative of any function represents the slope of the tangent to the graph of the function at every value of x at which the derivative exists. The slope of the tangent at a particular value of x can be approximated by the slope of a secant line drawn through two points close to that particular value of x.

If $y' = 2x \cos x$ is the derivative of $y = x^2 \sin x$, then each value of $y' = 2x \cos x$ will represent the slope of the tangent to $y = x^2 \sin x$ at the corresponding value of x and can be approximated by the slope of a secant line. Figure 48.1 shows a graph of the function and of a secant drawn through two points near $x = 2$.[1]

Consider the x-coordinates of points A and B, 1.93 and 2.07 (rounded to two decimal places). Because they are close to the value of 2, the slope of secant line AB should give a good approximation for the derivative of $y = x^2 \sin x$ at $x = 2$. The slope of this secant line is 1.96 and therefore, if $y' = 2x \cos x$ were the derivative of $y = x^2 \sin x$, then the value of the derivative at $x = 2$ should be approximately 1.96. However, when substituting $x = 2$, y' evaluated at $x = 2$ is equal to $2 \cdot 2 \cdot \cos(2) \approx -1.665$. This value and the slope value calculated from the secant line in Figure 48.1 do not

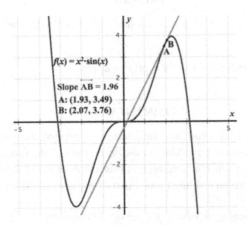

FIGURE 48.1. Graph of the function with rule $f(x) = x^2 \sin x$ and of a secant drawn through two points near $x = 2$.

agree, even with respect to their signs. Upon further inspection, it is apparent that the function is increasing on the interval around 2, and a negative derivative at the point $x = 2$ makes no sense. Hence, $2x \cos x$ cannot be the derivative of $x^2 \sin x$.

Mathematical Focus 2

The product rule for differentiation necessarily follows from the definition of the derivative of a function.

The derivative of a function $y = f(x)$ is defined as a limit of the slopes of secant lines drawn through the point $(x, f(x))$ and a point whose x-value approaches the x-value of the desired point on the graph of the function. Thus, the derivative of a function $f(x)$ at x is given by $f'(x) = \lim\limits_{h \to 0} \dfrac{f(x+h) - f(x)}{h}$. Substituting $x^2 \sin x$ for $f(x)$ yields:

$$f'(x) = \lim_{h \to 0} \frac{(x+h)^2 \sin(x+h) - x^2 \sin(x)}{h}.$$

Using trigonometric identities and algebraic manipulation produces an equivalent expression for $f'(x)$:

$$f'(x) = \lim_{h \to 0} \left[\frac{x^2 [\sin(x+h) - \sin(x)]}{h} + 2x \sin(x+h) + h[\sin(x+h)] \right].$$

It follows that:

$$f'(x) = x^2 \lim_{h \to 0} \left[\frac{[\sin(x+h) - \sin(x)]}{h} \right] + \lim_{h \to 0} \left[2x \sin(x+h) \right] + \lim_{h \to 0} h[\sin(x+h)]$$

$$= x^2 \frac{d}{dx}(\sin(x)) + 2x \sin(x) + 0$$

$$= x^2 \cos(x) + 2x \sin(x).$$

To demonstrate that $2x \cos x \neq x^2 \cos x + 2x \sin x$, one can evaluate both expressions for a particular value of x, say $x = \pi$, and get different results. When $x = \pi$,

$$2\pi \cos \pi \neq \pi^2 \cos \pi + 2\pi \sin \pi,$$

because the left member (i.e., $2\pi \cos \pi$) simplifies to -2π and the right member (i.e., $\pi^2 \cos \pi + 2\pi \sin \pi$) simplifies to $-\pi^2$. Because $2x \cos x \neq x^2 \cos x + 2x \sin x$, $y' = 2x \cos x$ cannot possibly be the derivative of $y = x^2 \sin x$.

In general, the definition of the derivative of a function implies the product rule. The function $k(x) = x^2 \sin x$ is a product of two functions, $f(x) = x^2$ and

$g(x) = \sin x$. Therefore, $k(x) = x^2 \sin x$ can be written as $k(x) = f(x) \cdot g(x)$. Using the definition of the derivative, it follows that:

$$k'(x) = \lim_{h \to 0} \frac{f(x+h) \cdot g(x+h) - f(x) \cdot g(x)}{h}$$

After algebraic manipulation, the result is,

$$k'(x) = \lim_{h \to 0} \left[\frac{f(x+h) \cdot g(x+h) - f(x+h)g(x) + g(x) \cdot f(x+h) - g(x) \cdot f(x)}{h} \right]$$

$$= \lim_{h \to 0} \left[f(x+h) \frac{g(x+h) - g(x)}{h} + g(x) \frac{f(x+h) - f(x)}{h} \right] ,$$

and by using the definition of the derivative and limit properties, the limit expression simplifies to:

$$k'(x) = f(x) \cdot g'(x) + g(x) \cdot f'(x).$$

This general result is known as the *product rule for differentiation*. The product rule applies to the problem in the Prompt because the function $y = x^2 \sin x$ is a product of two functions $f(x) = x^2$ and $g(x) = \sin x$, for which $f'(x) = 2x$ and $g'(x) = \cos x$. The derivative of $y = x^2 \sin x$ given by the product rule is $y' = x^2 \cdot \cos x + (\sin x) \cdot 2x$, the same as was developed directly from the definition of the derivative.

Mathematical Focus 3

For a function whose derivative is not identically 0, the function and its derivative cannot both be even, nor can they both be odd.

Functions that satisfy $f(-x) = -f(x)$ are called *odd* functions and functions satisfying $f(-x) = f(x)$ are called *even* functions. Odd functions with rules of the form $y = f(x)$ have graphs that are symmetric about the origin, whereas even functions with rules of the form $y = f(x)$ have graphs that are symmetric about the y-axis. Figure 48.2 shows odd functions, $f(x) = \sin x$ and $g(x) = 2x$, and even functions, $h(x) = \cos x$ and $k(x) = x^2$.

If a nonconstant polynomial function is even, then the function takes on positive values if the leading coefficient is positive (or negative values if the leading coefficient is negative), both for values of x approaching negative infinity and for values of x approaching positive infinity. The derivative of the function gives the slope of its graph at each value in its domain, and so the derivative must be negative when x is very small (i.e., when x approaches $-\infty$) and positive when x is large

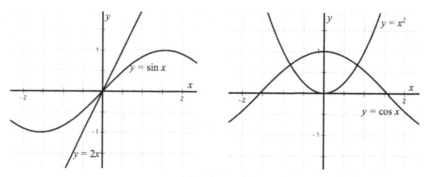

FIGURE 48.2. Graphs of odd functions, $f(x) = \sin x$ and $g(x) = 2x$, and even functions $h(x) = \cos x$ and $k(x) = x^2$.

(i.e., when x approaches $+\infty$), or vice versa if the leading term of the polynomial is negative. This means that the derivative of an even nonconstant polynomial function cannot be even. A similar argument shows that the derivative of an odd nonconstant polynomial function cannot be odd.

A symbolic argument that the derivative of any odd function is an even function and the derivative of any even function is an odd function follows.

To prove: The derivative of any even function is odd.

Proof: Let f be an even function. Then

$$f(x) = f(-x) \ \forall x \in \text{Domain}(f).$$

$$f'(-x) = \lim_{\Delta x \to 0} \frac{f(-x + \Delta x) - f(-x)}{\Delta x}$$

$$= \lim_{\Delta x \to 0} \frac{f(-(x - \Delta x)) - f(-x)}{\Delta x}$$

$$= \lim_{\Delta x \to 0} \frac{f(x - \Delta x) - f(x)}{\Delta x}$$

$$= \lim_{\Delta x \to 0} \frac{f(x - \Delta x) - f(x)}{-(-\Delta x)}$$

$$= -\lim_{\Delta x \to 0} \frac{f(x - \Delta x) - f(x)}{-\Delta x}$$

$$= -\lim_{-\Delta x \to 0} \frac{f(x + (-\Delta x)) - f(x)}{-\Delta x}$$

$$= -f'(x)$$

So f' is an odd function.

To prove: The derivative of any odd function is even.
Proof: Let f be an odd function. Then

$$f(-x) = -f(x) \ \forall x \in \text{Domain}(f)$$

$$f'(-x) = \lim_{\Delta x \to 0} \frac{f(-x+\Delta x) - f(-x)}{\Delta x}$$

$$= \lim_{\Delta x \to 0} \frac{f(-(x-\Delta x)) + f(x)}{\Delta x}$$

$$= \lim_{\Delta x \to 0} \frac{-f(x-\Delta x) + f(x)}{\Delta x}$$

$$= -\lim_{\Delta x \to 0} \frac{f(x-\Delta x) - f(x)}{\Delta x}$$

$$= -\lim_{\Delta x \to 0} \frac{f(x+(-\Delta x)) - f(x)}{-(-\Delta x)}$$

$$= -\left(-\lim_{-\Delta x \to 0} \frac{f(x+(-\Delta x)) - f(x)}{-\Delta x} \right)$$

$$= \lim_{-\Delta x \to 0} \frac{f(x+(-\Delta x)) - f(x)}{-\Delta x}$$

$$= f'(x).$$

So f' is an even function.

Because the only function that is both odd and even is the 0 function, and both $x^2 \sin x$ and its purported derivative $2x \cos x$ are odd, it must be the case that either $2x \cos x$ is the 0 function or it is not the derivative of $x^2 \sin x$.

The trigonometric functions do not exhibit such straightforward end-behavior, but the same relationship between a function and its derivative holds true: Both cannot be even nor can both be odd. Consider the graph of the even function f, defined by $f(x) = \cos x$ on the interval $(-\pi, \pi)$. It has positive slope on $(-\pi, 0)$ but negative slope on $(0, \pi)$; therefore, the derivative of f cannot be even. The same argument works for any odd or even trigonometric function. This property of odd and even functions and their derivatives can be applied to show that the derivative of $y = x^2 \sin x$ is not $y' = 2x \cos x$. By inspecting the graphs of both functions (see Figure 48.3), it is apparent that both are odd. Thus, one cannot be the derivative of the other.

Indeed, an even easier argument does not rely on graphing or inspection. Because $y = \sin(x)$ and $y = 2x$ are odd functions and $y = \cos x$ and $y = x^2$ are even functions, and because the product of an even function and an odd function is

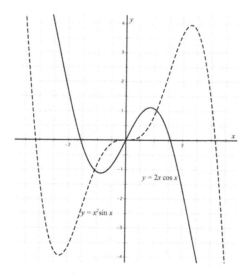

FIGURE 48.3. Graphs of $y = x^2 \sin x$ and $y = 2x \cos x$ that suggest that both functions are odd.

always an odd function, both $f(x) = x^2 \sin x$ and $g(x) = 2x \cos x$ are odd functions. Thus, one cannot be the derivative of the other.

NOTE

1.   In Figure 48.1, the displayed slope value, 1.96, and the displayed coordinates of the points A and B, (1.93, 3.49) and (2.07, 3.76) respectively, are accurate only to two decimal digits. Due to rounding, the displayed slope—calculated using more than two decimal digits—differs from the slope value, 1.93, that one would obtain using the rounded values.

FIGURE 16.7

CHAPTER 49

PROOF BY MATHEMATICAL INDUCTION

Situation 43 From the MACMTL–CPTM Situations Project

Erik Tillema, Jeremy Kilpatrick, Heather Johnson, Maureen Grady, Svet-lana Konnova, and M. Kathleen Heid

PROMPT

A teacher of a calculus course gave her students the opportunity to earn extra credit by proving various algebraic formulas by mathematical induction. For example, one of the formulas was the following:

$$1^2 + 2^2 + 3^2 + \cdots + n^2 = \frac{n(n+1)(2n+1)}{6}.$$

Only one student in the class was able to prove any of the formulas. After the student had presented his three proofs to the class on 3 consecutive days, another student complained: "I don't get what he is proving. And besides that, how do you get the algebraic formulas to start with?"

Mathematical Understanding for Secondary Teaching: A Framework and Classroom-Based Situations, pages 433–442.

COMMENTARY

Given that it is a deductive process, *mathematical induction* is an unfortunate term. George Pólya used to say that a better term would be *mathematical complement to induction*. Although an inductive process can be used to provide evidence for a conjecture involving *n* for any finite *n*, mathematical induction can be used to prove the conjecture for all natural numbers *n*.

The Foci consider several important aspects of developing proofs by mathematical induction that may be difficult or confusing, as well as some ways of generating conjectures that have an inductive proof. Focus 1 looks at a common image for what proof by mathematical induction accomplishes. Focus 2 examines a problem context in which a recursion relationship might be formed and used to formulate an algebraic statement that can be proved by mathematical induction. Focus 3 gives several examples of how figurate representation of numbers can lead to conjectures about algebraic formulas and subsequent proofs by induction. Focus 4 and Focus 5 analyze the structure of inductive proof by providing some examples in which one of the two conditions for induction fails.

MATHEMATICAL FOCI

Mathematical Focus 1

> One form of a proof by mathematical induction can be portrayed with an image of any process involving a basis (usually k = 1) and an inductive step (if the conjecture holds for n = k, it holds for n = k + 1). A second form of the principle of mathematical induction is also sometimes useful: If the assumption that a statement is true for all R < m implies it must be true for m, then the statement is true for all n.

It may be useful to provide some images of what the first form of mathematical induction does. One good image might be of a ladder that is infinitely tall (with the rungs corresponding to the set of positive integers). In doing a proof by induction, it is necessary to check whether one can get on the first rung of the ladder (which corresponds to checking whether the base case of the proposition holds true). Once one can get on the ladder, it is necessary to check whether if one is on any step of the ladder, one can proceed to the next step; namely, being on the kth rung, one can move to the $(k + 1)^{st}$ rung (which corresponds to showing that if the proposition holds for k it will hold for $k + 1$). By proving these two things, using the principle of mathematical induction, one can get to any rung of the ladder (which corresponds to proving that the proposition is true for all natural numbers, k).

An alternate image used to illustrate proof by mathematical induction involves the domino effect. Imagine an infinite row of dominoes standing on end. If one knows that the first domino will fall, and if one knows that when domino k falls, its next neighbor $(k + 1)$ will fall, then, by the principle of mathematical induction, all the dominos will fall.

Mathematical Focus 2

Proof by mathematical induction can be motivated by a recursion relationship using a graphic display.

Many recursion formulas create problem-solving opportunities that might lead to an algebraic formulation of a situation and a subsequent proof by mathematical induction. One such problem situation is the following.

Suppose one has a $2^n \times 2^n$ chessboard, an example of which is illustrated in Figure 49.1. Show that if one removes one square (shown in black), one can cover the remaining squares on the board with L-triominoes (the gray squares).

In Figure 49.1, the chessboard is an 8×8 chessboard. It is easy to show that it is possible to find a covering of the chessboard, but it is not always easy to see a pattern in a particular covering of a chessboard of this size. This difficulty provides a good opportunity to use Polya's heuristic strategy of looking at a related problem that is simpler. Consider a 2×2 and a 4×4 chessboard. In the 2×2 case, one can clearly cover the chessboard without any difficulty. That is, one can place a triomino as one pleases, and exactly one space will be left over, as in Figure 49.2.

Next consider a 4×4 board. Notice that it is composed of exactly four 2×2 chessboards (and further that there are four 4×4 chessboards in an 8×8 chessboard). In making this observation, one can establish a recursion relationship between successive chessboards. Also, one might observe that the 4×4 board will have exactly one 2×2 chessboard with the space taken out of it and three 2×2

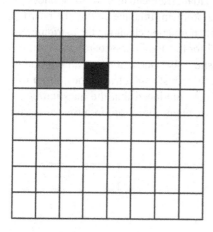

FIGURE 49.1. An 8×8 chessboard—as an example of a $2^n \times 2^n$ chessboard—with one square removed.

FIGURE 49.2. A 2×2 chessboard with one square removed.

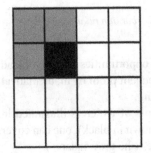 

FIGURE 49.3.   A 4 × 4 chessboard with one square removed.

chessboards with exactly one space not covered. One can cover those three spaces with exactly one triomino (shown on the right in Figure 49.3).

Having made these observations, one can set up a recursion relationship between successive boards and make a more formal proof by mathematical induction. The sequence of observations previously described provides a sufficient basis for developing a proof by mathematical induction.

On the other hand, one might use this problem to show that $\dfrac{2^{2n}-1}{3}$ is a whole number for all $n > 1$. To think of this algebraic formulation, notice that a $2^n \times 2^n$ chessboard contains 2^{2n} squares and that removing one of the squares leaves $2^{2n} - 1$ squares remaining to be covered. Also, each triomino contains 3 squares. Most importantly, one needs to recognize that the spatial arrangement of the triominoes on the board does not affect the solution of the problem. Imagine cutting the board at each row and putting the rows end to end. Then imagine changing the L-triominoes to straight triominoes (shown in Figure 49.4 for the 4 × 4 chessboard).

Once again, a recursive argument can show that one can cover all but one of the squares of the transformed board with the transformed triominoes. From this spatial configuration, one may become more convinced that the question being asked is a question of division. That is, how many threes are in a row of length $2^{2n} - 1$? The answer is *a whole number*.

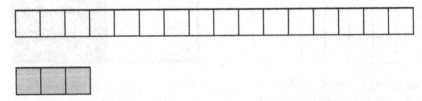

FIGURE 49.4.   End-to-end transformation of the squares of a 4 × 4 chessboard and a triomino.

Mathematical Focus 3

Figurate numbers can also be used to generate identities that can then be proved by mathematical induction.

Consider a figurate representation for the square numbers that uses dots. In that configuration, one might observe the pattern in Figure 49.5.

The configuration suggests that $n^2 = 1 + 3 + \ldots + (2n - 1)$. Having made a conjecture about the formula, one can then try to prove the conjecture using mathematical induction. (A related question is, how might a formula for the sum of the first n even numbers be found and proved?)

A second problem that uses figurate numbers and that extends Situation 4 (see Chapter 10), which demonstrates several ways to develop an intuition for why the sum of the first n integers equals $\frac{n(n+1)}{2}$, is to find the sum, S_k, of the first k sums of the first n integers. Algebraically, if $T_1 = 1$, $T_2 = 1 + 2$, $\ldots$, $T_n = 1 + 2 + \ldots + n$, then what is sought is $S_k = T_1 + T_2 + \ldots + T_k$. One can introduce this problem by finding both an additive and multiplicative solution to the following problem. Suppose five points are selected on a circle, as shown in the leftmost circle of Figure 49.6. How many possible triangles can be made using sets of three of those points?

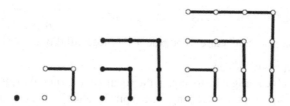

FIGURE 49.5.　A pattern in the first four square numbers represented using dots.

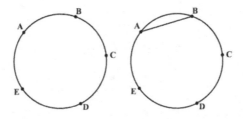

FIGURE 49.6.　A circle with five designated points and a circle with a segment connecting two of those points considered as the base for a triangle.

The problem can be structured additively by considering, for each of the designated points on the circle, the number of triangles that have that point as a vertex. Pick a point, say, A. There are six triangles involving A (ABC, ABD, ABE, ACD, ACE, and ADE), three more triangles involving B but not A (BCD, BCE, and BDE), and one more involving C but neither A nor B (CDE) or $T_1 + T_2 + T_3$. Multiplicatively, one can think about choosing sets of three points from five or $\binom{5}{3}$. After solving the problem for several numbers of points, one might conjecture that $T_1 + T_2 + \ldots + T_n = \binom{n+2}{3} = \frac{1}{6}(n+2)(n+1)(n)$, and use the problem context to prove the potential identity by mathematical induction. (An extension of the problem would be to find an additive and multiplicative way of thinking about the number of diagonals that can be drawn in a convex n-gon).

One might then consider using this identity to find an identity for $1^2 + 2^2 + \ldots + n^2$. Again, make a figurate representation of the problem to help illuminate what the identity might be. The first four square numbers are shown in Figure 49.7.

Using a formula for the sum of the first n triangular numbers, the figure can be broken into configurations representable by triangular numbers (see Figure 49.8).

From the example of the sum of the first four square numbers, one might observe that the sum of the first n square numbers is $2T_1 + 2T_2 + \ldots + 2T_{n-1} + T_n = \binom{n+2}{3} + \binom{n+1}{3}$.

Algebraic manipulation yields $1^2 + 2^2 + \ldots + n^2 = \frac{(2n+1)(n+1)(n)}{6}$, another formula that one can conjecture to be true and whose validity might be established by mathematical induction.

A final problem might be to look at the square of the sum of the first n integers: $(1 + 2 + \ldots + n)^2$. Again, make a figurate representation of the case for $n = 4$ (see Figure 49.9). How many groups of four are in Figure 49.9? There are $\binom{5}{2}$ verti-

FIGURE 49.7. Dot representation of $1^2 + 2^2 + \ldots + n^2$ for $n = 4$.

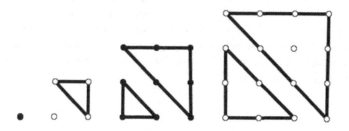

FIGURE 49.8.   Dot representation of $1^2 + 2^2 + \ldots + n^2$ for $n = 4$ partitioned into triangular number arrays.

cal groups (the sum of the first four integers), and there are $\binom{4}{2}$ horizontal groups (the sum of the first three integers), giving a total of $4\left(\binom{5}{2} + \binom{4}{2}\right)$ dots. Algebraic simplification shows that there are 4^3 dots in this outermost region of the figure. A similar process shows that computing the dots in groups of threes gives 3^3 dots, and so on. That yields a conjecture for the following identity: $(1 + 2 + \ldots + n)^2 = 1^3 + 2^3 + \ldots + n^3$. The formula can then be proved by mathematical induction.

Mathematical Focus 4

One form of proof by mathematical induction requires both a basis step and an inductive step; it fails if either is lacking.

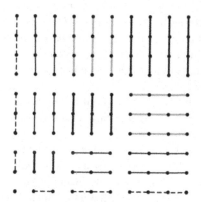

FIGURE 49.9.   Figurate representation of $(1 + 2 + \ldots + n)^2$ for $n = 4$.

Many of the examples in which one of the steps fails are somewhat trivial, but they can nonetheless give some feeling for the importance of both. For instance, consider the inequality $5(2^n) > 6^n$. It holds true for $n = 1$ (one can get on the ladder). However, the induction step cannot be proved. Assume the inequality is true for $n = k$ and attempt to show that it is true for $n = k + 1$. One needs to show that $5(2^{k+1}) > 6^{k+1}$, given that $5(2^k) > 6^k$. But

$$5(2^{k+1}) > 6^{k+1}$$
$$\Leftrightarrow 5(2^k)(2) > 6 \times 6^k$$
$$\Leftrightarrow 5(2^k)(2) > 6 \times 2^k \times 3^k$$
$$\Leftrightarrow 5(2) > 6 \times 3^k$$
$$\Leftrightarrow \frac{5}{3} > 3^k$$
$$\Leftrightarrow 5 \times 3^{-1} > 3^k$$
$$\Leftrightarrow 5 > 3^{k+1}$$
$$\Leftrightarrow k + 1 < \frac{\ln(5)}{\ln(3)} \doteq 1.46497$$

So $5(2^{k+1}) > 6^{k+1}$ follows only for certain values of $k + 1$. That is, the induction step cannot be proved. It is interesting to note that for $n = 2$, $5(2^n) > 6^n$ implies that $20 > 36$, which is false. The restriction on the value of $k + 1$ is consistent with the result that $5(2^k) > 6^k$ for $k + 1 = 1$ but false for $k + 1 = 2$. In this case, it was possible to get on the first step of the ladder but it was not possible to move from rung to rung.[1]

Now consider the inequality $3^n > 5(3^n)$. Suppose one begins by assuming that it is true for $n = k$ and tries to prove it for $n = k + 1$. One finds that $3^{k+1} = 3(3^k) > 3(5(3^k)) = 5(3^{k+1})$, and the inequality holds true based on the inductive hypothesis (being able to move from rung to rung). However, in the case of $n = 1$ (the base case) the inequality does not hold. One is unable to get on the ladder even though if one could, one could move from rung to rung. These examples illustrate that both conditions need to be met for a proof by mathematical induction to work. They also suggest that it is worth checking several cases when proving something by induction.

Checking several cases, however, is not a substitute for a proof,[2] as can be seen by the purported inequality $3^n > n!$. This inequality holds for the first six cases, but it is false when $n = 7$. For any a, the inequality $a^n > n!$ will eventually become false, even though one can make the n for which it becomes false arbitrarily large. This example demonstrates that proof by mathematical induction is, in fact, deductive. That is, one is not drawing a conclusion based solely on repeated observations (inductive reasoning), instead, one is establishing two premises and stating that if both premises hold true one can assert the conclusion (deductive reasoning). Taken together, the three examples in this Focus suggest the relationship of the premises to the conclusion as well as the role that observation plays in a proof by induction.

Mathematical Focus 5

One form of proof by mathematical induction requires one to consider the dependence of the basis step on the nature of the operations involved. If one wishes to prove a statement defined from a binary operation, then the basis step has to contain the n = 2 *case.*

In the middle of the 20th century, George Pólya posed the following: Find the error in the following argument, which proposes to prove that all horses are of the same color using the method of mathematical induction.

Basis: If there is only one horse, there is only one color.
Induction Step: Assume as induction hypothesis that within any set of n horses, there is only one color. Now look at any set of $n + 1$ horses. Number them: 1, 2, 3, ... , n, $n + 1$. Consider the sets {1, 2, 3, ... , n} and {2, 3, 4, ... , $n + 1$}. Each is a set of only n horses; therefore within each there is only one color. But the two sets overlap, so there must be only one color among all $n + 1$ horses.

Analysis of the use of the method of mathematical induction is necessary for the preceding example. For the basis step, the $n = 1$ case is trivial because any horse is the same color as itself. However, the logic of the inductive step is incorrect when $n = 2$. In case $n = 2$, two sets {1} and {2} need to be considered, and they do not overlap. Therefore, even though the inductive step is correct in all cases when $n > 2$, the logic of the inductive step is incorrect when $n = 2$. Consequently, the method of mathematical induction cannot be applied to prove that that all horses are of the same color, because the basis step for this example, in which a binary operation is involved, has to consider the $n = 2$ case, and it fails.

In this case, the operation being considered is a type of binary operation (i.e., forming the union of two different sets, {1, 2, 3, ..., n} and {2, 3, 4, ..., $n + 1$)}. A basis step that takes only one set into consideration (the set {1}) cannot serve as the basis step for mathematical induction in this case, and indeed it leads to patently false results in the example of the horses.

NOTES

1. Note that one could more easily prove that $5(2^n) > 6^n$ is false by finding a counterexample (e.g., $n = 2$). The point of this example, however, was to illustrate ways in which both the basis step and the induction step are needed in proving statements by mathematical induction.

2. A finite list of examples never proves that a pattern holds in an infinite number of cases. Lagrange polynomials allow one to produce infinitely many different sequences of numbers all of which have the same arbitrarily chosen sequence of n numbers at the beginning.

CHAPTER 50

SITUATING AND REFLECTING ON OUR SITUATIONS WORK

M. Kathleen Heid

Developing a framework for mathematical understanding for secondary teaching was an ambitious project—one that called for the attention of many, including six mathematics education faculty and several dozen mathematics education graduate students from two universities over several years. It required identifying mathematical opportunities that arise in secondary classrooms and countless hours of debating over myriad visions of the mathematics that a teacher could productively use in the context of those opportunities. As we developed the Situations and the resulting MUST Framework, we conducted three national conferences of mathematicians, mathematics educators, school mathematics leaders, and secondary mathematics teachers to provide feedback on our ideas and products. We are grateful for the generous, insightful, useful, and provocative feedback these professionals offered. Bringing the Situations and the Mathematical Understanding for Secondary Teaching Framework to fruition in their current forms took several additional years, with debates about their content and structure both vigorous and wide-ranging. Realizing that the debate can and must continue indefinitely, we decided to share the work at this particular juncture via this volume. This volume describes both the process of development of the Situations and the MUST Framework and the products themselves. We hope that the myriad ideas that have

Mathematical Understanding for Secondary Teaching: A Framework and Classroom-Based Situations, pages 443–452.

arisen in the context of conferences, mathematics education classes, and research studies that have involved our Situations and the MUST Framework inform the field regarding mathematical understanding for secondary teaching. We also hope that the field of mathematics education will view our work as a beginning, and will continue serious conversations about teachers' understandings of mathematics that will best serve secondary mathematics teaching.

TRENDS IN DOCUMENTS ON THE MATHEMATICAL PREPARATION OF TEACHERS

Others have contributed in various ways to the field's perception of mathematical understanding, mathematical knowledge, and/or formal mathematics background for secondary teaching (e.g., Ball, Lubienski, & Mewborn, 2001; Ferrini-Mundy & Findell, 2010). A history of documents on the mathematical preparation of teachers reveals a continuing national interest in this area. A brief journey into these documents reveals an evolving role for mathematical processes and a gradually deepening appreciation for the complex nature of mathematical understanding.

DOCUMENTS ON THE MATHEMATICAL PREPARATION OF TEACHERS

Early reports from the committees of the Mathematical Association of America (MAA) focused on programs of mathematics courses recommended for teachers as well as the general content of those courses (MAA, 1935, 1961a, 1961b, 1971a, 1971b, 1983). The 1935 MAA report on the training of teachers of mathematics recommended specific undergraduate courses (trigonometry, college algebra, calculus, analytic geometry, Euclidean geometry, and theory of equations) without elaborating on the content of those courses, explaining by stating that the content for mathematics courses was standardized and did not need elaboration. The document proposing this set of recommendations also described master's degree programs in mathematics and observed that "such a degree [master's in mathematics] by itself is most useful if it implies certification of exceptionally good preparation for teaching secondary mathematics" (p. 270). Almost three decades later, the MAA recommendations for high school teachers were still described in terms of courses, but those recommendations increased the mathematical rigor of the program and called for three courses in analysis (with analytic geometry), two courses in abstract algebra, two courses in geometry beyond analytic geometry, two courses in probability and statistics and three additional upper-division electives, including an introduction to the language of logic and sets (Mathematical Association of America, 1961b).

The 1971 revision of the MAA recommendations, in addition to supporting the 1961 MAA course recommendations (MAA, 1961a), further increased the rigor of the mathematics by recommending a course in real analysis, experience with computing, and a course in applications of mathematics. The 1971 (MAA, 1971a, 1971b) document echoed the 1935 recommendations in its adherence to graduate-school-level mathematics standards for secondary mathematics teachers. The 1971 document was written with "a view to maintaining, as far as possible, comparability of standards between prospective teachers and prospective entrants to a graduate school with a major in a mathematical science" (p. 170). Although the 1971 recommendations, like those preceding them, relied primarily on course titles and listed courses with little description of content, the 1971 document made minor reference to processes as well as content, noting the importance of communication and pointing out that the development of skills in reading, writing, speaking, and listening were essential to the preparation of teachers. A little more than a decade later, the MAA issued an updated document describing recommended mathematical preparation of teachers (MAA, 1983). This document recommended a somewhat broader program of courses, including, in addition to the course recommendations of earlier documents, courses in discrete mathematics, number theory, history and appreciation of mathematics, and mathematical modeling and applications. Unlike previous documents, the 1983 document specified for each course not only a list of course topics but also a list of learning objectives. The learning objectives were largely focused on development of specific skills and exposure to mathematical ideas and did not elaborate on mathematical processes.

It was not until the 1989 *Curriculum and Evaluation Standards for School Mathematics* (NCTM, 1989a) and the *Post-Baccalaureate Education of Teachers of Mathematics* (NCTM, 1989b) developed by the National Council of Teachers of Mathematics (NCTM) that discussion of the importance of mathematical processes came to prominence. The former document included overt discussion of processes for student learning, and the latter document noted that "problem solving, reasoning, and communication are key ingredients in the guidelines" for mathematics teachers' education (p. 3). NCTM's *Professional Standards for Teaching Mathematics* (NCTM, 1991) and its successor, *Mathematics Teaching Today: Improving Practice, Improving Student Learning* also recognized the importance of teachers' mathematical knowledge with such goals as "deep knowledge of sound and significant mathematics" (NCTM, 2007, p. 19). In 1991, the MAA Committee of the Mathematical Education of Teachers, following the lead of the NCTM's *Curriculum and Evaluation Standards for School Mathematics*, broke ground in its *Call for Change* (1991) by accentuating processes as well as content. In *Call for Change*, authors insisted that how mathematics was taught to teachers was as important as what was taught, and included the following process standards common to the preparation of teachers at the elementary, middle school, and high school levels: learning mathematical ideas, connecting mathematical ideas, communicating mathematical ideas, building mathematical models, using technology, and developing perspectives on the history and culture of mathematics.

This turn toward elaborating on how mathematics should be taught to teachers was developed more fully in *The Mathematical Education of Teachers* (MET)

(CBMS, 2001), the 2001 document published by the Conference Board of the Mathematical Sciences (CBMS). Along with recommending the equivalent of a mathematics major for prospective high school teachers of mathematics, MET recommendations included wording such as: "Prospective mathematics teachers need mathematics courses that develop a deep understanding of the mathematics they will teach" (p. 7), and "Courses on fundamental ideas of school mathematics should focus on a thorough development of basic mathematical ideas" (p. 8), and "Along with building mathematical knowledge, mathematics courses for prospective teachers should develop the habits of mind of a mathematical thinker" (p. 8). In its extensively developed chapters, the MET report situated mathematical processes as well as content squarely at the center of recommendations for the mathematical preparation of secondary teachers. The revision of the MET report, MET II (CBMS, 2012), delivered similar messages and gave an even more prominent role to processes, exemplifying the "habits of mind of a mathematical thinker" to which both MET reports referred with processes such as "reasoning and explaining, modeling, seeing structure, and generalizing" (p. 19). One direction for teachers' mathematical preparation suggested in the MET II document was an increased emphasis on the connections between mathematics at various levels: "[T]he mathematical topics in courses for prospective high school teachers and in professional development for practicing teachers should be tailored to the work of teaching, examining connections between middle grades and high school mathematics as well as those between high school and college" (p. 54). Further accentuating the importance of mathematical processes, the MET II authors posited that "teachers need opportunities for the full range of mathematical experience themselves: struggling with hard problems, discovering their own solutions, reasoning mathematically, modeling with mathematics, and developing mathematical habits of mind" (p. 54).

SITUATING THE SITUATIONS WORK IN THE CONTEXT OF TRENDS IN RECOMMENDATION DOCUMENTS

Our work on the Situations Project can be viewed in light of this history of documents on the mathematical preparation of secondary teachers. Unlike the early documents, we did not focus on the program of courses or mathematical topics that secondary teachers may need. Recognizing the importance of mathematical processes and practices as well as mathematical content, we included mathematical processes as essential features of mathematical understanding for secondary teaching and developed Mathematical Foci for Situations that highlighted those processes as well as content. Most of the documents we reviewed that related to the mathematical preparation of secondary teachers were written by mathematicians or mathematics educators with experience or knowledge of secondary teaching but were not based on the documentation of actual events of practice. Acknowledging the importance of connecting teacher knowledge to practice, our work centered on deriving our products from events and mathematical opportunities that we witnessed in the practice of teaching secondary mathematics. Recognizing the importance of teachers developing a solid understanding of the mathematical ideas that are the basis for secondary mathematics, we analyzed mathematical underpinnings of these events and mathematical opportunities. Our attention was continuously focused on the nature of the mathematical understandings that teachers could productively use in the context of their practice. The Situations we developed and the MUST Framework, thus, represent a next step in developing an understanding of the mathematics that would best serve a secondary mathematics teacher—a step that is intimately connected to practice and highly attentive to mathematical content and processes.

Our work is based on the recognition that there is considerable intellectual substance underpinning school mathematics—a position that was also articulated in both of the MET documents. This intellectual substance is particularly important in today's student-centered classrooms. In classrooms that thrive on open discourse among teachers and students, teachers need the capacity to react to and respond to a large variety of mathematical settings and circumstances. As educators with considerable mathematics teaching experience, we believed that it was imperative to start with practice-based mathematical incidents, and as mathematics educators, we opted to focus on the mathematical aspects of those incidents. An examination of the mathematics that a teacher could productively use in those settings gave us further evidence of the depth of mathematical understanding on which school mathematics is based. Fluency in school mathematics is not enough for a teacher to capitalize on the mathematical opportunities that arise in secondary mathematics classrooms. Teachers need the types of mathematical capacity suggested in the Mathematical Proficiency Perspectives of the MUST Framework. They need to be adept at carrying out the mathematical actions suggested in the Mathematical Activity Perspective of the MUST Framework, and they need the understanding and experience that gives them the capacity to act effectively in the settings of practice suggested in the Mathematical Context of the MUST Framework. Not only do they need facility as suggested in each of these Perspectives, but they also need to coordinate and integrate the three Perspectives as they encounter mathematical opportunities in their practice. Doing so requires the agility present in the best of teachers that allows them to move swiftly among Perspectives as called for with each circumstance as well as the ability to coordinate the simultaneous use of all three lenses. This tripartite focus and the need to apply it in order to help others understand mathematics is something that mathematics teachers need so that they can respond swiftly and insightfully to both anticipated and unanticipated circumstances. Teachers' responsibility to help others understand mathematics results in a critical core of mathematical understanding that differentiates it from other mathematical professions.

THE PROMISE AND LIMITATIONS OF OUR APPROACH

We have attempted to develop a way to think about mathematical knowledge for secondary teaching that accounts for the need for that knowledge to continue to grow throughout a teacher's career as well as the need to help others understand mathematics. We have referred to that knowledge as dynamic, meaning that its development does not stop with the completion of academic programs or with a fixed number of years of teaching experience. We have characterized it by the multifaceted mathematical proficiency that can productively and continually be called upon daily in teaching. We have described it as the recognition of mathematical entities and structures, the reasoning about those structures and entities in closed systems, and the creative application of mathematical capabilities to the development of new understandings and representations of existing understandings. In addition, we have emphasized the importance of teachers recognizing and analyzing situations in which their mathematical knowledge can enable them to navigate the difficult terrain of helping others develop flexible and viable mathematical understanding.

As carefully and intensely as we have worked on developing the MUST Framework, we recognize that it is not a finished product. We have offered numerous examples of classroom incidents that we have witnessed in which teachers could productively call on mathematical knowledge. We have drawn on our backgrounds in mathematics and in observing, studying, and teaching secondary mathematics to capture the nature of that mathematical knowledge in our framework. In addition to documenting events that occurred in the classrooms that we observed, we specifically solicited from observers of classes using reform-curricula descriptions of events that seemed to foster mathematical opportunities. Although we deliberately observed a range of classrooms, those settings were limited. Very few of our Situations directly involved the use of technology, most of our observations occurred in algebra classrooms, and very few occurred in classes that were using reform-curricula. Although this set of observations seemed to reflect current content emphases in schools, it is not necessarily representative of mathematics that could or should be taught at the secondary level. As the set of Situations grows to include Prompts arising from a broader range of settings and content, the Framework may need to be expanded to accommodate the additional ways of mathematical thinking and understanding that arise from analysis of those Situations.

WORKING TOWARD A COMMON UNDERSTANDING

Our development of the Situations and MUST Framework depended on the incidents we witnessed in the practice of teaching secondary mathematics. Although this approach guaranteed a certain authenticity to the results, it also was limited to those incidents that the faculty and graduate students working on the project had witnessed over a period of years. Others have taken different approaches to developing frameworks, including reflecting on experiences of researchers in their roles as expert teachers of mathematics, analyzing theoretical and empirical literature on mathematics learning, viewing available videos of mathematics instruction, and interviewing secondary mathematics teachers. Some of that development came in conjunction with developing assessments of knowledge of mathematics for teaching, through modification of items used in large-scale international assessments, through item-construction conferences, and through sessions at relevant conferences. Others have described programs that have focused on developing these frameworks (e.g., Ball, 2003; Ball & Bass, 2002; Hill, Rowan, & Ball, 2005; Krauss et al., 2008; Stacey, 2008). Given the growing interest in mathematical knowledge, understanding, and proficiency for teaching and the range of approaches to the problem of characterizing or measuring that knowledge, understanding, and proficiency, it would be informative to compare what can be learned from each of those approaches and whether there is promise in consolidating some of those approaches.

CHALLENGES AHEAD

Having developed the MUST Framework based on practice-based Situations is only the beginning. A major challenge to the MUST Framework is whether it can serve as the basis for assessment of secondary teachers' mathematical understanding. An effective assessment would need to help the assessor describe each of the sublevels of the three perspectives: the strands of the Mathematical Proficiency Perspective; the noticing, reasoning, and creating actions that comprise the Mathematical Activity Perspective; and the capacity to act in such contexts as probing mathematical ideas, understanding and assessing students' mathematical knowledge and thinking, and knowing and using mathematics curricula. Moreover, there would be a need to assess the teacher's ability to integrate these capabilities—to select and use their proficiencies in the execution of mathematical actions, and to recognize and act on settings in which those proficiencies and actions could productively be used. The challenge is daunting, and not one that our current work has broached.

As we look ahead beyond the work we have done in creating the Situations and the MUST Framework, we anticipate a range of uses of the Situations and the MUST Framework, including many of the uses reported or designed by attendees at the Users Conference reported in Chapter 6. By providing a characterization of the mathematical understanding that could productively be used by secondary mathematics teachers, the Framework could provide guidance regarding mathematics content to those engaged in the mathematical preparation of secondary mathematics teachers as well as those engaged in professional development of secondary mathematics teachers: curriculum writers, professional developers, and mathematics instructors. The Situations, either in part or in their entirety, could be used as content for engaging teachers in thinking about mathematics that underpins secondary mathematics. The Framework and/or the Situations could also be used as research tools to shape the study of instructional practices in secondary mathematics classrooms. As the Framework and Situations are used in a range of contexts, it is our hope that such uses will be studied in a scholarly and systematic fashion to generate insight into the mathematics of secondary teaching.

BROADENED OPPORTUNITIES

Today's technological world has broadened opportunities for engagement in mathematics. Informal settings for mathematics learning abound. Motion probes make roller coasters in amusement parks fair game for mathematical investigations, mathematics museums bring the wonders of mathematics within the reach of the general public, and the order and structure apparent in electronically enhanced or internet-available art exhibitions fascinate the mathematically inclined. Technology also enables structural changes in formal education, both in school and out of school. As mathematics education escapes the confines of the traditional classroom and emerges in such settings as gaming and flipped classrooms as well as museums, amusement parks, and art collections, teacher roles evolve, the nature of viable Situations will expand, and the MUST Framework will need to be revisited.

The Perspectives in the MUST Framework bring out the activity that underpins mathematical thinking, the range of ways to think about how we understand and use mathematics, and the specific contexts in which teachers need to enact their mathematical knowledge and understanding. As the field's vision of teacher understanding evolves to focus on actions and a broad range of proficiencies and learning contexts, more learner-friendly classrooms may emerge. Mathematics that is broadly conceived is likely to be accessible to a broader range of students.

REFERENCES

Ball, D. L. (2003, February). *What mathematical knowledge is needed for teaching mathematics*. Paper presented at Secretary's Summit on Mathematics, U.S. Department of Education, Washington, DC.

Ball, D. L., & Bass, H. (2002, May). Toward a practice-based theory of mathematical knowledge for teaching. In *Proceedings of the 2002 annual meeting of the Canadian Mathematics Education Study Group* (pp. 3–14). Edmonton, AB, Canada: CMESG/GCEDM.

Ball, D. L., Lubienski, S., & Mewborn, D. (2001). Research on teaching mathematics: The unsolved problem of teachers' mathematical knowledge. In V. Richardson (Ed.), *Handbook of research on teaching* (4th ed.). New York, NY: Macmillan.

Conference Board of the Mathematical Sciences. (2001). *Issues in Mathematics Education: Vol. 11. The mathematical education of teachers*. Providence, RI: American Mathematical Society in cooperation with Mathematical Association of America.

Conference Board of the Mathematical Sciences. (2012). *Issues in Mathematics Education: Vol. 17. The mathematical education of teachers II*. Providence, RI: American Mathematical Society in cooperation with Mathematical Association of America.

Ferrini-Mundy, J., & Findell, B. (2010). The mathematical education of prospective teachers of secondary school mathematics: Old assumptions, new challenges. In Committee on the Undergraduate Program in Mathematics (Ed.), *CUPM discussion papers about mathematics and the mathematical sciences in 2010*, (pp. 31–41). Washington, DC: Mathematical Association of America.

Hill, H. C., Rowan, B., & Ball, D. L. (2005). Effects of teachers' mathematical knowledge for teaching on student achievement. *American Educational Research Journal, 42*, 371–406. doi:10.3102/00028312042002371

Krauss, S., Brunner, M., Kunter, M., Baumert, J., Blum, W., Neubrand, M., & Jordan, A. (2008). Pedagogical content knowledge and content knowledge of secondary mathematics teachers. *Journal of Educational Psychology, 100*, 716–725. doi:10.1037/0022-0663.100.3.716

Mathematical Association of America. (1935). Report from the Commission on the Training and Utilization of Advanced Students in Mathematics on the training of teachers of mathematics. *American Mathematical Monthly, 42*, 263–277. doi:10.2307/2302225

Mathematical Association of America. (1961a). *Course guides for the training of teachers of junior high and high school mathematics*. Report from the Committee on the Undergraduate Program in Mathematics. Washington, DC: Author.

Mathematical Association of America. (1961b). *Recommendations for the training of teachers of mathematics*. Report from the Committee on the Undergraduate Program in Mathematics, Panel on Teacher Training. Washington, DC: Author.

Mathematical Association of America. (1971a) *Recommendations on course content for the training of teachers of mathematics*. Report from the Committee on the Undergraduate Program in Mathematics, Panel on Teacher Training, Mathematical Association of America. In *A compendium of CUPM recommendations: Studies, discussions, and recommendations by the Committee on the Undergraduate Program in Mathematics of the Mathematical Association of America*, (pp. 158–202). Washington, DC: Author.

Mathematical Association of America. (1971b). *Recommendations on course content for the training of teachers of mathematics*. Report from the Committee on the Undergraduate Program in Mathematics. Washington, DC: Author.

Mathematical Association of America. (1983). MAA Notes and Reports Series: Vol. 2. Recommendations on the mathematical preparation of teachers. Report from the Committee on the Undergraduate Program in Mathematics, Panel on Teacher Training. Washington, DC: Author.

Mathematical Association of America. (1991). *A call for change: Recommendations for the mathematical preparation of teachers of mathematics*. Report from the Committee on the Mathematical Education of Teachers of The Mathematical Association of America. Washington, DC: Author.

National Council of Teachers of Mathematics. (1989a). *Curriculum and evaluation standards for school mathematics*. Reston, VA: Author.

National Council of Teachers of Mathematics. (1989b). *Guidelines for the post-baccalaureate education of teachers of mathematics*. Reston, VA: Author.

National Council of Teachers of Mathematics. (1991). *Professional standards for teaching mathematics*. Reston, VA: National Council of Teachers of Mathematics.

National Council of Teachers of Mathematics. (2007). *Mathematics teaching today: Improving practice, improving student learning*. T. Martin (Ed.). Reston, VA: Author.

Stacey, K. (2008). Mathematics for secondary teaching: Four components of discipline knowledge for a changing teacher workforce. In P. Sullivan & T. Wood (Eds.), *The handbook of mathematics teacher education: Knowledge and beliefs in mathematics teaching and teaching development*. Rotterdam, The Netherlands: Sense Publishers.

APPENDIX

CONFERENCE PARTICIPANTS

The following individuals participated in one or more of the three conferences sponsored by the Mid-Atlantic Center for Mathematics Teaching and Learning and the Center for Proficiency in Teaching Mathematics that addressed the Mathematical Understanding of Secondary Teachers Situations and Framework.

Name	Affiliation
Bob Allen	Liberty High School (Brentwood, CA)
Thomas Banchoff	Brown University (emeritus) (RI)
Jane Barnard	Armstrong State University (GA)
Charlene Beckmann	Grand Valley State University (MI)
Sybilla Beckmann	University of Georgia
Steve Benson	Education Development Center (MA)
Glendon Blume	The Pennsylvania State University (emeritus)
Jerry Bona	University of Illinois at Chicago
Tracy Boone	Bald Eagle Area School District (PA)
Hilda Borko	Stanford University (CA)
David Bressoud	Macalester College (MN)
Lynn Breyfogle	Bucknell University (PA)

Mathematical Understanding for Secondary Teaching: A Framework and Classroom-Based Situations, pages 453–456.
Copyright © 2015 by Information Age Publishing
453

Name	Affiliation
Diane Briars	President, National Council of Teachers of Mathematics
Shawn Broderick	Keene State College (NH)
Maurice Burke	Montana State University (emeritus)
Bernard Camou	Scuola Italiana de Montevideo High School (Montevideo, Uruguay)
Herb Clemens	The Ohio State University
Tanya Cofer	Northeastern Illinois University
AnnaMarie Conner	University of Georgia
Al Cuoco	Education Development Center (MA)
Ed Dickey	University of South Carolina
Fred Dillon	Strongsville High School (OH)
Helen Doerr	Syracuse University (NY)
Sarah Donaldson	Covenant College (GA)
Kanita DuCloux	Western Kentucky University
Kelly Edenfield	Carnegie Learning
Thomas Evitts	Shippensburg University (PA)
James Fey	University of Maryland (emeritus)
Ryan Fox	Belmont University (TN)
Christine Franklin	University of Georgia
Brian Gleason	Nevada State College
Deborah Gober	Columbus State University (GA)
Maureen Grady	East Carolina University (NC)
Karen Graham	University of New Hampshire
Duane Graysay	The Pennsylvania State University
Henrique Guimaraes	University of Lisbon, Portugal
Linda Haffly	The Pennsylvania State University
William Harrington	State College Area High School (PA)
Larry Hatfield	University of Georgia (emeritus)
M. Kathleen Heid	The Pennsylvania State University
Sherry Hix	Jefferson High School (GA)
Mark Hoover-Thames	University of Michigan
Erik Jacobson	Indiana University
Darshan Jain	Adlai Stevenson High School (IL)
Natalie Jakucyn	Glenbrook South High School (IL)
Kathy Ivy Jaqua	Western Carolina University (NC)
Heather Johnson	University of Colorado Denver
Shiv Karunkaran	Washington State University
Signe Kastberg	Purdue University (IN)

Name	Affiliation
Jeremy Kilpatrick	University of Georgia
Hee Jung Kim	University of Georgia
Jake Klerlein	Amplify Learning (NY)
Svetlana Konnova	Prince George's Community College (MD)
John Lannin	University of Missouri
Glenda Lappan	Michigan State University
Hollylynne Lee	North Carolina State University
HyoenMi Lee	Gwang Moon High School (Korea)
Mark Levi	The Pennsylvania State University
Shlomo Libeskind	University of Oregon
Dennis Linn	Pella Community Schools (IA)
James Lynn	University of Illinois at Chicago
Karen Marrongelle	Portland State University
William McCallum	University of Arizona
Evan McClintock	University of Colorado Denver
Raven McCrory	Michigan State University
Ezetta Myers	Keenan High School (SC)
Sharon O'Kelley	Francis Marion University (SC)
Ira Papick	University of Nebraska–Lincoln
David Perkinson	Episcopal High School (LA)
Elizabeth Phillips	Michigan State University
Neil Portnoy	University of New Hampshire
David C. Royster	University of Kentucky
Tracy Scala	The Pennsylvania State University
Nanette Seago	WestEd (CA)
Walter Seaman	University of Iowa
Sharon Senk	Michigan State University
Lisa Sheehy	East Hall High School (GA)
Jeanne Shimizu	The College at Old Westbury, State University of New York
Ed Silver	University of Michigan
Jan Sinopoli	Pittsburgh Public Schools (PA)
Roy Smith	University of Georgia
Glenn Stevens	Boston University (MA)
David Stone	Georgia Southern University (emeritus)
Patrick Sullivan	Missouri State University
Dana TeCroney	Ohana Institute (FL)
Patrick Thompson	Arizona State University

Name	Affiliation
Zalman Usiskin	University of Chicago (IL) (emeritus)
Steve Viktora	New Trier High School (IL)
James Wilson	University of Georgia
Patricia Wilson	University of Georgia
Matt Winking	Phoenix High School (GA)
Matthew Winsor	Illinois State University
Rose Mary Zbiek	The Pennsylvania State University

Printed in the United States
By Bookmasters